Computational
Fluid Dynamics

CHAPMAN & HALL/CRC
Numerical Analysis and Scientific Computing

Aims and scope:

Scientific computing and numerical analysis provide invaluable tools for the sciences and engineering. This series aims to capture new developments and summarize state-of-the-art methods over the whole spectrum of these fields. It will include a broad range of textbooks, monographs, and handbooks. Volumes in theory, including discretisation techniques, numerical algorithms, multiscale techniques, parallel and distributed algorithms, as well as applications of these methods in multi-disciplinary fields, are welcome. The inclusion of concrete real-world examples is highly encouraged. This series is meant to appeal to students and researchers in mathematics, engineering, and computational science.

Proposals for the series should be submitted to one of the series editors above or directly to:
CRC Press, Taylor & Francis Group
4th, Floor, Albert House
1-4 Singer Street
London EC2A 4BQ
UK

Published Titles

Classical and Modern Numerical Analysis: Theory, Methods and Practice
*Azmy S. Ackleh, Edward James Allen, Ralph Baker Kearfott,
    and Padmanabhan Seshaiyer*

Computational Fluid Dynamics
Frédéric Magoulès

A Concise Introduction to Image Processing using C++
Meiqing Wang and Choi-Hong Lai

**Decomposition Methods for Differential Equations:
    Theory and Applications**
Juergen Geiser

**Discrete Variational Derivative Method: A Structure-Preserving Numerical
    Method for Partial Differential Equations**
Daisuke Furihata and Takayasu Matsuo

**Grid Resource Management: Toward Virtual and Services Compliant Grid
Computing**
Frédéric Magoulès, Thi-Mai-Huong Nguyen, and Lei Yu

Fundamentals of Grid Computing: Theory, Algorithms and Technologies
Frédéric Magoulès

Handbook of Sinc Numerical Methods
Frank Stenger

Introduction to Grid Computing
Frédéric Magoulès, Jie Pan, Kiat-An Tan, and Abhinit Kumar

**Mathematical Objects in C++: Computational Tools in a Unified Object-
Oriented Approach**
Yair Shapira

Numerical Linear Approximation in C
Nabih N. Abdelmalek and William A. Malek

Numerical Techniques for Direct and Large-Eddy Simulations
Xi Jiang and Choi-Hong Lai

Parallel Algorithms
Henri Casanova, Arnaud Legrand, and Yves Robert

Parallel Iterative Algorithms: From Sequential to Grid Computing
Jacques M. Bahi, Sylvain Contassot-Vivier, and Raphael Couturier

Computational Fluid Dynamics

Edited by
Frédéric Magoulès
Ecole Centrale Paris
Châtenay Malabry, France

CRC Press
Taylor & Francis Group
Boca Raton London New York

CRC Press is an imprint of the
Taylor & Francis Group, an **informa** business
A CHAPMAN & HALL BOOK

CRC Press
Taylor & Francis Group
6000 Broken Sound Parkway NW, Suite 300
Boca Raton, FL 33487-2742

First issued in paperback 2018

ISBN-13: 978-1-4398-5661-1 (hbk)
ISBN-13: 978-1-138-38209-1 (pbk)

Library of Congress Cataloging-in-Publication Data

Computational fluid dynamics / [edited by] Frédéric Magoulès.
    p. cm. -- (Chapman and hall/crc numerical analysis and scientific computation series ; 14)
    Summary: "This reference concentrates on advanced techniques of computational fluid dynamics. It offers illustrations of new developments of classical methods as well as recent methods that appear in the field. Each chapter takes a tutorial approach and covers a different method or application. Topics discussed include finite volumes, weighted residuals, spectral methods, smoothed-particle hydrodynamics (SPH), application of SPH methods to conservation equations, finite volume particle methods (FVPM), and numerical algorithms for unstructured meshes. The authors offer theory, algorithms, and applications for each topic"-- Provided by publisher.
    Summary: "This book concentrates on the numerical of computational fluid mechanics (including mathematical models in computational fluid mechanics, numerical methods in computational fluid mechanics, finite volume, finite difference, finite element, spectral methods, smoothed particle hydrodynamics methods, mixed-element-volume methods, free surface flow) followed by some focus of new development of classical methods, and to the recent methods appearing in this field. The topics covered in this book are wide ranging and demonstrate the extensive use in computational fluid mechanics. The book opens with a presentation of the basis of finite volume methods, weighted residual methods and spectral methods. These specific approaches are particularly important in the context of fluid mechanics, where they cover complementary domains of application. A unified point of view is introduced, based on the weighted residuals description. Chapter 1 presents the finite volume method. Chapter 2 describes the principles of weighted residuals methods. Chapter 3 introduces the spectral method. Chapter 4 presents computational fluid dynamics based on the smoothed particle hydrodynamics (SPH) method. Chapter 5 focuses on an improved SPH method based on an arbitrary Lagrange Euler (ALE) formalism. Chapter 6, using the similarity with the finite volumes method, introduces high order flux schemes between interacting points. Chapter 7 presents some numerical methods for compressible computational fluid dynamics. Chapter 8 deals with the prediction of turbulent complex flows as occur. Chapter 9 discusses the modeling and numerical simulation of free surface flows"-- Provided by publisher.
    Includes bibliographical references and index.
    ISBN 978-1-4398-5661-1 (hardback)
    1. Fluid dynamics--Mathematics. 2. Numerical analysis. I. Magoulès, F. (Frédéric) II. Title.

QA911.C622 2011
532.00285--dc22   2010041672

Visit the Taylor & Francis Web site at
http://www.taylorandfrancis.com

and the CRC Press Web site at
http://www.crcpress.com

Contents

List of Figures

List of Tables

Preface

Computational fluid dynamics has been a hot topic of research these last forty years, and several books have been published on this topics. This book concentrates on the numerical of Computational Fluid Mechanics, followed by some focus of new development of classical methods, and to the recent methods appearing in this field.

The present volume presents in nine chapters a selection of some numerical methods used in computational fluid mechanics including: mathematical models in computational fluid mechanics, numerical methods in computational fluid mechanics, finite volume, finite difference, finite element, spectral methods, smoothed particle hydrodynamics methods, mixed-element-volume methods, free surface flow. The topics covered in this book are wide ranging and demonstrate the extensive use in computational fluid mechanics.

The book opens with a presentation of the basis of finite volume methods, weighted residual methods and spectral methods. These specific approaches are particularly important in the context of fluid mechanics, where they cover complementary domains of application. A unified point of view is introduced, based on the weighted residuals description.

Chapter 1 presents the finite volume method. This approach is widely used in the context of industrial flows, for aerodynamic design for example. This comes from its natural property of conservativity [Hirsch, 2007], which is discussed at the beginning. Then, the integration on the control volumes is presented, constituting the first discretization step of the finite volume method. The second discretization step, the interpolation of the fluxes, is then discussed in a general frame. Some basic elements of time-marching resolution are briefly presented, on specific examples of finite volume discretization. This brings the question of upwinding, particularly important for aerodynamic design for example. This is discussed by focusing on the well-known scheme proposed by Roe [Roe, 1981]. Finally, the particular case of structured grids is presented, for which higher order schemes are easier to develop [Jameson et al., 1981].

Chapter 2 describes the principles of weighted residuals methods. This class of methods was developed mostly in the early 20th century, before the advent of the silicium era, to obtain accurate solutions of continuum mechanics differential equations by hand computations (as in the famous Rayleigh-Ritz method). In weighted residuals methods, an approximation function is expressed in a functional series. Constraints on the development coefficients come from the residual minimization of the initial differential problem applied to the approximation function. Usually, very few coefficients are enough to achieve good accuracy of the solution. Weighted residuals methods make a bridge between different methods, including finite volume and spectral methods.

Chapter 3 introduces the spectral method [Fletcher, 1991], widely used in academic research for its high accuracy and low dispersivity. Spectral methods are a class of methods that rely on the approximation of the solution of partial differential equations, and are thus completely defined from the functional base used for developing the solution, along with the projection base for minimizing the residual. Among these choices, wide classes of methods appear, using trigonometric polynomials—Fourier, Chebyshev—and different handlings of the weighting functions, that yield collocation—Galerkin—or tau-type methods [Canuto et al., 1987]. This chapter reviews these concepts, and illustrates them using simple cases of partial differential equations. The authors also present several issues raised by spectral methods, among which the treatment of nonlinear terms, the aliasing problem and the Gibbs phenomenon. Some practical details of the algorithm implemented for a spectral method are provided, especially the all-important fast Fourier transform, and its characteristics. A short discussion is finally proposed on the physical interpretation of wavespace representation and spectra in the Fourier space.

In Chapter 4, the authors present a novel approach of computational fluid dynamics based on the mesh-free technique named the smoothed particle hydrodynamics (SPH) method [Monaghan, 1992]. Mesh-free numerical methods have been much less investigated than mesh-based techniques although they can bring clear advantages on specific flow configurations like interfacial flows. The major benefit of using a mesh-free technique resides in the simplified management of calculation points, whose spatial distribution is not constrained by connectivity. In particular this allows large deformation of the initial point distribution and consequently the use of a Lagrangian description of the flow.

In Chapter 5, the authors focus on an improved SPH method based on an arbitrary Lagrange Euler (ALE) formalism [Vila, 1999]. This approach bridges the gap between the pure Lagrangian description used in the standard SPH method and the classical Eulerian description encountered in most of conventional computational fluid dynamics methods. The resulting framework is applied to the weak form of conservation laws and can be linked to a finite volume formalism. A generic treatment of boundary conditions is introduced by taking profit of the dual description, bringing an efficient answer to one of the main shortage of SPH [Marongiu et al., 2007c].

The similarity with the finite volumes method, in Chapter 6, is also used to introduce high order flux schemes between interacting points, which opens a way to increase the accuracy of the method. The underlying integration scheme which serves to sum flux contributions coming from neighboring points is at the heart of the mesh-free nature of SPH-ALE. It uses a regularizing function called the kernel function. This function has a great influence on the numerical behavior of SPH method and has been an important field of research. The convergence of the integration is not always ensured in practical configurations and in order to minimize the errors, the integration domain has to hold a bigger number of points than a classical numerical stencil in finite volumes, which is the reason for the relatively higher cost of mesh-free techniques. Some correction techniques are presented [Bonet and Lock, 1999], with a special emphasis on the finite volume particle method (FVPM) [Warnecke, 2005]. They aim at modifying locally the mesh-free integration procedure to improve its accuracy. Finally applications of the SPH-ALE method to simulate free surface flows are presented, demonstrating the ability of this method to represent properly highly distorded interfaces and highly dynamic flows.

In Chapter 7, the authors present some numerical methods for compressible computational fluid dynamics. This topic is a very animated one, since at least four main mathematical theories, finite differences, finite volumes, finite elements and spectral approximation, have been interacting in order to produce a large and sophisticated panoply of methods for the discretization of the compressible Navier-Stokes equations.

- The finite difference method remains the best reference for analyzing discretization CFD schemes. Most time-advancing schemes are of finite difference type and analyzed as such. Finite difference inspired notions like dissipation and dispersion which was well clarified with the modified or equivalent equation theory [Hirt, 1968], [Lerat and Peyret, 1974] and which remain the reference properties for advection phenomena.

- The finite volume method brought a revolution in nonlinear hyperbolics with the relation to weak solutions by Lax and Wendroff [Lax and Wendroff, 1972] and the introduction of Riemann solver by Godunov [Godunov, 1959].

- The finite element method was early applied to incompressible CFD, but entered through the back door for compressible CFD. There are now many reasons to think that finite elements will provide the best way to rise the order of accuracy for the simulation of compressible flows on unstructured meshes, through, for example, the discontinuous Galerkin formulations [Arnold et al., 2002].

This chapter deals with a low order scheme, the mixed-element-volume method, which combines features of finite differences, finite volumes and finite elements. On this basis several important problems, arising for any scheme but easier to analyze with the mixed-element-volume method, are addressed. Methods are proposed for addressing unstructured meshes, moving meshes, mesh adaptation, nonlinear stability and positiveness, and control of dissipation errors.

Chapter 8 deals with the prediction of turbulent complex flows, as occur in most environmental problems or engineering applications. The prediction of turbulent flows for such practical applications is one of the most difficult modeling problems. After centuries of investigations, the prediction relies on the combination of different methodologies, involving in first places turbulence modeling and numerical simulation.

- A first approach is direct numerical simulation (DNS), which assumes that the Navier-Stokes equations are an accurate model for a turbulent flow, if all the relevant scales of turbulence are resolved. DNS requires thus a very high accuracy in terms of number of unknowns and high-order numerical methods. A typical pioneering work has been published in [Kim ct al., 1987]. DNS is very useful for the study of physical phenomena related to turbulence, since it is able to provide a large amount of information on the flow dynamics, most of which are very difficult or impossible to be obtained experimentally. However, it is clear that this approach, because of the huge computational requirements, is limited to academic flows, characterized by low Reynolds numbers and very simple geometry.

- The large majority of flows of practical interest are still nowadays simulated by numerical discretization of the Reynolds-averaged Navier-Stokes equations (RANS). The aim of this approach is to provide the time-averaged flow variables. A closure of the RANS equations is needed, the so-called turbulence model, and this is the most critical issue in the RANS approach, since all information on turbulence fluctuations is contained in the model. A huge variety of closure models for the RANS equations has been proposed up to now, see references [Baldwin and Lomax, 1978], [Jones and Launder, 1972] for example of such models. However, most of them are strongly specialized to a particular class of flows and it is almost impossible to find a model of general validity. On the other hand, although the RANS approach produces rather stiff differential systems, the advantage of only simulating the averaged flow field is that this leads to moderate computational requirements also for complex flow configurations and high Reynolds numbers. Moreover, a high level of accuracy of the numerical method is not required and this makes the handling of complex geometries easier. As a consequence, industrial simulation tools were identified to RANS modeling, combined with low-order accurate approximations and slightly more recently with unstructured meshes.

- A third possible approach is large eddy simulation (LES), in which small scales of the flow are damped and their effect on the large-scale motion is provided by a closure model, the so-called subgrid-scale (SGS) model. LES is able to provide more detailed information than RANS, since part of the turbulent fluctuations are resolved, but this also brings computational costs much higher than RANS. Although LES was born for the simulation of very high Reynolds number atmospheric flows [Smagorinsky, 1963], the widest and best-established use of LES up to now is as a kind of extension of DNS, with strong accuracy requirements for the numerics, and still severe Reynolds number and geometry limitations. It is however expected that LES could be a well suited approach for some particular (very) high Reynolds complex flows for which RANS do not give accurate predictions. Paradigmatic examples are flows characterized by massive separation, such as bluff-body flows. In the last decade, the development of computational resources has indeed enhanced the tendency to apply LES-methodologies to turbulent flow problems of significant complexity, such as arise in various applications in technology and in many natural flows. However, in order that LES becomes a completely reliable tool for such applications, some still open issues, related to both modeling and numerics, must be solved. Just to give one example, in order to obtain accurate LES predictions, an extremely fine resolution in attached boundary layers is needed, which increases with the Reynolds number and leads to prohibitive computational costs for Reynolds numbers typical of practical applications. This motivated the building of RANS-LES hybrid models, which try to combine RANS and LES in the computational domain. A well-known example is the detached eddy simulation model of Spalart et al. [Spalart et al., 1997].

In this chapter, the coupling of a numerical solver, originally designed for RANS simulation of complex compressible flows on unstructured grids and described in Chapter 7, with first, a LES approach, and, then, a RANS/LES hybrid method, is described. The main numerical and modeling issues, which are typical of LES of complex engineering problems, are discussed and some solutions are proposed. Illustrative applications to massively separated flows are finally presented.

In Chapter 9, the authors discuss the modeling and numerical simulation of free surface flows, and with applications to liquids and ice. Free surface flows are ubiquitous in nature, and arise in many fields of sciences and engineering. A wide range of behaviors can be exhibited when considering various constitutive laws (Newtonian vs. non-Newtonian flows) and regimes (small vs. large Reynolds numbers). In this chapter, the authors consider two main examples, and gather them into one mathematical and numerical framework. The first situation arises in mold casting, when a liquid is injected at high velocity into a mold [Maronnier et al., 2003], [Caboussat et al., 2005], [Caboussat, 2006]. This Newtonian flow is usually turbulent, involves high Reynolds numbers in complex geometries, and frequent topological changes. The second situation arises in glacier modeling, in order to predict the evolution of alpine glaciers. This non-Newtonian ice flow contain large viscous effects [Jouvet et al., 2008]. The applications are numerous. Liquid flows in mold casting are used for instance in metallurgy, material sciences, chemistry, and environmental sciences. Ice flows allow the modeling and prediction of glaciers, which has huge implications on the future management of natural risks, the energy production through hydroelectric plants, the water supply for agriculture, the freshwater stocks, and tourism. Since glaciers are one of the main agents suffering from global warming, the numerical simulation and prediction of glaciers' behavior is of paramount importance for policy-makers and the electricity production. A numerical method for the simulation of such flows that relies on an Eulerian approach based on the volume-of-fluid formulation is advocated. A time splitting algorithm, together with a two-grids method, allows the various physical phenomena to be decoupled and computed accurately. Finite element techniques on an unstructured mesh of tetrahedra are used for the approximation of the diffusion phenomena, while the advection operators are discretized with a characteristics method on a grid of small cubic cells. This numerical framework allows a natural incorporation of additional effects, such as the addition of compressible gas, addition of mass, surface tension effects, treatment of a nonlinear viscosity, or various boundary conditions. Then the chapter investigates real-world applications. Numerical results in a wide range of applications (mold casting, sloshing problems, bubbles flow, and alpine glaciers) show the ability of the numerical framework to simulate with success very diverse free surface flows, at very large and very small Reynolds numbers. In particular, the prediction of the future behaviors of glaciers, following climate scenarios, shows the risks encountered due to a global temperature increase.

The various technology presented in this book demonstrates the wide aspects of interest in computational fluid mechanics, and the many possibilities and venues that exist in the research in this area. We are sure that this interest is only going to further evolve, and that many exciting developments are still awaiting us.

Frédéric Magoulès
Ecole Centrale Paris, France

Warranty

Every effort has been made to make this book as complete and as accurate as possible, but no warranty of fitness is implied. The information is provided on an as-is basis. The authors, editor and publisher shall have neither liability nor responsibility to any person or entity with respect to any loss or damages arising from the information contained in this book or from the use of the code published in it.

Chapter 1

Finite volume methods

Jérôme Boudet
Ecole Centrale de Lyon, LMFA
36 avenue Guy de Collongue
69134 Ecully Cedex, France

1.1   Introduction

The finite volume method is a very popular approach for the computation of industrial flows. Domains of application include aeronautics, for the simulation of external or internal aerodynamics (see Figure 1.1). The popularity of this approach comes from the particular attention paid to conservativity. Indeed, the flux balance is controlled on the discrete level, the first discretization step consisting in the integration of the equations on elementary control volumes. The following presentation of the finite volume method only constitutes an overall introduction to the strategy of discretization. More details can be found in books such as [Hirsch, 2007], [Peyret, 1996], [Versteeg and Malalasekera, 1995], and references cited in the text.

1

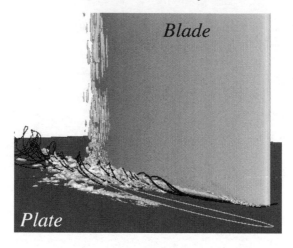

FIGURE 1.1: Large eddy simulation of the tip-clearance vortex and turbulent structures developing from the gap between a blade and a plate. Vortical structures are visualized by an iso-surface of Q-criterion.

1.2 Conservativity

The transport equations of a flow can be generally written in the following form:

$$\frac{\partial \mathbf{U}}{\partial t} + \nabla.\mathbf{F}(\mathbf{U}) = \mathbf{Q} \tag{1.1}$$

This expresses conservativity: the evolution of a scalar variable U_i is controlled by the flux $\mathbf{F}_i$ and the source Q_i. The form of the equation (1.1) is called the *conservative form*, and the associated variables U_i are the *conservative variables*.

For example, a viscous compressible flow is governed by the following set of transport equations:

$$\frac{\partial \rho}{\partial t} + \frac{\partial (\rho u_j)}{\partial x_j} = 0 \tag{1.2}$$

$$\frac{\partial (\rho u_i)}{\partial t} + \frac{\partial (\rho u_i u_j)}{\partial x_j} = -\frac{\partial p}{\partial x_i} + \frac{\partial \tau_{ij}}{\partial x_j} \tag{1.3}$$

$$\frac{\partial (\rho e_t)}{\partial t} + \frac{\partial}{\partial x_j} \left[(\rho e_t + p) u_j \right] = \frac{\partial (u_i \tau_{ij})}{\partial x_j} - \frac{\partial q_j}{\partial x_j} \tag{1.4}$$

with summation on repeated indices. ρ is the density, u_i ($i = 1..3$) are the components of velocity, p is the pressure, e_t is the specific total energy, τ_{ij} are the viscous constraints, and q_j is the heat conduction flux. In this case, the conservative form (1.1) is obtained with:

$$\mathbf{U} = \begin{pmatrix} \rho \\ \rho u_1 \\ \rho u_2 \\ \rho u_3 \\ \rho e_t \end{pmatrix}$$

$$\mathbf{F} = \begin{pmatrix} \rho u_1 & \rho u_2 & \rho u_3 \\ \rho u_1 u_1 + p - \tau_{11} & \rho u_1 u_2 - \tau_{12} & \rho u_1 u_3 - \tau_{13} \\ \rho u_2 u_1 - \tau_{21} & \rho u_2 u_2 + p - \tau_{22} & \rho u_2 u_3 - \tau_{23} \\ \rho u_3 u_1 - \tau_{31} & \rho u_3 u_2 - \tau_{32} & \rho u_3 u_3 + p - \tau_{33} \\ \{(\rho e_t + p)u_1 - u_i \tau_{i1} + q_1\} & \{(\rho e_t + p)u_2 - u_i \tau_{i2} + q_2\} & \{(\rho e_t + p)u_3 - u_i \tau_{i3} + q_3\} \end{pmatrix}$$

REMARK 1.1   For resolution purpose, with given initial and boundary conditions, the transport equations (1.2)-(1.4) must be completed with constitutive laws on τ_{ij} (e.g. Newtonian fluid hypothesis) and q_j (e.g. Fourier law), and equations of state (e.g. perfect gas). $\quad\square$

Returning to the general equation (1.1), integration on a volume Ω and application of the theorem of Reynolds yields:

$$\frac{d}{dt} \int_\Omega \mathbf{U} dv + \int_{\delta\Omega} \mathbf{F}(\mathbf{U}).d\mathbf{S} - \int_{\delta\Omega} \mathbf{U}(\mathbf{W}.d\mathbf{S}) = \int_\Omega \mathbf{Q} dv \qquad (1.5)$$

where $\delta\Omega$ represents the border of Ω, moving with local velocity $\mathbf{W}$ ($= \mathbf{0}$ for a fixed domain), and $d\mathbf{S}$ is the outward normal surface vector. Considering different domains Ω_i whose union is connected, the summation of the corresponding equations (1.5) results in the same equation for the union domain. For example, considering the four domains presented in Figure 1.2, the sum of the different equations is:

$$\frac{d}{dt} \int_{\Omega_1 \bigcup \Omega_2 \bigcup \Omega_3 \bigcup \Omega_4} \mathbf{U} dv + \int_{\delta\Omega_1 \bigcup \delta\Omega_2 \bigcup \delta\Omega_3 \bigcup \delta\Omega_4} \mathbf{F}(\mathbf{U}).d\mathbf{S}$$
$$- \int_{\delta\Omega_1 \bigcup \delta\Omega_2 \bigcup \delta\Omega_3 \bigcup \delta\Omega_4} \mathbf{U}(\mathbf{W}.d\mathbf{S}) = \int_{\Omega_1 \bigcup \Omega_2 \bigcup \Omega_3 \bigcup \Omega_4} \mathbf{Q} dv \qquad (1.6)$$

then:

$$\frac{d}{dt} \int_{\Omega_1 \bigcup \Omega_2 \bigcup \Omega_3 \bigcup \Omega_4} \mathbf{U} dv + \int_{\delta(\Omega_1 \bigcup \Omega_2 \bigcup \Omega_3 \bigcup \Omega_4)} \mathbf{F}(\mathbf{U}).d\mathbf{S}$$
$$- \int_{\delta(\Omega_1 \bigcup \Omega_2 \bigcup \Omega_3 \bigcup \Omega_4)} \mathbf{U}(\mathbf{W}.d\mathbf{S}) = \int_{\Omega_1 \bigcup \Omega_2 \bigcup \Omega_3 \bigcup \Omega_4} \mathbf{Q} dv \qquad (1.7)$$

which is indeed the equation (1.5) for the domain $\Omega_1 \bigcup \Omega_2 \bigcup \Omega_3 \bigcup \Omega_4$. The key point is the compensation of the fluxes on the inner boundaries, because of the opposite normal vectors (in the example, the surface of integration $\delta\Omega_1 \bigcup \delta\Omega_2 \bigcup \delta\Omega_3 \bigcup \delta\Omega_4 \to \delta(\Omega_1 \bigcup \Omega_2 \bigcup \Omega_3 \bigcup \Omega_4))$. This summation property of the integral equations is the expression of conservativity.

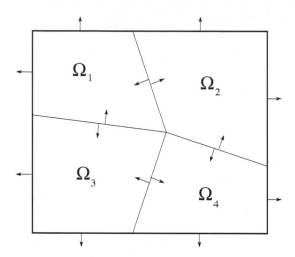

FIGURE 1.2: Conservativity: 2D illustration.

1.3 Control volume integration

The finite volume method uses integration on elementary control volumes as a first discretization step, in order to ensure conservativity at the discrete level. Thus, a preliminary step consists in the division of the computational domain into a finite number of elementary control volumes. Overlapping control volumes can be used, but internal boundaries (i.e. boundaries not on the external frontier of the domain) must be common to two control volumes, in order to allow conservativity by flux compensation on the internal boundaries. Nevertheless, following illustrations will consider non-overlapping control volumes, more generally encountered. From here, and for the following presentation, the control volumes will be supposed fixed, but formulation can be easily extended to moving computational grids. For a given control volume Ω_I (e.g. Figure 1.3), from equation (1.5), the first discretization step yields:

$$\frac{\partial}{\partial t}\left(\mathbf{U}_I\Omega_I\right) + \sum_{d\mathbf{S}_J \in \delta\Omega_I} \mathbf{F}_J.d\mathbf{S}_J = \mathbf{Q}_I\Omega_I \qquad (1.8)$$

where by definition:

$$\mathbf{U}_I = \frac{1}{\Omega_I} \int_{\Omega_I} \mathbf{U} dv$$

$$\mathbf{Q}_I = \frac{1}{\Omega_I} \int_{\Omega_I} \mathbf{Q} dv$$

$\mathbf{F}_J$ represent discrete estimates of the fluxes on the boundary faces. They must be interpolated from the values $\mathbf{U}_I$ on the neighboring control volumes, in order to complete discretization. This is discussed in a following section. It should just be noted that if the internal faces are common to two control volumes (as previously mentionned), and fluxes are interpolated with the same formula on each side, discrete conservativity is ensured. By explicitly considering the fluxes on the discrete level, the finite volume method yields a particularly convenient condition for conservativity.

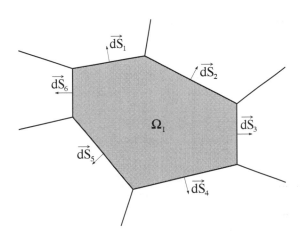

FIGURE 1.3: Example of 2D control volume for finite volume integration.

By definition, the variable $\mathbf{U}_I$ is the average of the conservative variable vector on the control volume Ω_I. For the interpolation of the fluxes on the faces, it could be convenient to have a point of application for $\mathbf{U}_I$. If $\mathbf{g}_I$ is the center of gravity of Ω_I, and $(\hat{x}_j, j = 1..3)$ the coordinates centered on $\mathbf{g}_I$, a Taylor series expansion gives:

$$\forall i : \ U_{Ii} = \frac{1}{\Omega_I} \int_{\Omega_I} U_i dv = \frac{1}{\Omega_I} \int_{\Omega_I} \left[U_i(\mathbf{g}_I) + \frac{\partial U_i}{\partial x_j}(\mathbf{g}_I)\hat{x}_j \right] dv + O(\Delta x^2)$$

$$U_{Ii} = U_i(\mathbf{g}_I) + O(\Delta x^2)$$

$\mathbf{U}_I$ can thus be identified with the vector of conservative variables at $\mathbf{g}_I$, at order 2. This gives a point of application for $\mathbf{U}_I$.

REMARK 1.2   However, it should not be prematurely concluded that finite volumes are limited to order 2. As will be seen in section 1.9.1 for example, higher order discretizations can be built, by a proper balancing of the discretization steps.   □

1.4   Grid

The grid is made of points and the edges between, covering the computational domain, and used for the definition of the control volumes.

A 3D *structured grid* is made of a topological parallelepiped, which means each point can be identified by a set of three indices. The grid lines, defined by a fixed couple of indices, can be curved for adaptation to the geometry of the computational domain. All the cells are hexahedral.

By opposition, a grid that is not made of a topological parallelepiped is called *unstructured*, and the cells can be tetrahedral, hexahedral, etc.

The structured grids are particularly effective for programming, the neighbors of a given point (used for flux interpolation, see section 1.5) being directly found by increment of the indices. Conversely, unstructured grids require specific procedures for the identification of the neighbors (link tables ...), but are easier to adapt to complex geometries. Figure 1.4 presents examples of 2D structured and unstructured grids. The finite volume method can be ap-

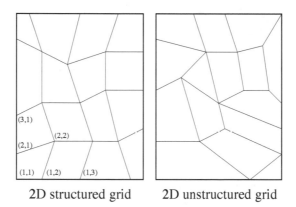

2D structured grid   2D unstructured grid

FIGURE 1.4: Examples of 2D structured and unstructured grids.

plied to both structured and unstructured grid, even with different kind of cells combined over the same computational domain (tetrahedral, hexahedral,

...). The control volumes are defined on the grid, and two major strategies are possible:

- Cell-centered: the control volumes are centered on the cells of the grid;

- Cell-vertex: the control volumes are centered on the points of the grid.

The control volumes can be defined over one or more cells or points, and it is also possible to use overlapping control volumes. Examples of control volumes are given in Figure 1.5.

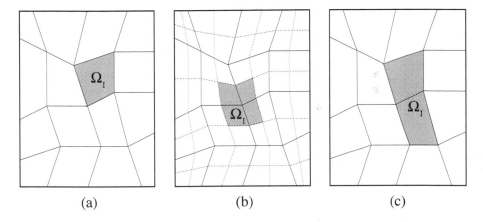

(a)   (b)   (c)

FIGURE 1.5: Examples of control volumes on a 2D structured grid. (a) cell-centered; (b) cell-vertex; (c) cell-centered on two cells.

1.5 General flux interpolation

Returning to equation (1.8), the discretization then requires the estimate of the fluxes $\mathbf{F}_J$ on the control volume boundaries (normal vectors: $d\mathbf{S}_J$). This is an interpolation step, the fluxes being functions of the conservative variables $\mathbf{U}_I$ that are defined on the control volumes.

In a general approach, different strategies are possible. For illustration, we can consider the 2D case presented in Figure 1.6, where the flux $\mathbf{F}_{PQ}$ on the edge PQ have to be interpolated. A basic interpolation can be obtained from the control volumes on each side, here Ω_1 and Ω_4. Two strategies are possible,

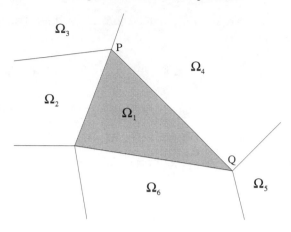

FIGURE 1.6: Flux interpolation: general case.

either interpolating the conservative variables:

$$\mathbf{F}_{PQ} = \mathbf{F}\left(\frac{\mathbf{U}_1.V_1 + \mathbf{U}_4.V_4}{V_1 + V_4}\right)$$

or interpolating the fluxes:

$$\mathbf{F}_{PQ} = \frac{\mathbf{F}(\mathbf{U}_1).V_1 + \mathbf{F}(\mathbf{U}_4).V_4}{V_1 + V_4}$$

where V_i denote the volumes. Interpolation can also be constructed from the end points:

$$\mathbf{F}_{PQ} = \frac{1}{2}\left(\mathbf{F}_P + \mathbf{F}_Q\right)$$

where:

$$\mathbf{F}_P = \mathbf{F}\left(\frac{\mathbf{U}_1.V_1 + \mathbf{U}_2.V_2 + \mathbf{U}_3.V_3 + \mathbf{U}_4.V_4}{V_1 + V_2 + V_3 + V_4}\right)$$

or:

$$\mathbf{F}_P = \frac{\mathbf{F}(\mathbf{U}_1).V_1 + \mathbf{F}(\mathbf{U}_2).V_2 + \mathbf{F}(\mathbf{U}_3).V_3 + \mathbf{F}(\mathbf{U}_4).V_4}{V_1 + V_2 + V_3 + V_4}$$

1.6   Resolution and time discretization

From equation (1.8), with interpolation of the fluxes from the neighboring control volumes, the semi-discrete finite volume equation can be expressed as:

$$\forall I : \quad \frac{\partial \mathbf{U}_I}{\partial t} = \mathbf{R}(..., \mathbf{U}_{I-1}, \mathbf{U}_I, \mathbf{U}_{I+1}, ...) \tag{1.9}$$

The residual **R** includes the fluxes and the sources, expressed as functions of the conservative variables on the neighboring control volumes. The last discretization step must address the time derivative.

1.6.1 Unsteady resolution

Classical strategies are employed for time discretization. For example, finite difference:

$$\frac{\mathbf{U}_I^{n+1} - \mathbf{U}_I^n}{\Delta t} \tag{1.10}$$

constitutes an estimate of $\partial \mathbf{U}_I / \partial t$ between time steps n and $n+1$. The fully discrete equation obtained for each contre volume should then relate the conservative variables ($\mathbf{U}_J^n$, $J = 1..J_{max}$) at any instant n, to the variables at the instant after $(n+1)$.

A scheme is said *explicit* when the conservative variable vector $\mathbf{U}_I^{n+1}$, for a given control volume I and instant $n+1$, can be directly expressed as a function of the flow field at instant n (i.e. $\mathbf{U}_J^n$, $J = 1..J_{max}$). For example, if discretization is carried out at instant n with formula (1.10):

$$\forall I : \quad \frac{\mathbf{U}_I^{n+1} - \mathbf{U}_I^n}{\Delta t} = \mathbf{R}(..., \mathbf{U}_{I-1}^n, \mathbf{U}_I^n, \mathbf{U}_{I+1}^n, ...)$$

thus:

$$\forall I : \quad \mathbf{U}_I^{n+1} = \mathbf{U}_I^n + \Delta t \times \mathbf{R}(..., \mathbf{U}_{I-1}^n, \mathbf{U}_I^n, \mathbf{U}_{I+1}^n, ...)$$

From the flow field ($\mathbf{U}_J^n$, $J = 1..J_{max}$) at instant n, each vector $\mathbf{U}_I^{n+1}$ can be directly calculated.

In contrast, with an *implicit* scheme, the conservative variable vector $\mathbf{U}_I^{n+1}$, at a given control volume I and instant $n+1$, is cross-dependant with other control volumes at the same instant, and requires a coupled resolution. For example, if time is discretized at instant $n+1$ with the same formula (1.10):

$$\forall I : \quad \frac{\mathbf{U}_I^{n+1} - \mathbf{U}_I^n}{\Delta t} = \mathbf{R}(..., \mathbf{U}_{I-1}^{n+1}, \mathbf{U}_I^{n+1}, \mathbf{U}_{I+1}^{n+1}, ...)$$

thus:

$$\forall I : \quad \mathbf{U}_I^{n+1} - \Delta t \times \mathbf{R}(..., \mathbf{U}_{I-1}^{n+1}, \mathbf{U}_I^{n+1}, \mathbf{U}_{I+1}^{n+1}, ...) = \mathbf{U}_I^n$$

The number of equations equals the number of unknowns ($\mathbf{U}_J^{n+1}$, $J = 1..J_{max}$), but the equations are cross-dependant between various control volumes at instant $n+1$. A coupled resolution must be set-up, computationally more expensive (requiring for example matrix inversions of linear systems), but such schemes benefit from particularly interesting stability properties that will be presented in section 1.7.

In should be noted that in the previous explicit and implicit discretizations, the choice of the discretization time only influences the variables of $\mathbf{R}$, the discrete expression of the time derivative being the same for the instants n and $n+1$. A general θ-scheme, resulting from the linear combination of the explicit scheme (coefficient $1 - \theta$) and the implicit scheme (coefficient θ), can be built:

$$\forall I: \quad \mathbf{U}_I^{n+1} = \mathbf{U}_I^n + \Delta t \times \left[(1 - \theta)\mathbf{R}(..., \mathbf{U}_{I-1}^n, \mathbf{U}_I^n, \mathbf{U}_{I+1}^n, ...) \right.$$
$$\left. + \theta\mathbf{R}(..., \mathbf{U}_{I-1}^{n+1}, \mathbf{U}_I^{n+1}, \mathbf{U}_{I+1}^{n+1}, ...) \right]$$

where $\theta \in [0, 1]$, $\theta = 0$ and 1 yielding the original explicit and implicit schemes respectively.

Fractional step methods can involve multiple explicit steps to approach implicit method properties, at moderate computational cost. For example, the 4-step Runge-Kutta scheme writes:

$$\mathbf{U}_I^{n+1} = \mathbf{U}_I^n + \frac{\Delta t}{6} \left[\mathbf{R}(\mathbf{U}_I^n) + 2\mathbf{R}(\mathbf{U}_I^{n+0.5*}) + 2\mathbf{R}(\mathbf{U}_I^{n+0.5**}) + \mathbf{R}(\mathbf{U}_I^{n+1*}) \right]$$

where the predictors are:

$$\mathbf{U}_I^{n+0.5*} = \mathbf{U}_I^n + 0.5\Delta t\mathbf{R}(\mathbf{U}_I^n)$$
$$\mathbf{U}_I^{n+0.5**} = \mathbf{U}_I^n + 0.5\Delta t\mathbf{R}(\mathbf{U}_I^{n+0.5*})$$
$$\mathbf{U}_I^{n+1*} = \mathbf{U}_I^n + \Delta t\mathbf{R}(\mathbf{U}_I^{n+0.5**})$$

Other variants are available, with different numbers of steps, and different properties (precision, stability conditions, see section 1.7).

In a purely unsteady simulation, the initial flow field is provided by the physical problem, and the temporal evolution is described iteratively by the discretized equations. In cases where a statistically steady solution is investigated, the initial flow field can be artificial, and a criterion must be designed to define the end of the transient regime and the beginning of the regime to investigate. For example, in a turbomachine unsteady computation (rotor/stator interaction), the criterion can consider the level of periodicity of a local or integral quantity.

1.6.2   Steady resolution

In this case:

$$\forall I: \quad \frac{\partial \mathbf{U}_I}{\partial t} = \mathbf{0}$$

and the equation writes for each control volume I:

$$\mathbf{R}(..., \mathbf{U}_{I-1}, \mathbf{U}_I, \mathbf{U}_{I+1}, ...) = \mathbf{0}$$

Supposing appropriate boundary conditions are defined, this equation is adapted for numerical resolution. It is generally non-linear (a basic feature in fluid mechanics), which calls for iterative methods of resolution. A conceptually simple approach, known as time-marching, consists in starting from an *artificial* flow field $\mathbf{U}_I^0$ and computing the unsteady transient regime up to the steady solution (from the complete unsteady equation (1.9)). This approach comes down to the unsteady resolution, previously described. However, two features distinguish the time marching resolution of a steady problem from the unsteady resolution:

- The initial flow field is *artificial*, not given by the physical problem.

- The physical constraints on the time resolution are alleviated. For example, local time stepping (i.e. spatially varying time steps) can be employed in order to increase the convergence speed.

Convergence is theoretically achieved when:

$$\forall I : \quad \frac{\partial \mathbf{U}_I}{\partial t} = \mathbf{0} \iff \mathbf{R}(..., \mathbf{U}_{I-1}, \mathbf{U}_I, \mathbf{U}_{I+1}, ...) = \mathbf{0}$$

Practically, because the computational encoding of numbers induce truncation errors, and because local flow instabilities often persist in apparently steady problems (e.g. local flow separations, ...), the computed steady solution is instead characterized by:

$$\forall I : \mathbf{R}(..., \mathbf{U}_{I-1}, \mathbf{U}_I, \mathbf{U}_{I+1}, ...) = \epsilon_I$$

with "low" magnitudes of ϵ_I. During a computation, in order to evaluate if a steady solution is reached and decide at which iteration the computation can be stopped, a measure of the residual magnitude must be carried out over the domain. At a given iteration n, and for a given scalar component j, a norm of the residual over the control volumes can be calculated: $\|R_j(..., \mathbf{U}_{I-1}^n, \mathbf{U}_I^n, \mathbf{U}_{I+1}^n, ...)\|$. Among the most popular norms:

$$L^2\text{-norm}: \quad \|R_j^n\|_2 = \sqrt{\sum_I \left(R_j(..., \mathbf{U}_{I-1}^n, \mathbf{U}_I^n, \mathbf{U}_{I+1}^n, ...) \right)^2}$$

$$L^\infty\text{-norm}: \quad \|R_j^n\|_\infty = \max_I \left| R_j(..., \mathbf{U}_{I-1}^n, \mathbf{U}_I^n, \mathbf{U}_{I+1}^n, ...) \right|$$

The L^2-norm measures the global convergence of the computation, whereas the L^∞-norm focus on the least converged point. An example of residual convergence is plotted in Figure 1.7. A logarithmic scale is used for the residual norm, which is normalized by the value at the initial time step ($n = 0$), yielding a reference level of 0 for the initial residual. These are common practices. A steady computation will be considered as converged if both conditions are achieved:

- The residuals reach a plateau. This means the solver reached its convergence limit.

- The plateau is at a sufficiently low level. A threshold must be defined, for example $\log 10 \left(||R_j^n||/||R_j^0|| \right) = -4$. This evaluates if the computed flow field can be considered as a steady solution. Indeed, $||R_j^n||/||R_j^0||$ evaluates the unsteadiness ratio $||\frac{\partial U_j^n}{\partial t}||/||\frac{\partial U_j^0}{\partial t}||$ between the current flow field and the initial flow field (considered as reference).

The value of the threshold must be defined with respect to the requested physical precision and the computational encoding precision of the numbers. In the example given in Figure 1.7, the solver has been compiled with a double precision option, which allows for very small residuals to be reached (nearly $\log 10 \left(||R_j^n||/||R_i^n|| \right) = -12$).

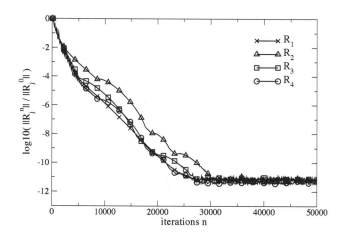

FIGURE 1.7: Example of convergence curves for a 2D compressible solver computation on a wing. The four residuals are respectively associated with the conservative variables: ρ, ρu_1, ρu_2 and ρe_t, and $||.||$ is the L^∞-norm.

Other convergence monitors can be used, such as the mass flow in a continuous flux system, and the force or torque induced by the flow on a structure. For these quantities, the steadiness of the flow will be evaluated by the stationarity of the values, and the identity of the mass flow between the inlet and outlet of the continuous flux system. Such monitors are interesting because of their physical meaning, and their role in the interpretation of the computation.

Divergence   Instabilities of the numerical schemes (see section 1.7), or physical unsteadiness (including physical instabilities), can prevent the solver from reaching the convergence criteria. Strong numerical instabilities will generally result in $\log 10 \left(\|R_j^n\| / \|R_j^0\| \right) >> 0$, and can be detected by un-physical flow quantities (negative pressure or density in compressible solvers, etc). Weak numerical instabilities or physical unsteadiness will generally result in insufficiently low values of $\log 10 \left(\|R_j^n\| / \|R_j^0\| \right)$, and possible oscillations.

Finally, it must be pointed out that time marching convergence can involve physical phenomena perturbing the transient convergence to the steady solution. In a compressible solver, acoustic waves can be generated (e.g. from the artificial initial flow) that should be evacuated properly by the boundary conditions. Also, convergence can pass through a supersonic state that prevents the upstream propagation of information, and blocks the convergence.

1.7 Consistency, stability, and convergence

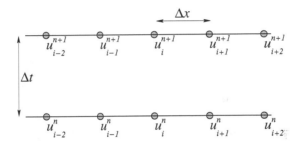

FIGURE 1.8: Discretization space.

For illustration purpose, we consider the 1D scalar equation:

$$\frac{\partial u}{\partial t} + \alpha \frac{\partial u}{\partial x} = 0 \tag{1.11}$$

discretized with cell-vertex finite volumes:

$$\Delta x \times \frac{\partial u_i}{\partial t} + \alpha u_{i+0.5} - \alpha u_{i-0.5} = 0$$

and explicit time discretization:

$$\frac{u_i^{n+1} - u_i^n}{\Delta t} + \alpha \frac{u_{i+0.5}^n - u_{i-0.5}^n}{\Delta x} = 0$$

Consistency A numerical scheme is *consistent* if the discrete equation tend to the continuous differential equation when grid and time steps tend to zero. In the present example, using a left-sided interpolation (i.e. $u_{i+0.5}^n = u_i^n$ and $u_{i-0.5}^n = u_{i-1}^n$):

$$\frac{u_i^{n+1} - u_i^n}{\Delta t} + \alpha \frac{u_i^n - u_{i-1}^n}{\Delta x} = 0 \tag{1.12}$$

Using Taylor series expansions with respect to point i and instant n, the discretized equation is shown equivalent to:

$$\frac{\partial u_i^n}{\partial t} + \frac{\Delta t}{2}\frac{\partial^2 u_i^n}{\partial t^2} + O(\Delta t^2) + \alpha\frac{\partial u_i^n}{\partial x} - \alpha\frac{\Delta x}{2}\frac{\partial^2 u_i^n}{\partial x^2} + O(\Delta x^2) = 0$$

The discrete equation actually tends to the continuous differential equation (at point i and instant n), with order 1 in time and order 1 in space. The orders can be increased by using a centered discretization. For example, using a centered interpolation of the fluxes:

$$u_{i+0.5} = \frac{u_i + u_{i+1}}{2} \quad \text{and:} \quad u_{i-0.5} = \frac{u_{i-1} + u_i}{2}$$

then discretization yields:

$$\frac{u_i^{n+1} - u_i^n}{\Delta t} + \alpha\frac{u_{i+1}^n - u_{i-1}^n}{2\Delta x} = 0 \tag{1.13}$$

$$\Longleftrightarrow \frac{\partial u_i^n}{\partial t} + \frac{\Delta t}{2}\frac{\partial^2 u_i^n}{\partial t^2} + O(\Delta t^2) + \alpha\frac{\partial u_i^n}{\partial x} + \alpha\frac{\Delta x^2}{6}\frac{\partial^3 u_i^n}{\partial x^3} + O(\Delta x^3) = 0$$

In this case, the discretization is 1st order in time and 2nd order in space. Centering the flux interpolation allowed increasing the order spatially. Considering again a left-sided interpolation, the order can also be increased by increasing the number of interpolation points. For example, a left-sided interpolation on 3 points can be designed:

$$u_{i+0.5} = \frac{-u_{i-1} + 5u_i + 2u_{i+1}}{6} \quad \text{and:} \quad u_{i-0.5} = \frac{-u_{i-2} + 5u_{i-1} + 2u_i}{6}$$

and the discrete equation writes:

$$\frac{u_i^{n+1} - u_i^n}{\Delta t} + \alpha\frac{u_{i-2}^n - 6u_{i-1}^n + 3u_i^n + 2u_{i+1}^n}{6\Delta x} = 0$$

$$\Longleftrightarrow \frac{\partial u_i^n}{\partial t} + \frac{\Delta t}{2}\frac{\partial^2 u_i^n}{\partial t^2} + O(\Delta t^2) + \alpha\frac{\partial u_i^n}{\partial x} + \alpha\frac{\Delta x^3}{12}\frac{\partial^4 u_i^n}{\partial x^4} + O(\Delta x^4) = 0$$

Stability A numerical scheme is *stable* if perturbations on the numerical solution are bounded. Among other methods, the von Neumann stability analysis consists in calculating the amplification factor $G = A^{n+1}/A^n$ of harmonic perturbations $w_i^n = A^n exp(\hat{i}\kappa i\Delta x)$, where $\hat{i} = -1$, imposed on the numerical solution $\overline{u_i}^n$. The total flow field writes: $u_i^n = \overline{u_i}^n + w_i^n$, and is controlled by the discrete equation. The scheme is stable if $\forall \kappa : |G| \leq 1$. For illustration, considering equation (1.11), the left-sided discretization introduced in (1.12) yields:

$$u_i^{n+1} = \left(1 - \frac{\alpha \Delta t}{\Delta x}\right) u_i^n + \frac{\alpha \Delta t}{\Delta x} u_{i-1}^n$$

and the amplification factor is:

$$G = \frac{A^{n+1}}{A^n} = \left(1 - \frac{\alpha \Delta t}{\Delta x}\right) + \frac{\alpha \Delta t}{\Delta x} \exp(-\hat{i}\kappa \Delta x)$$

In this example, the numerical scheme is stable for: $\alpha \geq 0$ and $\left|\frac{\alpha \Delta t}{\Delta x}\right| \leq 1$.

Alternative explicit schemes. For the same equation (1.11), the scheme constructed with the two-point centered interpolation (equation (1.13)) is shown always instable. The right-sided interpolation scheme:

$$\frac{u_i^{n+1} - u_i^n}{\Delta t} + \alpha \frac{u_{i+1}^n - u_i^n}{\Delta x} = 0 \tag{1.14}$$

is stable for: $\alpha \leq 0$ and $\left|\frac{\alpha \Delta t}{\Delta x}\right| \leq 1$. Considering the lagged schemes (1.12) and (1.14), beside the opposite condition on the sign of α, they both require $\left|\frac{\alpha \Delta t}{\Delta x}\right| \leq 1$. This is a CFL condition (CFL: Courant-Friedrichs-Lewy), commonly associated with explicit schemes. It limits the time step Δt for a given grid density Δx. The finer the grid density, the lower the time step, which increases the effective computational time.

Implicit scheme. The left-sided implicit scheme:

$$\Delta x \times \frac{u_i^{n+1} - u_i^n}{\Delta t} + \alpha(u_i^{n+1} - u_{i-1}^{n+1}) = 0$$

is stable for $\alpha \geq 0$. There is no CFL condition, and the time step only needs to be adapted to the temporal evolution rate of the flow. The implicit scheme is a pertinent alternative to the explicit methods: the much larger time steps allowed can compensate the higher computational effort required by the coupled resolution of all the grid points at each iteration.

Convergence A numerical scheme is said *convergent* if the discrete solution tend to the continuous solution, when the discretization steps (spatial and temporal) tend to zero. The equivalence theorem of Lax (cf. Hirsch [Hirsch, 2007]) states that for a well-posed initial-value linear problem, discretized in a consistent way, stability is an equivalent condition to convergence. Consequently, the practical analysis of linear numerical schemes will consist in a consistency and stability analysis.

1.8   Upwind interpolation

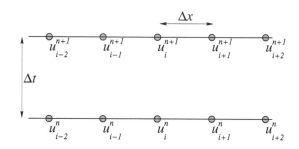

FIGURE 1.9: Discretization space.

The discretization analysis of equation (1.11) is particularly instructive. This equation is an elementary hyperbolic equation, describing advection at a constant velocity α. Value $\alpha > 0$ (resp. $\alpha < 0$) corresponds to propagation in the direction of positive x (resp. negative x). As a summary, the property analysis on three different explicit discretizations have shown:

- The explicit centrered scheme (1.13) is 2nd order spatially, but always instable.

- The explicit left-sided scheme (1.12) is 1st order spatially, and stable for: $\alpha \geq 0$ and $\left| \frac{\alpha \Delta t}{\Delta x} \right| \leq 1$.

- The explicit right-sided scheme (1.14) is 1st order spatially, and stable for: $\alpha \leq 0$ and $\left| \frac{\alpha \Delta t}{\Delta x} \right| \leq 1$.

For a given value of α imposed by the physical problem, among the three explicit schemes considered, only the discretization lagged in the upwind direction can be stable (left-sided for $\alpha > 0$, and right-sided for $\alpha < 0$). Since the flow equations are strongly influenced by hyperbolic phenomena, this stability property of the upwind schemes is particularly important. In practice, dealing with unsteady or time marching resolutions, the flow outside of the shear layers (boundary layers, ...) is mainly inviscid, and is governed by hyperbolic equations. Upwind schemes then provide stable methods, while the use of centered schemes requires artificial numerical viscosity.

The analysis of the canonical one-dimensional scalar hyperbolic equation is instructive, but generalization of upwinding to multidimensional non-linear vectorial equations is not straightforward. Different approaches have been proposed, and two of them are briefly presented below. First, the continuous equation (1.1) must be decomposed as:

$$\frac{\partial \mathbf{U}}{\partial t} + \nabla.\mathbf{F}_{visc.}(\mathbf{U}) + \nabla.\mathbf{F}_{inv.}(\mathbf{U}) = \mathbf{Q}$$

where $\mathbf{F}_{visc.}$ and $\mathbf{F}_{inv.}$ represents respectively the viscous and inviscid fluxes, with: $\mathbf{F} = \mathbf{F}_{visc.} + \mathbf{F}_{inv.}$. The upwinding only addresses the inviscid fluxes, associated with the hyperbolic tendency of the equations. In the following presentation of the approaches, only the inviscid part of the equation will be considered:

$$\frac{\partial \mathbf{U}}{\partial t} + \nabla.\mathbf{F}_{inv.}(\mathbf{U}) = 0$$

When decomposing the divergence, the equation writes:

$$\frac{\partial \mathbf{U}}{\partial t} + \frac{\partial \mathbf{F}}{\partial x} + \frac{\partial \mathbf{G}}{\partial y} + \frac{\partial \mathbf{H}}{\partial z} = 0$$

with $\mathbf{F}_{inv.} = (\mathbf{F}|\mathbf{G}|\mathbf{H})$. Moreover, only the x-component will be considered, the methods applying similarly to the other components. Consequently, for the presentation of the methods, the equation can be reduced to:

$$\frac{\partial \mathbf{U}}{\partial t} + \frac{\partial \mathbf{F}}{\partial x} = 0 \qquad (1.15)$$

where $\mathbf{F}$ is the x-component of the inviscid fluxes, generally non-linear.

1.8.1   Steger-Warming approach

This approach [Steger and Warming, 1981b] constitutes a rather direct generalization of the linear scalar case. First, the equation is written in a quasi-linear form:

$$\frac{\partial \mathbf{U}}{\partial t} + \mathbf{A}\frac{\partial \mathbf{U}}{\partial x} = 0$$

where $\mathbf{A}$ is the Jacobian matrix $\partial \mathbf{F}/\partial \mathbf{U}$. Verifying $\mathbf{F}(\mathbf{U})$ is a homogeneous function of degree one in $\mathbf{U}$ (i.e. $\forall \alpha : \mathbf{F}(\alpha \mathbf{U}) = \alpha \mathbf{F}(\mathbf{U})$), the Euler theorem yields: $\mathbf{F} = \mathbf{A}\mathbf{U}$. The equation then writes:

$$\frac{\partial \mathbf{U}}{\partial t} + \frac{\partial \mathbf{A}\mathbf{U}}{\partial x} = 0$$

where $\mathbf{A}$ can be diagonalized: $\mathbf{A} = \mathbf{L}^{-1}\mathbf{\Lambda}\mathbf{L}$. The diagonal matrix $\mathbf{\Lambda}$ is made of the real eigenvalues of $\mathbf{A}$ (the problem is hyperbolic). Splitting the matrix $\mathbf{\Lambda} = \mathbf{\Lambda}^+ + \mathbf{\Lambda}^-$, with $\mathbf{\Lambda}^+$ (resp. $\mathbf{\Lambda}^-$) made of the positive (resp. negative) eigenvalues, the flux writes:

$$\mathbf{F} = \mathbf{F}^| + \mathbf{F} = \mathbf{L}^{-1}\mathbf{\Lambda}^+\mathbf{L}\mathbf{U} + \mathbf{L}^{-1}\mathbf{\Lambda}^-\mathbf{L}\mathbf{U}$$

and each contribution $\mathbf{F}^{\pm}$ can be interpolated with an adequate upwind approach, i.e. a left-sided interpolation for $\mathbf{F}^{+}$ and a right-sided interpolation for $\mathbf{F}^{-}$.

Considering for example a uniform grid, as shown in Figure 1.9, with cell-vertex control volumes (width: Δx), the finite volume discretization of equation (1.15) writes:

$$\Delta x \frac{\partial \mathbf{U}_i}{\partial t} + \left(\mathbf{F}^{+}_{i+0.5} + \mathbf{F}^{-}_{i+0.5}\right) - \left(\mathbf{F}^{+}_{i-0.5} + \mathbf{F}^{-}_{i-0.5}\right) = \mathbf{0}$$

The flux decomposition guides the upwind interpolation (i.e. left-sided for $\mathbf{F}^{+}$, and right-sided for $\mathbf{F}^{-}$). For example, with a one-point interpolation:

$$\Delta x \frac{\partial \mathbf{U}_i}{\partial t} + \left(\mathbf{F}^{+}_{i} + \mathbf{F}^{-}_{i+1}\right) - \left(\mathbf{F}^{+}_{i-1} + \mathbf{F}^{-}_{i}\right) = \mathbf{0}$$

And the elementary explicit scheme yields:

$$\mathbf{U}^{n+1}_i = \mathbf{U}^{n}_i - \Delta t \frac{\mathbf{F}^{+n}_{i} - \mathbf{F}^{+n}_{i-1}}{\Delta x} - \Delta t \frac{\mathbf{F}^{-n}_{i+1} - \mathbf{F}^{-n}_{i}}{\Delta x}$$

Detailed examples of applications (e.g. mono-dimensional Euler equations) are given in [Steger and Warming, 1981b].

1.8.2   Roe scheme: approximate Riemann solver

A Riemann problem is an initial value problem where equation (1.15) applies to a piecewise constant initial field with a single discontinuity. Experimentally, such conditions can be obtained in shock tube configurations. Numerically, the flux interpolation at the boundary between two control volumes can be seen as a Riemann problem. Indeed, the evolution of the flux at the boundary is controlled by the conservative variables on each control volume, with a discontinuity at the interface in the general case. The analysis of such a problem requires to write the equation (1.15) in a quasi-linear form, as previously done in the Steger-Warming approach:

$$\frac{\partial \mathbf{U}}{\partial t} + \mathbf{A} \frac{\partial \mathbf{U}}{\partial x} = \mathbf{0}$$

where $\mathbf{A} = \partial \mathbf{F}/\partial \mathbf{U}$ is the Jacobian matrix. Instead of carrying the direct analysis of this problem (see Steger and Warming method in the previous paragraph), Roe [Roe, 1981] proposes to study an approximate problem:

$$\frac{\partial \mathbf{U}}{\partial t} + \hat{\mathbf{A}} \frac{\partial \mathbf{U}}{\partial x} = \mathbf{0}$$

where matrix $\hat{\mathbf{A}}(\mathbf{U}_L, \mathbf{U}_R)$, function of the variables on the left- (L) and right-hand side (R) of the discontinuity, must respect the following conditions:

- The eigenvalues of $\hat{\mathbf{A}}$ are real, and the eigenvectors are linearly independant.

- $\hat{\mathbf{A}}(\mathbf{U}_L, \mathbf{U}_R) \to \mathbf{A}(\mathbf{U}) = \partial \mathbf{F}/\partial \mathbf{U}$ when $\mathbf{U}_L \to \mathbf{U}$ and $\mathbf{U}_R \to \mathbf{U}$.

- $\forall \mathbf{U}_L, \mathbf{U}_R, \hat{\mathbf{A}}(\mathbf{U}_L, \mathbf{U}_R) \times (\mathbf{U}_R - \mathbf{U}_L) = \mathbf{F}_R - \mathbf{F}_L$.

The characteristic variable $\mathcal{U}$ is defined by:

$$\mathcal{U}_R - \mathcal{U}_L = \mathbf{L}\,(\mathbf{U}_R - \mathbf{U}_L)$$

where L is the left eigenvector matrix of $\hat{\mathbf{A}}$. Then, the third property above yields:

$$\mathbf{L}^{-1}\boldsymbol{\Lambda}\,(\mathcal{U}_R - \mathcal{U}_L) = \mathbf{F}_R - \mathbf{F}_L$$

Identifying the left (L) and right (R) states with generic indices i and $i+1$, the following interpolations are then proposed on the boundary:

$$\mathbf{F}_{i+0.5} = \mathbf{F}_i + \mathbf{L}^{-1}_{i+0.5}\boldsymbol{\Lambda}^{-}_{i+0.5}\,(\mathcal{U}_{i+1} - \mathcal{U}_i)$$

$$\mathbf{F}_{i+0.5} = \mathbf{F}_{i+1} - \mathbf{L}^{-1}_{i+0.5}\boldsymbol{\Lambda}^{+}_{i+0.5}\,(\mathcal{U}_{i+1} - \mathcal{U}_i)$$

An average of these interpolations makes the practical expression:

$$\mathbf{F}_{i+0.5} = \frac{1}{2}\,(\mathbf{F}_i + \mathbf{F}_{i+1}) - \frac{1}{2}\mathbf{L}^{-1}_{i+0.5}\left(\boldsymbol{\Lambda}^{+}_{i+0.5} - \boldsymbol{\Lambda}^{-}_{i+0.5}\right)(\mathcal{U}_{i+1} - \mathcal{U}_i)$$

$$\Longleftrightarrow \mathbf{F}_{i+0.5} = \frac{1}{2}\,(\mathbf{F}_i + \mathbf{F}_{i+1}) - \frac{|\hat{\mathbf{A}}_{i+0.5}|}{2}\,(\mathbf{U}_{i+1} - \mathbf{U}_i)$$

where: $|\hat{\mathbf{A}}_{i+0.5}| = \mathbf{L}^{-1}_{i+0.5}\left(\boldsymbol{\Lambda}^{+}_{i+0.5} - \boldsymbol{\Lambda}^{-}_{i+0.5}\right)\mathbf{L}_{i+0.5}$. As shown in [Roe, 1981] and [Peyret, 1996], a possible choice for $\hat{\mathbf{A}}$ is the Jacobian matrix $\mathbf{A}$ evaluated from the Roe's average variables, defined as:

$$\hat{\rho}_{i+0.5} = \sqrt{\rho_i \rho_{i+1}}$$

and for the other components:

$$\hat{U}_{i+0.5} = \frac{\sqrt{\rho_i}U_i + \sqrt{\rho_{i+1}}U_{i+1}}{\sqrt{\rho_i} + \sqrt{\rho_{i+1}}}$$

1.9 Particular case of structured grids

In this section, the special case of structured grids is considered. It allows to construct higher order schemes, in a rather simple way, by some analogy with the finite difference approach. The structured approach is popular in different disciplines, such as aerodynamics, because of its numerical efficiency. A first section is dedicated to the flux interpolation on regular grids. The following section then discusses the generalization to curvilinear grids.

1.9.1   Flux interpolation on regular grids

For illustration, the 2D control volume Ω_I, shown in Figure 1.10, is considered. The partial differential equation (1.1) yields in this case:

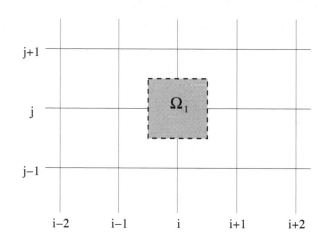

FIGURE 1.10: Flux interpolation on a 2D regular grid.

$$\frac{\partial \mathbf{U}}{\partial t} + \frac{\partial \mathbf{F}(\mathbf{U})}{\partial x} + \frac{\partial \mathbf{G}(\mathbf{U})}{\partial y} = \mathbf{Q} \tag{1.16}$$

and the integration on the control volume Ω_I gives:

$$\frac{\partial}{\partial t}(\mathbf{U}_I \Omega_I) + (\mathbf{F}_{i+0.5,j} - \mathbf{F}_{i-0.5,j})\Delta y + (\mathbf{G}_{i,j+0.5} - \mathbf{G}_{i,j-0.5})\Delta x = \mathbf{Q}_I \Omega_I$$

Setting $\mathbf{U}_I = \mathbf{U}_{i,j}$ and $\mathbf{Q}_I = \mathbf{Q}_{i,j}$, and dividing by $\Omega_I = \Delta x \times \Delta y$, the space discretization yields:

$$\frac{\partial \mathbf{U}_{i,j}}{\partial t} + \frac{\mathbf{F}_{i+0.5,j} - \mathbf{F}_{i-0.5,j}}{\Delta x} + \frac{\mathbf{G}_{i,j+0.5} - \mathbf{G}_{i,j\ 0.5}}{\Delta y} = \mathbf{Q}_{i,j} \tag{1.17}$$

As a first approach, the fluxes through the faces can be interpolated as:

$$\mathbf{F}_{i-0.5,j} = (\mathbf{F}_{i-1,j} + \mathbf{F}_{i,j})/2 \quad ; \quad \mathbf{F}_{i+0.5,j} = (\mathbf{F}_{i,j} + \mathbf{F}_{i+1,j})/2$$

$$\mathbf{G}_{i,j-0.5} = (\mathbf{G}_{i,j-1} + \mathbf{G}_{i,j})/2 \quad ; \quad \mathbf{G}_{i,j+0.5} = (\mathbf{G}_{i,j} + \mathbf{G}_{i,j+1})/2$$

This yields:

$$\frac{\partial \mathbf{U}_{i,j}}{\partial t} + \frac{\mathbf{F}_{i+1,j} - \mathbf{F}_{i-1,j}}{2\Delta x} + \frac{\mathbf{G}_{i,j+1} - \mathbf{G}_{i,j-1}}{2\Delta y} = \mathbf{Q}_{i,j} \tag{1.18}$$

The following Taylor series expansions can be drawn:

$$\frac{\mathbf{F}_{i+1,j} - \mathbf{F}_{i-1,j}}{2\Delta x} = \frac{\partial \mathbf{F}_{i,j}}{\partial x} + O(\Delta x^2) \quad ; \quad \frac{\mathbf{G}_{i,j+1} - \mathbf{G}_{i,j-1}}{2\Delta y} = \frac{\partial \mathbf{G}_{i,j}}{\partial y} + O(\Delta y^2)$$

This shows the semi-discrete equation (1.18) equals the partial differential equation (1.16), at point indexed (i,j), at order 2 in space.

In order to increase the discretisation precision up to order 4 in space, the interpolation must use the 4 neighboring points:

$$\mathbf{F}_{i-0.5,j} = \left(-\mathbf{F}_{i-2,j} + 7\mathbf{F}_{i-1,j} + 7\mathbf{F}_{i,j} - \mathbf{F}_{i+1,j}\right)/12 \qquad (1.19)$$

and similarly for the other fluxes in (1.17). This yields:

$$\frac{\partial \mathbf{U}_{i,j}}{\partial t} + \frac{\mathbf{F}_{i-2,j} - 8\mathbf{F}_{i-1,j} + 8\mathbf{F}_{i+1,j} - \mathbf{F}_{i+2,j}}{12\Delta x}$$
$$+ \frac{\mathbf{G}_{i,j-2} - 8\mathbf{G}_{i,j-1} + 8\mathbf{G}_{i,j+1} - \mathbf{G}_{i,j+2}}{12\Delta y} = \mathbf{Q}_{i,j} \qquad (1.20)$$

From the Taylor series expansions of the fluxes F and G at the different points, with respect to the point indexed (i,j), this semi-discrete equation is actually shown to be equivalent to:

$$\frac{\partial \mathbf{U}_{i,j}}{\partial t} + \frac{\partial \mathbf{F}_{i,j}}{\partial x} + O(\Delta x^4) + \frac{\partial \mathbf{G}_{i,j}}{\partial y} + O(\Delta y^4) = \mathbf{Q}_{i,j}$$

Which is the partial differential equation (1.16), at point indexed (i,j), with a spatial discretization error of order 4. It is important to note that formula (1.19) is not a 4th order interpolation of the flux F at point indexed $(i-0.5,j)$. The coefficients are designed to yield a 4th order discretization of the *complete equation* in which the interpolation is used.

1.9.2 Curvilinear grids

Regular grids are practically unsuitable for most of the flows. Indeed, common geometrical boundaries generally require the use of curvilinear grids. Discretization then differs from the formula obtained on regular grids, mainly concerning integration and flux interpolation. This is addressed below.

Integration Integration by substitution is a convenient tool to recover formulas similar to the Cartesian grid case. The integral equation on the control volume writes:

$$\frac{d}{dt}\int_\Omega \mathbf{U}dv + \int_{\delta\Omega} \mathbf{F}(\mathbf{U}).d\mathbf{S} = \int_\Omega \mathbf{Q}dv$$

then:

$$\frac{d}{dt}\int_{\Omega'} \mathbf{U}\sqrt{g}\prod_{j=1}^{3} d\xi_j \pm \sum_{i=1}^{3}\int_{S^{i\pm}} \mathbf{F}(\mathbf{U}).\mathbf{a}_i\sqrt{g}\prod_{j\neq i} d\xi_j = \int_{\Omega'} \mathbf{Q}\sqrt{g}\prod_{j=1}^{3} d\xi_j$$

where ξ_j $(j = 1..3)$ are the coordinates on an intermediate computational grid of Cartesian structure (as illustrated in Figure 1.11), $\mathbf{a}_i = grad\xi_i$ $(i = 1..3)$, and:

$$\sqrt{g} = \frac{D(x, y, z)}{D(\xi_1, \xi_2, \xi_3)}$$

is the Jacobian of the geometrical transformation. A possible set of intermediate coordinates (ξ_1, ξ_2, ξ_3) is provided by the point indices (i, j, k). Here, (ξ_1, ξ_2, ξ_3) will be supposed aligned with the indices (i, j, k), as illustrated in Figure 1.11. Then:

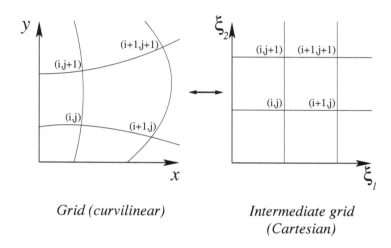

Grid (curvilinear)   *Intermediate grid (Cartesian)*

FIGURE 1.11: Curvilinear grid and intermediate Cartesian grid for integration by substitution.

$$\begin{pmatrix} \partial\xi_1/\partial x & \partial\xi_1/\partial y & \partial\xi_1/\partial z \\ \partial\xi_2/\partial x & \partial\xi_2/\partial y & \partial\xi_2/\partial z \\ \partial\xi_3/\partial x & \partial\xi_3/\partial y & \partial\xi_3/\partial z \end{pmatrix} = \begin{pmatrix} \partial x/\partial\xi_1 & \partial x/\partial\xi_2 & \partial x/\partial\xi_3 \\ \partial y/\partial\xi_1 & \partial y/\partial\xi_2 & \partial y/\partial\xi_3 \\ \partial z/\partial\xi_1 & \partial z/\partial\xi_2 & \partial z/\partial\xi_3 \end{pmatrix}^{-1}$$

The right-side matrix is easily computed at each point of the grid, using finite differences. For example:

$$\frac{\partial x}{\partial\xi_1} = \frac{x_{i+1,j,k} - x_{i-1,j,k}}{2\Delta\xi_1} + O(\Delta\xi_1^2)$$

Finally, inverting the matrix poses no difficulty (dimensions are 3×3), and when the grid is fixed, it only needs to be done at the beginning of the computation.

REMARK 1.3   This approach is not mandatory, it only constitutes a simplification when dealing with curvilinear structured grids. ⬜

Flux interpolation   As previously shown in section 1.9.1, higher orders of discretization can be achieved by the finite volume method on structured grids. In this approach, the integral variables on the control volume are assimilated with the values on the grid points, similarly to the finite difference method. As a consequence, when dealing with curvilinear grids, the interpolation will be influenced by the grid point locations, and significantly more complex formula are then obtained. Moreover, it must be kept in mind that the spatial orders of discretization are calculated on the complete equation, which must guide the optimization of the interpolation coefficients.

1.10   Boundary conditions

The interpolation strategies developed in the core of the computational domain are constrained on the external boundaries because of the absence of neighboring control volumes. Thus, specific interpolations must be set-up, in conjunction with the specification of the physical boundary conditions. As an illustration, the elementary hyperbolic equation (1.11) can be considered, with $\alpha > 0$ and a left-sided interpolation. The one-dimensional grid is shown in Figure 1.12 (points $i = 1..i_{max}$), together with a schematic representation of discretization (1.12). The discrete equation cannot apply directly on the first point ($i = 1$), because there is not neighbor on the left-hand side (i.e. u_0 is not defined).

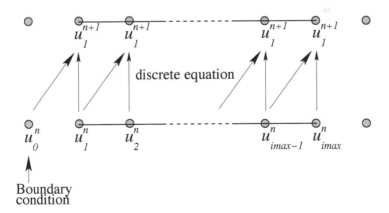

FIGURE 1.12: Numerical sketch of discretization (1.12).

Boundary conditions must be defined in agreement with the nature of the equations. Only physically ingoing quantities should be imposed. In the example above, the equation represents advection of u in the positive x direction ($\alpha > 0$). Consequently, the value of u can be imposed on the left-hand side boundary of the domain. This can be done by setting the value of u_0^n to the expected boundary condition. On the other end of the domain ($i = i_{max}$), no boundary condition can be imposed. Fortunately, the present interpolation naturally requires no data on this side. In a more general case, when no boundary condition can be imposed on a given frontier, a proper upwinding should be defined, using only points from the interior of the domain. Such a specific interpolation strategy can apply on the frontier only, while the interpolation remains unchanged in the core of the domain. More details about the analysis of the equations and the specification of the boundary conditions can be found in [Hirsch, 2007].

Chapter 2

Weighted residuals methods

Fabien Godeferd

Ecole Centrale de Lyon, LMFA
36 avenue Guy de Collongue
69134 Ecully Cedex, France

2.1 Introduction

Among the numerical methods used for the spatial resolution of partial differential equations (PDE), the finite differences method (FD) and the finite elements (FE) method use a *local* approximation for the functions; for instance, the FE are based on low-order polynomials, whose support—the subdomain over which the function is not zero—is very reduced with respect to the complete resolution domain. This allows to remove the strong coupling between the approximation functions over each finite element. The corresponding matrix systems are therefore more easily solved, their being very often block-diagonal and sparse. Finite volumes methods (FV) use the conservative formulation of the PDEs, and evaluate integrals over the elementary volumes, but eventually require FD approximations for the differential operators.

These methods use an approximation of the system of equations itself. Another possibility is to keep the system of PDEs unchanged, but specify an approximation of the *solution* in a specific given form. Most often, the approximation function is written as a series of analytically known functions chosen in a functional space of which they form a complete base, e.g. polynomials or other families of functions. The formal projection of the system of equations onto the reduced chosen functional space defines the essence of the method. The original PDEs are then transformed into equations for the *coefficients* of the development, and often become a simpler system of coupled ordinary differential equations (ODE). The weighted residuals method (WR) defines the condition for the approximation solution to converge to the exact solution when increasing the number of degrees of freedom, that is the number of coefficients in the series. These methods are well suited for simple problems, for rapidly achieving an accurate solution, and are described for instance in [Chung, 2002] or [Fletcher, 1991] among the other methods (FV, FE, FD) which are local.

Contrary to local methods, spectral methods presented in Chapter 3 are also a kind of *global methods*, in the sense that they use high-order polynomials or Fourier series. The main difference with a general WR method is the careful choice of the functional space such that the development functions are mutually orthogonal. When the context allows, the result is an unrivaled accuracy with respect to local methods, at equivalent resolution; in other terms a very fast convergence. Moreover, when the computation is done for very large domains, this added accuracy allows to limit the increase in resolution, which therefore also saves computational resources (computational memory, storage, and the related operations). In addition, the numerical dissipation, which is artificial with respect to physical dissipation terms in the solved equations, is limited to a very low value. This is a major advantage for the computation of high Reynolds number flows (roughly, high energy flows), since the Reynolds number is inversely proportional to the molecular viscosity: $\Re \propto 1/\nu$. Local methods such as finite differences require high-order discretization to avoid this artifact. An illustration of a turbulent flow computed with a Fourier-Fourier-Chebyshev collocation method is presented in Figure 3.1. Not only a spectral method introduces very little numerical dissipation, but it is also a low-dispersive technique, a property which is important for treating phenomena such as wave propagation. The propagation and reflection of inertial waves in a rotating channel is illustrated on Figure 3.2.

Two characteristics of the spectral methods prevent their use in the computation of complex flows such as industrial or aeronautic flows. The spectral methods "suffer" more than FD from the presence of solid boundaries with complex geometry, since they require to find functional bases adapted to each different geometry. Secondly, strong discontinuities in the solution are treated with difficulty by series expansions. This is the case e.g. in the presence of shocks in compressible flows. The treatment of complex geometries or discontinuities can be addressed by an extension of the spectral method which is called "spectral elements method", in which the domain is decomposed into sub-domains, or elements, over which the spectral method is applied, requiring a suitable formulation of the matching conditions at the boundary of the elements.

The spectral methods are generally classified in three categories: "Galerkin" (a denomination which is also found in FE), "tau" and "collocation" (also "pseudo-spectral") method. The Galerkin and tau techniques are based on the *coefficients* of the functional global development of the approximation solution, whereas the collocation method uses the *values* of this global development at suitable discretization points. This last method will be discussed more in the following in relation with the treatment of nonlinear terms appearing in PDEs. The spectral methods are very efficient by specific choices of the development functions, since they often allow to convert differential space operators into algebraic operators, and with a smart use of the properties of the development functions, which can be trigonometric polynomials such as the Fourier, Chebyshev, or Legendre polynomials.

A significant advantage of using trigonometric polynomials is the possibility to implement numerically the spectral method by using the algorithm of fast Fourier transforms (FFT) for performing the required discrete Fourier transformations. This algorithm was first proposed by Cooley & Tukey in 1965. Using the FFT reduces the numerical cost for evaluating a Fourier transform from $8N^2$ real arithmetic operations to about $5N \log_2 N$ when the resolution N is a power of two.

In the following, after the principles of the weighted residuals method are given, a presentation of the spectral methods is given, illustrated with examples of application. The reader will thus get an introductory point of view on the spectral methods, before entering more complex matter presented in further readings of highly specialized and exhaustive presentations such as that of [Canuto et al., 1987] or of [Boyd, 2001], the latter opus specializing in Fourier and Chebyshev spectral methods.

2.2   Principles of the weighted residuals method

The discretization methods based on weighted residuals are procedures that approach the solution of an ensemble of differential—or integral—equations of the form

$$Au = f \tag{2.1}$$

on the domain Ω, with given boundary conditions

$$Su = g \tag{2.2}$$

on the sub-manifold Γ of Ω. The unknown function u depends on the space variable $x = (x_1, x_2, \cdots, x_n)$. The function u as the exact solution of (2.1) and (2.2) is approached by an estimate u_N built in the form of a combination of base functions $\Phi(x)$:

$$u_N(x) = \sum_{k=1}^{N} \alpha_k \Phi_k(x) \tag{2.3}$$

in which the α_k are the N free parameters, and the functions $\Phi(x)$ form an independent functional base (chosen in a space yet to be defined). We impose that the functions $\Phi_k(x)$ satisfy the boundary conditions and are smooth enough and with the needed properties required by equation (2.1). By inserting the estimate (2.3) in (2.1), one obtains the error function, which is the departure from the exact solution, also called the *residual*

$$\varepsilon = Au_N - f \neq 0$$

The method consists in canceling ε in a certain sense, that is canceling the integral of the residual function weighted as follows:

$$\langle \varepsilon, W_i \rangle = \int_\Omega \varepsilon(x) W_i(x) \mathrm{d}x = 0, \qquad i = 1, \cdots, N \qquad (2.4)$$

in which the weighting functions are the $W_i(x)$'s. The various possibilities in choosing the weighting functions yield various kinds of methods. One could thus achieve a finite volume method, a finite elements method, a spectral method, and even recover finite differences. The set of equations (2.4) correspond to the projection of the residual onto the space generated by the weighting functions $W_i(x)$, $i = 1, \cdots, N$, using the inner product in function space $\langle f, g \rangle = \int_\Omega f(x), g(x) \mathrm{d}x$. In a sense, this corresponds to defining a projector over a functional space.

2.3   Collocation or pseudo-spectral method

As indicated by its name, in the collocation method one chooses to satisfy exactly equation (2.1) in a finite number of points chosen in advance, in the same amount as the unknowns α_k to be determined in the development of the approximation function. This corresponds to the specific choice

$$W_i(x) = \delta(x_i), \qquad \text{Dirac functions in } x_i$$

and one obtains N equations in the form

$$\int_\Omega \varepsilon(x) W_i(x) \mathrm{d}x = \varepsilon(x_i) = 0, \qquad i = 1, \cdots, N$$

2.4   Least squares method

This methods consists in minimizing the square of the norm of the residual function

$$F = \langle \varepsilon, \varepsilon \rangle = \int_\Omega \varepsilon^2 \mathrm{d}x$$

and if u is approached by $u_N(x) = \sum_{k=1}^N \alpha_k \Phi_k(x)$, then the minimum is obtained by solving the following equations in the unknowns α_k, $k = 1, \cdots, N$:

$$\frac{\partial F}{\partial \alpha_i} = 0 \qquad i = 1, \cdots, N$$

(in which it is still necessary to check that the extremum is indeed a minimum). This also gives

$$\frac{\partial}{\partial \alpha_i} \langle \varepsilon, \varepsilon \rangle = 0, \qquad i = 1, \cdots, N$$

whence for the general equation $Au - f = 0$, one obtains

$$\frac{\partial}{\partial \alpha_i} (\langle A \sum \alpha_k \Phi_k, A \sum \alpha_k \Phi_k \rangle - 2 \langle A \sum \alpha_k \Phi_k, f \rangle + \langle f, f \rangle = 0$$

In the case in which A is a linear operator, the equation becomes simply

$$\langle Au - f, A\Phi_i \rangle = 0, \qquad i = 1, \cdots, N$$

The weighting functions W_i, corresponding to the least squares method, are therefore in this case the images by A of the Φ_is. They are also automatically obtained as $W_i(x) = \frac{\partial \varepsilon(x)}{\partial \alpha_i}$.

2.5   Method of moments

As hinted by its name, in this method the weighting functions are chosen to be the elements of the suite $\{1, x, x^2, \cdots, x^N\}$, whence the equations

$$0 = \int_\Omega \varepsilon \mathrm{d}x = \int_\Omega \varepsilon x \mathrm{d}x = \cdots = \int_\Omega \varepsilon x^N \mathrm{d}x$$

corresponding to the cancelling of the N successive moments of the residual function $\varepsilon(x)$.

2.6   Galerkin approximation

In the celebrated method of Galerkin, the weighting functions W_i are *identical* to the base functions Φ_i, or also $W_i(x) = \frac{\partial u_N(x)}{\partial \alpha_i}$. One obtains the system of equations

$$\int_\Omega \varepsilon \Phi_i(x) dx = 0 \qquad i = 1, \cdots, N$$

REMARK 2.1   It is possible to provide a "physical" interpretation by defining a variation $\delta u = \Phi_1 \delta \alpha_1 + \Phi_2 \delta \alpha_2 + \cdots + \Phi_N \delta \alpha_N$ with arbitrary $\delta \alpha_i$s. The Galerkin method therefore implies

$$\int_\Omega (Au - f)\delta u dx = 0$$

where δu is a virtual variation compatible with the boundary conditions. With the background of continuum mechanics in mind, these variations can be seen as virtual displacements, and the Galerkin method therefore derives from the principle of virtual works.   ▯

2.7   Subdomains

In this method, the weighting functions are the characteristic functions of subdomains Ω_i that constitute a mapping of the initial resolution domain Ω, namely $W_i(x) = 1$ if $x \in \Omega_i$, 0 otherwise; this yields

$$\int_\Omega \varepsilon W_i(x) dx = \int_{\Omega_i} \varepsilon dx = 0 \qquad i = 1, \cdots, N.$$

Unlike the collocation method, in this case the integral is still continuous, and, provided the equations are written in conservative form, the resulting equations are similar to those obtained in the finite volumes method described in Chapter 1.

2.8   An example

We illustrate the way the WR method is applied on a very simple example of ordinary differential equation for a purely spatial linear problem. One seeks an approximate solution $u_N(x)$ of the following equation with the boundary conditions that prescribe a unique exact solution $u(x)$:

$$\frac{du}{dx} + u = 2x , \qquad x \in [0,1] , \tag{2.5}$$

$$u(1) = \frac{1}{2} .$$

The exact solution can be computed easily: $u(x) = 2x - 2 + \frac{1}{2}e^{1-x}$. We shall use an approximation function which satisfies the boundary conditions, and allow two degrees of freedom in the development space, in the form:

$$u_N(x) = \alpha_1(x - x^2) + \alpha_2(x - 1) + \frac{1}{2}x \tag{2.6}$$

where the first two coefficients of the functional development are α_1 and α_2, and the first two functions are the linearly independent polynomials $x - x^2$ and $x - 1$. The third term $\frac{1}{2}x$ ensures the boundary conditions without having to carry an additional explicit constraint on the approximation function. Upon injecting this approximation in equation (2.5), one obtains the residual

$$\varepsilon(x) = \alpha_1(1 - 2x) + \alpha_2 + 1/2 + \alpha_1(x - x^2) + \alpha_2(x - 1) + x/2 - 2x^2$$

which is not necessarily zero since u_N has no reason to be a solution of the system (2.5) at this stage. The Galerkin method suggests to use the weighting functions defined from $W_i(x) = \frac{du_N}{d\alpha_i}$, which yields $W_1(x) = x - x^2$, $W_2(x) = x - 1$. And by canceling the integrals of the weighted residual $I_i = \int_0^1 W_i(x)\varepsilon(x)\,dx$, one obtains the system of equations in terms of the unknown α_is

$$\frac{1}{30}\alpha_1 + \frac{1}{12}\alpha_2 = -\frac{1}{40} \tag{2.7}$$

$$-\frac{1}{4}\alpha_1 - \frac{1}{6}\alpha_2 = \frac{1}{6} , \tag{2.8}$$

which provides the values for the coefficients of the Galerkin approximation, which writes $u_N^{\text{galerkine}}(x) = -\frac{7}{11}(x - x^2) - \frac{1}{22}(x - 1) + \frac{1}{2}x$. In a similar way, the approximation by the least squares method is obtained, with slightly modified values for the coefficients. The approximation by the method of subdomains depends on the choice of the intervals of the resolution domain; we choose for instance the weighting functions such that $w_1(x) = 1$ on $[0, 1/2]$ and $w_1(x) = 0$ on $]1/2, 1]$, and $w_2(x) = 0$ on $[0, 1/2]$ and $w_2(x) = 1$ on $]1/2, 1]$. The resulting approximating solutions are compared for the values obtained at $x = \frac{9}{10}$, given in Table 2.1, as well as using a Galerkin approximation with an additional degree of freedom. One observes on the table that a good accuracy is obtained already with only two coefficients in the development, but an order of magnitude in the accuracy is obtained when using an additional coefficient, illustrating the fast convergence of the WR methods.

	exact	Galerkin	least squares	subdomains	Galerkin (three coefficients)
$u_N(\frac{9}{10})$	0.352	0.397	0.377	0.378	0.358

Table 2.1: Solutions of problem (2.5) evaluated at $x = \frac{9}{10}$ for the different choices of the WR method. The last column corresponds to the Galerkin method with three degrees of freedom using $u_N(x) = \alpha_1(x - x^2) + \alpha_2(x - 1) + \alpha_3(x^2 - x^3) + \frac{1}{2}x$.

Chapter 3

Spectral methods

Fabien Godeferd
Ecole Centrale de Lyon, LMFA
36 avenue Guy de Collongue
69134 Ecully Cedex, France

3.1 Introduction

We present in this chapter the spectral methods, which are a kind of weighted residuals methods with a specific choice of orthogonal development functions. We start by providing the background principles underlying the three main spectral methods applied to a linear problem. Then, we specialize the presentation to the case of Burgers equation and of Helmholtz-type equations, giving progressively more details in the spectral technique for the case of Fourier and Chebyshev polynomials. The implicit time-discretization of the heat equation is presented exhaustively, so that the practical implementation of the Chebyshev-tau method for this particular case can be done without particular difficulty.

3.2 Linear problem: Galerkin, tau, and collocation methods

The problem to be solved mixes the partial differential equation with prescribed initial conditions and boundary conditions. We choose in a first time to consider the linear problem for simplicity, the treatment of nonlinear terms is postponed to a further section. Denoting f the possible external forcing

FIGURE 3.1: Iso-surfaces of longitudinal velocity for the fully-developed high
Reynolds number turbulent flow in a plane channel. A spectral Fourier-
Chebyshev collocation method with 256^3 mesh points is used. A simulation
with finite differences at the same resolution would require a sixth-order spatial scheme to achieve an equivalent accuracy.

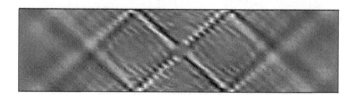

FIGURE 3.2: Inertial waves propagating in a fluid at rest within a two-
dimensional plane channel rotating in a direction normal to the solid walls.
A Fourier-Chebyshev collocation technique with 64^2 mesh points is used.

term,

$$\frac{\partial u}{\partial t}(x,t) = L(x,t)u(x,t) + f(x,t) \qquad x \in \mathcal{D}, t \geq 0 \qquad (3.1)$$

$$B(x)u(x,t) = 0 \qquad x \in \partial\mathcal{D}, t > 0 \qquad (3.2)$$

$$u(x,0) = g(x) \qquad x \in \mathcal{D} . \qquad (3.3)$$

The resolution of this system is performed over the spatial domain $\mathcal{D}$ with boundaries $\partial\mathcal{D}$, $L(x,t)$ is a linear differential operator in space, and $B(x)$ represents the boundary conditions linear operator, independent of time. The unknown function u here is a scalar for the sake of simplicity, however the extension to a multi-variable problem is straightforward. In the short present introduction, we limit the scope to a linear problem with homogeneous boundary conditions. Since the spectral space of coefficients is discretized, although not the physical space, the *semi-discrete* approximation of (3.1) is

$$\frac{\partial u_N}{\partial t}(x,t) = L_N u_N(x,t) + f_N(x,t) \qquad (3.4)$$

in which at each time t the approximation function $u_N(x,t)$ belongs to a N-dimensional subspace $\mathcal{B}_N$ of the functional space $\mathcal{B}$ of functions that verify the boundary conditions (3.2) (within the initial Hilbert space $\mathcal{H}$), and L_N is a linear operator acting over $\mathcal{H}$ and resulting in $\mathcal{B}_N$ of the form

$$L_N = P_N L P_N$$

The operator P_N is therefore the projector from $\mathcal{H}$ to $\mathcal{B}_N$ such that for each function f, the associated discrete function is $f_N = P_N f$. The choice of a spectral method thus goes through the choice of an approximation space $\mathcal{B}_N$ and of a projection operator P_N, in the same way we have chosen a set of development functions and a set of weighting functions in Chapter 2.

In the following, we describe the corresponding spectral methods that are more commonly used.

3.2.1 Galerkin approximation

Let us describe shortly the way one sets up a Galerkin approximation for the problem (3.1)-(3.3). As done in the general weighted residual method in Chapter 2, the approximation solution u_N is sought in the form of a truncated series

$$u_N(x,t) = \sum_{m=1}^{N} \alpha_m(t)\Phi_m(x) \qquad (3.5)$$

where the time-dependent functions Φ_n are linearly independent and belong to $\mathcal{B}_N$. In this way, u_N immediately satisfies the boundary conditions. The unsteady coefficients α_n of the development are obtained by solving the Galerkin equation that are written as

$$\frac{d}{dt}\left(\Phi_n, u_N\right) = \left(\Phi_n, L u_N\right) + \left(\Phi_n, f\right) \qquad n = 1, \ldots, N , \qquad (3.6)$$

or with (3.5)

$$\sum_{m=1}^{N} \left(\Phi_n, \Phi_m\right) \frac{d\alpha_m}{dt} = \sum_{m=1}^{N} \alpha_m \left(\Phi_n, L\Phi_m\right) + \left(\Phi_n, f\right) . \qquad (3.7)$$

The inner product over $\mathcal{B}_N$ is denoted (,) as in Chapter 2. One transforms the previous equations, implicit in terms of the development coefficient, into an explicit form by applying the projector P_N such that

$$P_N u(x) = \sum_{n=1}^{N} \sum_{m=1}^{N} p_{nm} \left(\Phi_m, u\right) \Phi_n(x)$$

in which the coefficients p_{nm} are the components of the inverse matrix of order $N \times N$ whose components are (Φ_n, Φ_m).

In terms of the minimization of the weighted residual obtained for equation (3.4), this is therefore equivalent to the application of the orthogonal projection operator P_N on the residual, the problem may then be recast as:

$$u_N \in \mathcal{B}_N$$

$$P_N \left(\frac{\partial u_N}{\partial t}(x,t) - L_N(x,t) u_N(x,t) - f_N(x,t) \right) = 0 \qquad (3.8)$$

or, from a variational point of view,

$$u_N \in \mathcal{B}_N$$

$$\left(\frac{\partial u_N}{\partial t}(x,t) - L_N(x,t) u_N(x,t) - f_N(x,t), v \right) = 0 \qquad \forall v \in \mathcal{B}_N \qquad (3.9)$$

Equations (3.7) provide a system of ODEs in terms of the unknowns α_is. The time-evolving solution of the problem is obtained by:

- (a) projecting the initial conditions onto the development functions for obtaining the $\alpha_i(t = 0)$;

- (b) time-marching the α_is by finite differences of the time derivative $\frac{\partial}{\partial t}$;

- (c) obtain the solution at each desired time-step and space point by computing the series (3.5).

3.2.2 Tau method

This method was first introduced by Lanczos as soon as 1938. The base functions for the series development are the functions Φ_n $(n = 1, 2 \ldots)$, elements of a complete set of orthonormal functions. The solution $u_N(x, t)$ is therefore expressed as

$$u_N(x, t) = \sum_{n=1}^{N+k} \alpha_n(t) \Phi_n(x) . \tag{3.10}$$

The integer k is the number of boundary conditions applied to the approximation function, expressed as $Bu_N = 0$. The most significant difference between (3.10) for the tau approximation and (3.5) for the Galerkin approximation, lies in the fact that the functions Φ_n, in the tau method do not individually verify the boundary conditions (3.2). The boundary constraints are, here, explicit:

$$\sum_{n=1}^{N+k} \alpha_n B\Phi_n = 0 , \tag{3.11}$$

corresponding to as many as k equations, and are specified directly as constraints applied to the development coefficients a_n of the functions, such that they belong to $\mathcal{B}_N$. The projection operator is therefore defined as

$$P_N \left(\sum_{n=1}^{\infty} A_n \Phi_n \right) = \sum_{n=1}^{N} A_n \Phi_n + \sum_{m=1}^{k} b_m \Phi_{N+m} \tag{3.12}$$

relation in which the b_m $(m = 1, \ldots, k)$ are chosen so that they satisfy the boundary conditions in the form: $BP_N u = 0, \forall u \in \mathcal{H}$. The tau approximation of the problem (3.1)-(3.3) is fully defined by the relations (3.10), (3.11) and the N equations

$$\frac{d\alpha_n}{dt} = (\Phi_n, Lu_N) + (\Phi_n, f) \qquad n = 1, \ldots, N . \tag{3.13}$$

The name of this method, "tau", comes from the fact that the approximation function u_N thus obtained is also the solution of a modified problem of type

$$\frac{\partial u_N}{\partial t} = Lu_N + f + \sum_{p=1}^{\infty} \tau_p(t) \Phi_{N+p}(x)$$

where the coefficients τ_p have to be determined such that $u_N \in \mathcal{B}_N$.

3.2.3 Collocation method

This method is also known as the pseudo-spectral method, especially when applied to the Navier-Stokes conservation equations for fluid flows. In the collocation method, the projection operator P_N is determined after the definition of N points $x_1, x_2, \ldots, x_N$ in the interior of the domain $\mathcal{D}$ over which the solution is sought. These points are the "collocation points". Again, we use the functions Φ_n $(n = 1, \ldots, N)$ that constitute a base in the approximation space $\mathcal{B}_N$ such that $\det \Phi_n(x_m) \neq 0$. Then, for each $u \in \mathcal{H}$,

$$P_N u = \sum_{n=1}^{N} \alpha_n \Phi_n(x) \tag{3.14}$$

with coefficients α_n that are solutions of the equations

$$\sum_{n=1}^{N} \alpha_n \Phi_n(x_i) = u(x_i) \qquad i = 1, \ldots, N \ . \tag{3.15}$$

The collocation is thus characterized by the conditions $P_N u(x_i) = u(x_i)$ for $i = 1, \ldots, N$ and $P_N u \in \mathcal{B}_N$. It appears that the results of the collocation depend both on the choice of the points x_n and of the functions $\Phi_n(x)$ for $n = 1, \ldots, N$.

3.3   Applications: Fourier

We assume here that the domain has no solid boundary, that is the spatial domain is assumed to be infinite or periodic. In the latter case, the spectral methods are extremely efficient. One example of periodic conditions is found in the global circulation models for geophysical flows, used by meteorologists for the atmosphere around the Earth. Such models are spectral, as opposed to predictions obtained by finite differences formulas for local meteorology.

3.3.1   Fourier Galerkin approximation for the Burgers equation

We seek here the solution of a problem constituted of the nonlinear Burgers equation with given boundary conditions. We recall that Burgers' equation corresponds to a "pressureless" fluid:

$$\frac{\partial u}{\partial t} + u \frac{\partial u}{\partial x} - \nu \frac{\partial^2 u}{\partial x^2} = 0 \tag{3.16}$$
$$u(x, 0) = u_0(x)$$

in which ν is a positive constant that represents the molecular viscosity in the fluid. The domain is assumed to be periodic in space over the interval $[0, 2\pi]$. The functional space which is chosen is the set of trigonometric polynomials of order $\leq N/2$. The approximate solution function u_N with N degrees of freedom then writes

$$u_N(x,t) = \sum_{k=-N/2}^{N/2-1} \hat{u}_k(t)e^{ikx} . \qquad (3.17)$$

The unknowns of the problem are the spectral coefficients $\hat{u}_k(t)$ with $k = -N/2, \ldots, N/2 - 1$, renamed with respect to the previous notation (α_is) to conform to the classical notation for Fourier spectral coefficients; the functions e^{ikx} correspond to the development functions $\Phi_n(x)$. The ordinary differential equations for their evolution are obtained upon writing that the residual is orthogonal to each of the chosen weighting functions, identical as the development functions in the case of the Galerkin method. This implies that

$$\int_0^{2\pi} \left(\frac{\partial u_N}{\partial t} + u_N \frac{\partial u_N}{\partial x} - \nu \frac{\partial^2 u_N}{\partial x^2} \right) e^{-ikx} \, dx = 0 \qquad k = -\frac{N}{2}, \ldots, \frac{N}{2} - 1 ,$$
$$(3.18)$$

in which the inner product of complex-valued functions is defined as $(f, g) = \int_0^{2\pi} f(x)g^*(x)dx$. Since by construction the test functions and the development functions are orthogonal, one thus obtains

$$\frac{\partial \hat{u}_k}{\partial t} + \left(\widehat{u_N \frac{\partial u_N}{\partial x}} \right)_k + k^2 \nu \hat{u}_k = 0 \qquad k = -\frac{N}{2}, \ldots, \frac{N}{2} - 1 , \qquad (3.19)$$

with the following definition of the transform of the nonlinear term:

$$\left(\widehat{u_N \frac{\partial u_N}{\partial x}} \right)_k = \frac{1}{2\pi} \int_0^{2\pi} u_N \frac{\partial u_N}{\partial x} e^{-ikx} \, dx . \qquad (3.20)$$

The initial conditions are readily obtained in spectral space by

$$\hat{u}_k(0) = \frac{1}{2\pi} \int_0^{2\pi} u(x,0)e^{-ikx} \, dx . \qquad (3.21)$$

We postpone to section 3.6.2 the description of the evaluation of the nonlinear convolution term (3.20). In equation (3.19), we note an interesting feature of the spectral Fourier transform, in that it converts spatial differential operators, e.g. here $\nu \frac{\partial^2 u}{\partial x^2}$, into algebraic operators, here $k^2 \nu \hat{u}$. This is what allows to convert PDEs in physical space into ODEs in spectral space in terms of the development coefficients. Provided one knows how to compute the nonlinear term (3.20), explained in section 3.6, the resolution of (3.19), starting from the initial conditions (3.21), only involves time-marching by discretizing the time derivative with a classical FD scheme; e.g. using a time-step Δt

$$\frac{\partial \hat{u}}{\partial t} = \frac{\hat{u}_k^{n+1} - \hat{u}_k^n}{\Delta t} + O(\Delta t) .$$

3.3.2   Fourier collocation for Burgers' equation

Let us consider the previous example (3.16) for which the approximation function u_N is this time represented by its values at the spatial points $x_j = 2\pi j/N$, $j = 0, \ldots, N-1$, which constitute the chosen mesh points. It is of course possible to switch from the values of u_N at the mesh points to the Fourier coefficients by the relations (3.77) and (3.79). The collocation method requires the verification of equation (3.16) at each of the mesh point, that is

$$\left[\frac{\partial u_N}{\partial t} + u_N \frac{\partial u_N}{\partial x} - \nu \frac{\partial^2 u_N}{\partial x^2} \right]_{x=x_j} = 0 \qquad j = 0, 1, \ldots, N-1 . \qquad (3.22)$$

The initial conditions are obviously $u_N(x_j, 0) = u_0(x_j)$.

Let us note that the orthogonal projection operator is already included in equation (3.22), with respect to the following discrete inner product:

$$(u, v)_N = \frac{2\pi}{N} \sum_{j=0}^{N-1} u(x_j) v^*(x_j) .$$

(* represents the complex conjugate.) In this method, the derivative $\partial u_N/\partial x$ if efficiently evaluated by the spectral discrete differentiation operator $\mathcal{D}_N$ defined as:

$$(\mathcal{D}_N u)_l = \sum_{k=-N/2}^{N/2-1} \alpha_k e^{2ikl\pi/N} \qquad l = 0, 1, \ldots, N-1 \qquad (3.23)$$

$$\alpha_k = \frac{ik}{N} \sum_{j=0}^{N-1} u_j e^{-2ikj\pi/N} . \qquad (3.24)$$

This differentiation operation may be represented by a matrix such that:

$$(\mathcal{D}_N u)_l = \sum_{j=0}^{N-1} (\mathcal{D}_N)_{lj} u_j \qquad (3.25)$$

with matrix coefficients explicitly obtained with the following formula:

$$(\mathcal{D}_N)_{lj} = \begin{cases} \frac{1}{2}(-1)^{l+j} \cot\left[\frac{(l-j)\pi}{N} \right] & l \neq j \\ 0 & l = j . \end{cases} \qquad (3.26)$$

The resulting matrix is therefore skew-symmetric, and its eigenvalues are the iks, $k = -N/2 + 1, \ldots, N/2 - 1$, where 0 is an eigenvalue with order of multiplicity 2, corresponding to the eigenfunctions 1 and $\cos(Nx/2)$. (The case of an odd number of collocation points is slightly different, but is very seldom used.)

Let us consider again equation (3.16) by rewriting Burgers equation in an equivalent form, although slightly different, with a conservative expression of the nonlinear term:

$$\frac{\partial u}{\partial t} + \frac{1}{2}\frac{\partial}{\partial x}(u^2) - \nu\frac{\partial^2 u}{\partial x^2} = 0 \ . \tag{3.27}$$

The nonlinear operator is therefore approximated here by the discretized operator $(1/2)\mathcal{D}_N\left[(u^2)\right]$. The collocation discretization of the problem therefore becomes

$$\frac{\partial u_N}{\partial t} + \frac{1}{2}\mathcal{D}_N(u_N \boxtimes u_N) - \nu\mathcal{D}_N^2 u_N = 0 \tag{3.28}$$

whereas it amounted to

$$\frac{\partial u_N}{\partial t} + u_N \boxtimes \mathcal{D}_N u_N - \nu\mathcal{D}_N^2 u_N = 0 \tag{3.29}$$

in the previous case. The operator $\boxtimes$ stands for the point-wise product in discretized space. In equation (3.28), the nonlinear term is evaluated by a square power (point-wise, that is to say), then by the differentiation. The resulting system of equations (3.28) is therefore *not* equivalent to equations (3.29), since the operators $\boxtimes$ and $\mathcal{D}_N$ do not necessarily commute. This is not the case for the Galerkin method which provides the same system to be solved numerically, independently of the precise way one writes the partial differential equations of the problem.

3.4 Applications: Chebyshev

The Chebyshev polynomials are a kind of trigonometric polynomials with uneven variations in space (see Figure 3.3) and limit values that prove useful in spectral methods (see the illustrations of the computation of fluid motion in a channel flow 3.2 and 3.1). A Chebyshev polynomial of order n is the function $T_n(x) = \cos(n \arccos x)$ for $x \in [-1, 1]$; for example $T_0(x) = 1$, $T_1(x) = x$, $T_2(x) = 2x^2 - 1$, $\ldots$, with particularly useful recurrence relations, e.g. $T_{n+1}(x) = 2xT_n(x) - T_{n-1}(x)$.

3.4.1 Computation of derivatives

Among many nice properties of the Chebyshev polynomials and the series decomposition of functions on their base, the recurrence relations can be an efficient way for obtaining the coefficients of the development of derivatives of a function $f(x)$ by using the coefficients of the development of $f(x)$ itself.

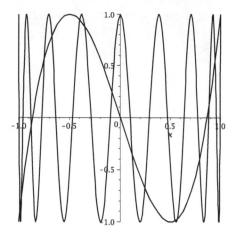

FIGURE 3.3: Two Chebyshev polynomials of order $k = 3$ and order $k = 16$ showing the uneven distribution of extrema and zeros, upon which can be defined grid points for an efficient Chebyshev collocation method. In addition, non-zero values at the boundaries of the interval allow to specify Dirichlet boundary conditions to the problem whenever necessary, in the context of a Chebyshev-tau method.

For instance, in the case of the Chebyshev polynomials, the coefficients of the development of the derivative are

$$\alpha_n^{(1)} = \frac{2}{c_n} \sum_{\substack{p = n + 1 \\ p + n = \text{odd}}}^{N} p \alpha_p \tag{3.30}$$

(see below the details on the coefficients c_n) and they verify the recurrence relation

$$c_n \alpha_n^{(1)} = \alpha_{n+2}^{(1)} + 2(n + 1)\alpha_{n+1} \tag{3.31}$$

with the understanding that, due to the truncation of the series, $\alpha_N^{(1)} = \alpha(1)_{N+1} = 0$. In this way, $\alpha_n^{(1)}$ can be evaluated for each n in N operations. The existence of the recurrence relation (3.31) is guaranteed by the properties of the Chebyshev polynomials, namely the relation

$$2T_n = \frac{T'_{n+1}}{n + 1} - \frac{T'_{n-1}}{n - 1} \qquad \text{for } n > 1 \ .$$

By writing analogous recurrence relations, on can also efficiently evaluate the coefficients of arbitrary order derivatives.

3.4.2 Chebyshev tau approximation for Burgers' equation

The solution of equation (3.16) is here sought over the interval $[-1, 1]$ such that it satisfies Dirichlet boundary conditions

$$u(-1, t) = u(1, t) = 0$$

The test functions, i.e. the set of functions over which one looks for the solution, are polynomials of degree $\leq N$, a space which we denote $\mathbb{P}_N$, which take a zero value at $x = \pm 1$. The discrete solution of the problem is expressed in terms of a Chebyshev series:

$$u_N(x, t) = \sum_{k=0}^{N} \hat{u}_k(t) T_k(x) \tag{3.32}$$

One ensures that Burgers equation is verified by specifying that the residual be orthogonal to the weighting (or test) functions, chosen within the set $\mathbb{P}_{N-2}$ (accounting for the two boundary conditions). This yields

$$\int_{-1}^{1} \left(\frac{\partial u_N}{\partial t} + u_N \frac{\partial u_N}{\partial x} - \nu \frac{\partial^2 u_N}{\partial x^2} \right) T_k(x)(1-x^2)^{-1/2} \, dx = 0 \qquad k = 0, \dots, N-2 \tag{3.33}$$

We note that the multiplying factors functions $w(x) = (1-x^2)^{-1/2}$ ensure the orthogonality of the Chebyshev polynomials with one another, thus requiring a slightly different inner product. Moreover, the boundary conditions impose the two additional relations:

$$u_N(-1, t) = u_N(1, t) = 0 \ .$$

Equation (3.33) reduces to

$$\frac{\partial \hat{u}_k}{\partial t} + \left(\widehat{u_N \frac{\partial u_N}{\partial x}} \right)_k - \nu \hat{u}_k^{(2)} = 0 \qquad k = 0, 1, \dots, N-2 \tag{3.34}$$

where the second order derivative coefficients $\hat{u}_k^{(2)}$ are obtained from the $\hat{u}_k$ by using the known formulas for differentiating Chebyshev polynomials, that is

$$\hat{u}_k^{(2)} = \frac{1}{c_k} \sum_{\substack{p = k+2 \\ p+k \text{ even}}}^{\infty} p(p^2 - k^2) \hat{u}_p$$

with

$$c_k = \begin{cases} 2 & \text{if } k = 0 \\ 1 & \text{if } k \geq 1 \end{cases} \ ;$$

and the nonlinear term is formally expressed as

$$\left(\widehat{u_N \frac{\partial u_N}{\partial x}} \right)_k = \frac{?}{\pi c_k} \int_{-1}^{1} \left(u_N \frac{\partial u_N}{\partial x} \right) T_k(x)(1 - x^2)^{-1/2} \, dx \ . \tag{3.35}$$

Note that, when implementing the method, it is interesting to define a *vector* of coefficients c_k and use its components, instead of having to test on the value of k, thus saving computational time. The boundary conditions are stated through the relation $T_k(\pm 1) = (\pm 1)^k$ as follows:

$$\sum_{k=0}^{N} \hat{u}_k = 0 \qquad (3.36)$$

$$\sum_{k=0}^{N} (-1)^k \hat{u}_k = 0 \ ,$$

and the initial conditions are given by

$$\hat{u}_k(0) = \frac{2}{\pi c_k} \int_{-1}^{1} u_0(x) T_k(x)(1 - x^2)^{-1/2} \, dx \qquad k = 0, 1, \ldots, N \ .$$

This set of equations constitutes the complete system of ordinary differential equations to be solved for obtaining the solution approximation by the Chebyshev tau spectral method.

From the point of view of the projection of the equations, on observes that equation (3.33) is the variational expression of an orthogonal projection P_N. The latter operator is defined with respect to the continuous inner product associated with the weighting functions $w(x)$ as a transformation within $L_w^2(-1, 1)$—the set of functions with integrable square with respect to this inner product—within $\mathbb{P}_{N-2}$.

REMARK 3.1   Note that the nonlinear term (3.35) is a specific case of the transform of a product:

$$\widehat{(uv)}_k = \frac{2}{\pi c_k} \int_{-1}^{1} uv T_k(x)(1 - x^2)^{-1/2} \, dx \ , \qquad (3.37)$$

which also amounts to

$$\widehat{(uv)}_k = \frac{1}{2} \sum_{p+q+k} \hat{u}_p \hat{v}_q + \sum |p - q| = k \hat{u}_p \hat{v}_q \ . \qquad (3.38)$$

in discretized space.   ⧠

A typical time scheme employs the explicit evaluation of the nonlinear term and an implicit evaluation for the linear term. Using homogeneous boundary conditions of the Neumann type (flux conditions) transforms the conditions (3.36) into

$$\sum_{k=0}^{N} k^2 \hat{u}_k = 0 \tag{3.39}$$

$$\sum_{k=0}^{N} (-1)^k k^2 \hat{u}_k = 0 . \tag{3.40}$$

3.4.3 Chebyshev collocation

In the collocation method for the Dirichlet problem, the set of base functions for developing the approximation function is identical to the one chosen in the preceding section, and the solution is represented by its values over the mesh points $x_j = \cos(\pi j/N)$, $j = 0, 1, \ldots, N$. The values of u_N on the grid are linked to the Chebyshev coefficients by the matrices obtained explicitly by writing the values of the Chebyshev series at the grid points, namely

$$C_{kj} = \frac{2}{N \bar{c}_j \bar{c}_k} \cos \frac{\pi j k}{N}$$

for the transformation from physical space to the Chebyshev coefficient space, and

$$\left(C^{-1}\right)_{jk} = \cos \frac{\pi j k}{N}$$

for the inverse transformation. The coefficients $\bar{c}_j$ are given by

$$\bar{c}_j = \begin{cases} 2 & j = 0, N \\ 1 & 1 \leq j \leq N-1 . \end{cases}$$

These two transforms can be handled efficiently by using Fast Fourier Transforms (FFT).

The discretization of the partial differential equation therefore is

$$\frac{\partial u_N}{\partial t} + u_N \frac{\partial u_N}{\partial x} - \nu \left[\frac{\partial^2 u_N}{\partial x^2} \right]_{x=x_j} = 0 \qquad j = 1, \ldots, N-1 , \tag{3.41}$$

with

$$u_N(-1, t) = u_N(1, t) = 0 \tag{3.42}$$

$$u_N(x_j, 0) = u_0(x_j) \qquad j = 0, \ldots, N . \tag{3.43}$$

It is thus possible to identify the linear operator $L_N u_N$ which amounts to $-\nu \mathcal{D}_N^2 u_N$, in which $\mathcal{D}_N$ is the differentiation operator of the Chebyshev collocation, and the nonlinear term is $u_N \mathcal{D}_N u_N$. The orthogonal projection is therefore expressed by the relations (3.41)-(3.43), that transform $\mathcal{C}^0([-1, 1])$ into the space of polynomials of degree $\leq N$ which verify the boundary conditions (with respect to the previously introduced inner product).

3.5   Implicit equations

Often, in stationary problems, spectral methods appear to be very useful
in the case of implicit equations. Moreover, many unsteady problems can
be solved by using implicit or semi-implicit algorithms in time. We therefore
illustrate here the treatment of such implicit problems by considering elliptical
linear equations of Helmholtz type

$$\Delta u - \lambda u = f \tag{3.44}$$

where f represents a function of the space coordinate x and λ is a real positive
constant. This kind of equations gathers such problems as potential stationary
incompressible flows, and implicit time discretization of the heat equation.

The spectral collocation approximation transforms the problem in a linear
system of type

$$LU = F \tag{3.45}$$

in which U and F are vectors of values over the mesh points for u and f
respectively, including the boundary conditions as well. The matrix L is the
matrix of a tensorial product in two dimensions in space, or more. For the
Galerkin and tau methods, U and F are the vectors gathering the coefficients
of the development of u and f, as well as the boundary conditions data, and
L is the corresponding matrix in spectral transform space. The linear system
is generally full, and may be solved in several ways, e.g. by direct inversion
methods or iterative methods.

3.5.1   Fourier approximation

In the Fourier approximation, we study the simple case of a one-dimensional
equation with constant coefficients for a space-periodic domain $x \in [-1, 1]$:

$$\frac{d^2u}{dx^2} - \lambda u = f \ . \tag{3.46}$$

The Galerkin approximation gives

$$-k^2\hat{u}_k - \lambda\hat{u}_k = \hat{f}_k \qquad k = -\frac{N}{2}, \dots, \frac{N}{2} - 1 \tag{3.47}$$

where the Fourier coefficients $\hat{u}_k$ have been defined previously as well as the
corresponding truncated Fourier series. The solution of this equation is triv-
ially obtained with the simple algebraic operation

$$\hat{u}_k = -\hat{f}_k/(k^2 + \lambda) \qquad k = -\frac{N}{2}, \dots, \frac{N}{2} - 1 \tag{3.48}$$

where $\hat{u}_0$ is arbitrarily chosen if $\lambda = 0$.

The collocation method itself provides the following equations

$$\left[\frac{d^2u}{dx^2} - \lambda u - f\right]_{x=x_j} = 0 \qquad j = 0,\ldots,N-1 . \tag{3.49}$$

These may be solved by using the discrete Fourier transform $\hat{u}_k = \frac{1}{N}\sum_{j=0}^{N-1} u(x_j)e^{-ikx_j}$ in order to diagonalize (3.49):

$$-k^2\hat{u}_k - \lambda\hat{u}_k = \hat{f}_k \qquad k = -\frac{N}{2},\ldots,\frac{N}{2}-1 . \tag{3.50}$$

The solution of the system is obtained in terms of the $\hat{u}_k$, and the final solution is obtained by transforming again using the inverse transform, so as to get $u_j = u(x_j)$, $j = 0,\ldots,N-1$. The cost of this solution method is of course reduced by using Fast Fourier Transforms.

REMARK 3.2 Again, one notes that even if the collocation method is stated in physical space through (3.49), the efficiency of the resolution requires the passage to Fourier space. $\quad\Box$

3.5.2 Chebyshev tau approximation

We consider here a classical problem of one-dimensional heat diffusion

$$\frac{\partial u}{\partial t}(x,t) - \beta^2\frac{\partial^2 u}{\partial x^2}(x,t) = F(x,t) \qquad -1 \le x \le 1 \tag{3.51}$$

with Dirichlet boundary conditions

$$u(-1,t) = u_0 = 2T_0 \tag{3.52}$$
$$u(1,t) = u_1 = 0 \tag{3.53}$$

and the initial condition $u(x, t = 0) = u^0$. The inhomogeneous term F can be a forcing, either source or sink. The heat diffusion coefficient is β^2. The implicit time discretization of equation (3.51) is:

$$\frac{u^{n+1} - u^n}{\Delta t} = \beta^2 \left(\frac{\partial^2 u}{\partial x^2}\right)^{n+1} + F(x,t) , \tag{3.54}$$

where Δt is the time-step. This requires the resolution at every time-step of a problem of Helmholtz kind:

$$\lambda u^{n+1} \quad \left(\frac{\partial^2 u}{\partial x^2}\right)^{n+1} - \frac{1}{\beta^2}\left(\frac{u^n}{\Delta t} + F(x,t)\right) \tag{3.55}$$

with $\lambda = 1/(\alpha^2 \Delta t)$. We shall adopt a resolution method of type Chebyshev tau for the resolution of this problem, by developing the approximation solution in a series of Chebyshev polynomials $T_n(x)$ truncated at order N:

$$u_N(x) = \sum_{n=0}^{N} \alpha_n T_n(x) \; . \tag{3.56}$$

The right term in the equation is denoted

$$G(x,t) = \frac{1}{\beta^2} \left(\frac{u^n}{\Delta t} + F(x,t) \right) \; , \tag{3.57}$$

and is developed in the following manner:

$$G_N = \sum_{n=0}^{N} \hat{g}_n T_n(x) \; . \tag{3.58}$$

The spectral method consists in considering the residual of the equation, that is

$$\varepsilon_N = -\frac{\partial^2 u_N}{\partial x^2}(x) + \lambda u_N(x) - G_N(x) \tag{3.59}$$

which can be also written using (3.58) and (3.56):

$$\varepsilon_N = \sum_{n=0}^{N} \left[-\alpha_n^{(2)} + \lambda \alpha_n - \hat{g}_n \right] T_n(x) \; . \tag{3.60}$$

We require the canceling of the projection of the residual onto the base of Chebyshev polynomials:

$$(\varepsilon_N, T_i) = \int_{-1}^{1} \frac{\varepsilon_N T_i}{\sqrt{1 - x^2}} \, dx \qquad i = 0, 1, \ldots, N - 2 \tag{3.61}$$

or again

$$\left(\sum_{n=0}^{N} \left[-\alpha_n^{(2)} + \lambda a_n - \hat{g}_n \right] T_n, T_i \right) = 0 \qquad i = 0, 1, \ldots, N - 2 \tag{3.62}$$

and by using the orthogonality property of the development polynomials, one obtains $N - 1$ equations

$$-\alpha_n^{(2)} + \lambda a_n - \hat{g}_n = 0 \qquad n = 0, 1, \ldots, N - 2 \; . \tag{3.63}$$

The unknowns are the $N + 1$ coefficients $(\alpha_k)_{0 \le k \le N}$, knowing that the development coefficients of the second-order derivative may be computed as

$$\alpha_k^{(2)} = \frac{1}{c_k} \sum_{\substack{p = k + 2 \\ p + k \text{ even}}} p(p^2 - k^2)\alpha_p \; .$$

One needs to complete the previous system with the two boundary conditions

$$u_N(-1) = \sum_{k=0}^{N} (-1)^k a_k = u_0 \tag{3.64}$$

$$u_N(1) = \sum_{k=0}^{N} a_k = u_1 . \tag{3.65}$$

Upon writing this system of $N + 1$ equations in matrix form, it is equivalent to $LU = G$, in which the matrix L is upper triangular, whose resolution requires N^2 operations, and G is the vector of coefficients $\hat{g}_n$. A better resolution method is obtained by recasting the equations (3.63), each of which being denoted E_k, using the recurrence relation for spectral coefficients of the second-order derivative, known in the general form

$$P_k \alpha_{k-2}^{(2)} + Q_k \alpha_k^{(2)} + S_k \alpha_{k+2}^{(2)} = \alpha_k \qquad 2 \leq k \leq N . \tag{3.66}$$

The recurrence coefficients are

$$P_k = \frac{c_{k-2}}{4k(k-1)} \qquad Q_k = \frac{-e_{k+2}}{2(k^2-1)} \qquad S_k = \frac{e_{k+4}}{4k(k+1)} \tag{3.67}$$

with

$$c_n = \begin{cases} 0 & n < 0 \\ 2 & n = 0 \\ 1 & n > 0 \end{cases} \qquad e_n = \begin{cases} 1 & n \leq N \\ 0 & n > N \end{cases} . \tag{3.68}$$

The following linear combination of equations

$$P_k E_{k-2} + Q_k E_k + S_k E_{k+2} = 0 \tag{3.69}$$

then provides, after very little algebra,

$$\lambda P_k a_{k-2} + (\lambda Q_k - 1) a_k + \lambda S_k a_{k+2} = P_k \hat{g}_{k-2} + Q_k \hat{g}_k + S_k \hat{g}_{k+2} \qquad k = 2, \ldots, N . \tag{3.70}$$

Let us denote h_k the right-hand side of this equation. Equation (3.70) leads to an algebraic system in which the even-order coefficients and the odd-order coefficients are decoupled. Accordingly, the system is separated in two linear systems, whose unknowns are respectively the $\alpha_0, \alpha_2, \ldots$ and the $\alpha_1, \alpha_3, \ldots$. The boundary conditions write

$$\alpha_0 - \alpha_1 + \alpha_2 + \ldots + \alpha_N = u_0 \tag{3.71}$$
$$\alpha_0 + \alpha_1 + \alpha_2 + \ldots + \alpha_N = u_1 \tag{3.72}$$

and yield, by sum and difference,

$$\alpha_0 + \alpha_2 + \ldots + \alpha_N = \frac{u_0 + u_1}{2} \tag{3.73}$$

$$\alpha_1 + \alpha_3 + \ldots + \alpha_{N-1} = \frac{u_1 - u_0}{2} . \tag{3.74}$$

One now needs to solve two linear systems:

$$M_p A_p = F_p \qquad M_I A_I = F_I \, .$$

The unknowns are the vectors

$$A_p = \begin{bmatrix} \alpha_0 \\ \alpha_2 \\ \vdots \\ \alpha_{N-2} \\ \alpha_N \end{bmatrix} \qquad A_I = \begin{bmatrix} \alpha_1 \\ \alpha_3 \\ \vdots \\ \alpha_{N-1} \end{bmatrix}$$

and the matrices are quasi-tridiagonal, of the form:

$$M_p = \begin{bmatrix} \lambda P_2 & \lambda Q_2 - 1 & \lambda S_2 & 0 & & \cdots & \\ 0 & \lambda P_4 & \lambda Q_4 - 1 & \lambda S_4 & & & \\ & & \ddots & \ddots & \ddots & & \\ & & & \lambda P_{N-2} & \lambda Q_{N-2} - 1 & 0 \\ & & & 0 & \lambda P_N & -1 \\ 1 & 1 & \cdots & & & & 1 \end{bmatrix}$$

and

$$M_p = \begin{bmatrix} \lambda P_3 & \lambda Q_3 - 1 & \lambda S_3 & 0 & & \cdots & \\ 0 & \lambda P_5 & \lambda Q_5 - 1 & \lambda S_5 & & & \\ & & \ddots & \ddots & \ddots & & \\ & & & \lambda P_{N-3} & \lambda Q_{N-3} - 1 & 0 \\ & & & 0 & \lambda P_{N-1} & \lambda Q_{N-1} - 1 \\ 1 & 1 & \cdots & & 1 & 1 \end{bmatrix}$$

with right-hand terms

$$F_p = \begin{bmatrix} h_2 \\ h_4 \\ \vdots \\ h_{N-2} \\ h_N \\ \frac{u_0 + u_1}{2} \end{bmatrix} \qquad F_I = \begin{bmatrix} h_3 \\ h_5 \\ \vdots \\ h_{N-1} \\ h_N \\ \frac{u_1 - u_0}{2} \end{bmatrix} \, .$$

REMARK 3.3 Note that these two linear systems are well-conditioned, that is with dominant diagonal, whereas it was not the case of the system in the initial formulation. ⬜

The inversion method by double-sweeping allows to solve the two systems in terms of the unknowns $(\alpha_k)_{0 \le k \le N}$ with a cost proportional to N. We recall that the unknowns are the coefficients of the Chebyshev series development of the solution function u_N at time t^{n+1}. The complete solution is therefore

$$u_N^{n+1}(x) = \sum_{n=0}^{N} \alpha_k^{n+1} T_k(x) \ ,$$

which can be evaluated at each desired point x in space. Let us stress that at each step, the vectors F_p and F_I change, since they are computed from the solution u_N^n at the preceding timestep. It is therefore necessary to perform the resolution of the linear systems at every timestep.

3.6 Evaluation of nonlinear terms

We have up to now considered linear equations as in (3.1), or have delayed the explicit explanation on how to treat nonlinear terms in the considered problems. Such terms appear in many problems of fluid mechanics due to the convection operator $(u \cdot \nabla)u$ in the Navier-Stokes equations, and can be optimally computed in spectral numerical schemes by using transformation methods. The basic idea is to apply the spectral transform (FFT or other) to switch from the spectral space representation of the nonlinear terms to their equivalent in physical space, instead of computing the explicit convolution product which is required when spectrally transforming a nonlinear term. In the case of Chebyshev series, fast Fourier transforms allow the computation of the coefficients of arbitrary nonlinear terms with a computational cost proportional to $N \log N$ arithmetic operations (in each direction of space). Generally speaking, collocation methods are often more efficient for the computation of the coefficients of nonlinear terms than the Galerkin or the tau methods. It is therefore recommended for nonlinear problems. For PDEs with quadratic nonlinearity, e.g. the Navier-Stokes equations for an incompressible fluid, as mentioned, the cost of the Galerkin or tau method is at least double that of the collocation method. However, the Galerkin approximation may be used for its conservative character, when exact energy conservation properties are required.

For instance, the hyperbolic problem

$$\frac{\partial u}{\partial t} + e^{u+x} \frac{\partial u}{\partial x} = f(x,t) \qquad x \in [-1,1], \qquad t > 0 \qquad (3.75)$$
$$u(-1,t) = 0 \ ,$$

is more complex to treat with a Galerkin or tau approach, but in a more straightforward way with the collocation technique. Let us explain how the time-stepping is performed for equation (3.75) approximated by a Chebyshev collocation method. The $N + 1$ collocation points are $x_j = \cos(\pi j / N)$ for $j = 0, \ldots, N$, such that the solution $u_j = u_N(x_j)$ is known in the form

$$u_j = \sum_{n=0}^{N} a_n \cos \frac{\pi n j}{N} \ . \tag{3.76}$$

The application of the inverse fast Fourier transform to (3.76) allows to obtain the coefficients a_n for $n = 0, \ldots, N$ and thus to compute the coefficients of $\frac{\partial u}{\partial x}$ $a_n^{(1)}$ by using the relationship (3.31). It is then necessary to evaluate

$$\left. \frac{\partial u}{\partial x} \right|_{x=x_j} = \sum_{n=0}^{N} a_n^{(1)} \cos \frac{\pi n j}{N}$$

by using the fast Fourier transform. Finally, one can also compute $e^{u_j + x_j \frac{\partial u}{\partial x}}$ at each mesh point x_j and use the result for the time advancement of the solution over one timestep. However, when dealing with nonlinear terms in the equations, another procedure is required (described in section 3.6.1), which raises the problem of aliasing.

3.6.1 Problem of aliasing

The back and forth transforms between physical and spectral space carries a drawback when manipulating truncated series of the functions. This phenomenon called *aliasing* creates spurious spectral coefficients that are not physically present; all the more when the amplitudes of the spectral coefficients do not decrease fast enough when increasing their order. A special treatment is therefore required, which is called "de-aliasing".

We illustrate the case of the discrete Fourier transform, based on a set of grid points in physical space

$$x_j = \frac{2\pi j}{N} \qquad \text{for } j = 0, \ldots, N - 1 \ ,$$

where N is a positive integer. The discrete Fourier coefficients of a complex valued function $u(x)$ for $x \in [0, 2\pi]$ using this physical space discretization are

$$\hat{u}_k = \frac{1}{N} \sum_{j=0}^{N-1} u(x_j) e^{-ikx_j} \qquad k \in \left[\frac{-N}{2}, \frac{N}{2} \right] \ . \tag{3.77}$$

Using the orthogonality of the Fourier polynomials, we have

$$\frac{1}{N} \sum_{j=0}^{N-1} e^{ipx_j} = \begin{cases} 1 \text{ if } p = Nm, \ m = 0, \pm 1, \pm 2, \ldots \\ 0 \qquad\qquad\qquad \text{otherwise} \end{cases} \tag{3.78}$$

which allows to inverse the formula, which then writes

$$u(x_j) = \sum_{k=-N/2}^{N/2-1} \hat{u}_k e^{ikx_j} \qquad j = 0, \dots, N-1 . \qquad (3.79)$$

In this relation, one identifies the polynomial

$$I_N^u(x) = \sum_{k=-N/2}^{N/2-1} \hat{u}_k e^{ikx}$$

as a trigonometric interpolation polynomial of order $N/2$ of the function u using the grid points x_j. It is therefore the discrete Fourier series of u.

As shown by (3.78), the issue of aliasing lies in the fact that a spectral coefficient of order k is not distinguishable from a spectral coefficient of order $k + mN$, where m is an integer, when the frequencies are confined to vary within the interval $\left[\frac{-N}{2}, \frac{N}{2}\right]$; since, in the transform (3.77), the exponential term $e^{-ikx_j} = e^{-i(k+mN)x_j}$ when $x_j = \frac{2\pi j}{N}$. This is illustrated on Figure 3.4.

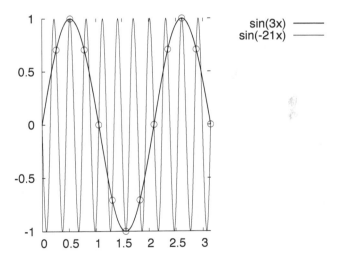

FIGURE 3.4: Illustration of aliasing: two sine functions with frequency $k = 3$ and $k = -21$ are indistinguishable when represented on a physical grid with $N = 24$ points (symbols), due to the coarseness of the grid. For increasing frequencies, the required number N of grid points required to avoid aliasing increases dramatically.

3.6.2   Convolution sums

In a pseudo-spectral method, aliasing errors may arise whenever a FFT or an inverse FFT transform occurs. In the Galerkin context, the treatment of nonlinear terms, or terms in the equations in which the coefficients are non-uniform and depend on space, introduce convolution products that need to be computed.

Let us consider a term $w(x)$ in the equation that needs to be solved, as a product of the functions $u(x)$ and $v(x)$ in physical space:

$$w(x) = u(x)v(x) \ . \tag{3.80}$$

In the context of developing terms in infinite series, one recovers the infinite sum that corresponds to the convolution, in the form of

$$\hat{w}_k = \sum_{m+n=k} \hat{u}_m \hat{v}_m \tag{3.81}$$

in which

$$u(x) = \sum_{m=-\infty}^{\infty} \hat{u}_m e^{imx} \tag{3.82}$$

$$v(x) = \sum_{n=-\infty}^{\infty} \hat{v}_n e^{inx} \tag{3.83}$$

and

$$\hat{w}_k = \frac{1}{2\pi} \int_0^{2\pi} w(x) e^{-ikx} \, dx \ .$$

When u, v and w are approximated by their respective Fourier series truncated at order $N/2$, the approximation of the convolution product becomes

$$\hat{w}(k) = \sum_{\substack{m+n=k \\ |m|,\,|n|\,\leq\,N/2}} \hat{u}_m \hat{v}_n \tag{3.84}$$

in which $|k| \leq N/2$. The direct computation of this sum involves $\mathcal{O}(N^2)$ operations, with an exaggerated cost compared to the $\mathcal{O}(N)$ cost for evaluating this same term in a finite differences technique (these estimates correspond to a one-dimensional problem).

The use of a transformation method allows to reduce this cost, which becomes exaggerated for computations with a large number of nodes, to a cost of order $\mathcal{O}(N \log_2 N)$ (again in a one-dimensional setting).

3.6.3   Numerical evaluation by a pseudo-spectral transformation method

The pseudo-spectral method uses the inverse discrete Fourier transforms of $\hat{u}_m$ and $\hat{v}_n$, since the functions u and v are known by their spectral coefficients, either as the result of a previous timestep, or from initial conditions. Instead of computing the nonlinear term by a convolution product in transformation space, the product is therefore computed in physical space as in (3.80) by a point-wise multiplication, that is at each grid point, and the result is Fourier transformed again to obtain $\hat{w}_k$.

The discrete Fourier transforms are denoted

$$U_j = \sum_{k=-N/2}^{N/2-1} \hat{u}_k e^{ikx_j}$$

$$V_j = \sum_{k=-N/2}^{N/2-1} \hat{v}_k e^{ikx_j} \qquad \text{for } j = 0, 1, \dots, N-1 \tag{3.85}$$

and one defines

$$W_j = U_j V_j \qquad j = 0, 1, \dots, N-1 \tag{3.86}$$

and

$$\hat{W}_k = \frac{1}{N} \sum_{j=0}^{N-1} W_j e^{-ikx_j} \qquad k = -N/2, \dots, N/2-1 \tag{3.87}$$

where the mesh points coordinates in physical space are $x_j = 2\pi j/N$, and U_j, V_j, W_j are the respective values of $u(x_j)$, $v(x_j)$, $w(x_j)$. Using the orthogonal property of the Fourier exponential with respect to the inner product, one gets

$$\hat{W}_k = \underbrace{\sum_{m+n=k} \hat{u}_m \hat{v}_n}_{\hat{w}_k} + \sum_{m+n=k\pm N} \hat{u}_m \hat{v}_n . \tag{3.88}$$

In the latter formula, the required convolution product $\hat{w}_k$ is identified, in addition to an undesired term which corresponds to the aliasing error. As explained in section 3.6.1, modes of order $k + N$ are misinterpreted as modes of order k in the second sum of the right-hand-side of equation (3.88).

The cost of evaluating the nonlinear term in such a way can be estimated to $(15/2)N \log_2 N$ multiplications (corresponding to three fast Fourier transforms and N multiplications). The extension of the method to three-dimensional space is straightforward. The objective at this point is to get rid of the undesired term in (3.88), which can be realized using two techniques. Each technique has a slightly different memory usage and computational cost, and the choice of one or the other depends on the available computer.

3.6.4 De-aliasing by the 3/2 rule

The underlying idea of this method is to use Fourier series with a larger dimension $M \geq 3N/2$ instead of N in (3.87).

Using Fourier series of size M (denoted by $\tilde{\ }$) transform the equation (3.88) in

$$\tilde{W}_k = \underbrace{\sum_{m+n=k} \tilde{u}_m \tilde{v}_n}_{\tilde{w}_k} + \sum_{m+n=k\pm M} \tilde{u}_m \tilde{v}_n \ . \tag{3.89}$$

It is important to understand that in the context of a spectral methods with N degrees of freedom, one needs to compute the convolution product $\tilde{W}_k$ for $|k| \leq N/2$ *only*. Thus, for being able to remove the contribution of aliased coefficients in (3.89), one needs to force them to appear only beyond the interval $[-N/2, N/2]$ by an adequate choice of M. It can be shown that if $M \geq \frac{3N}{2} - 1$, the desired effect is obtained, that is the aliasing only appears for the largest non useful coefficients. At each timestep after the M-size transformation of the convolution product, it is necessary to set to zero all the coefficients that lie beyond the interval $[-N/2, N/2]$.

The additional cost associated with the transforms of size M points is 50%, and the memory usage is of course more important. One must add that a pseudo-spectral method based on Chebyshev polynomials is also subject to aliasing, which also can be removed exactly with the 3/2 rule.

3.6.5 De-aliasing by phase-shifting

This method for de-aliasing consists in evaluating twice the transformation product (3.87) using two different meshes: one based on the grid points x_j and another mesh which is shifted with grid points locations $x_j + \Delta$. In the latter case, the convolution product is denoted $\hat{W}'_k$. One can show that the unaliased convolution product is the result of the linear combination of the two estimates:

$$\hat{w}_k = \frac{1}{2} \left[\hat{W}_k + \hat{W}'_k \right] \ ,$$

provided $\Delta = \pi/N$. The cost of this technique is slightly higher than that of the 3/2 method. It can still be valuable in that it does not involve Fourier transforms of augmented size, an increase that may be impossible when the number of grid points reaches the limits of the memory size of the computer.

3.6.6 Errors and convergence

As in all methods for the numerical solution of PDEs, the discretization with a finite number of points raises the questions about the accuracy of the method, the errors, and its convergence when increasing the number of degrees of freedom. As already mentioned in Introduction, the convergence of spectral methods is very fast with respect to the geometric rate of convergence of DF, EF or VF methods. In this respect the convergence of spectral methods is super-geometric.

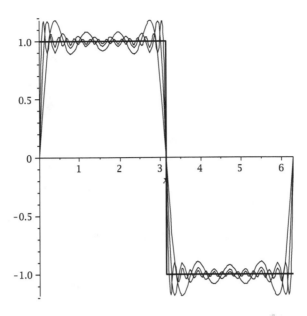

FIGURE 3.5: The truncated Fourier series representation of the step function over the interval $[0, 2\pi]$, using $N = 8$, 16 and 32 spectral coefficients.

We have also seen that the truncated series can be seen as interpolation polynomials for the developed functions. In the collocation method, the use of trigonometric interpolation over equidistant points raises a problem known as the *Gibbs phenomenon* discovered at the beginning of the 20th century in wave physics. This phenomenon refers to the typical oscillations of a truncated Fourier series or a discrete Fourier series, for a bounded function close to a discontinuity. More generally, steep gradients in the developed functions require a very large number of spectral coefficients. A step function is the worst case, in which an infinite number of frequencies are contained.

In general, all the truncated Fourier series present an oscillating behavior around the exact function. These oscillations are often negligible, however, for instance for the step function (and generally in the neighborhood of strong gradients), these oscillations have specific properties. The maximal amplitude of the oscillations close to the discontinuity—the overshoot—has a finite limit, and the location of this overshoot gets closer to the discontinuity when the retained number of spectral coefficients in the series increases. This is clearly illustrated on Figure 3.5. In a formal way, if p_N is the interpolation polynomial of the analytical function f at order N, one can show that the error $||f - p_N|| = \mathcal{O}(1)$ when $N \to \infty$. The methods to overcome this problem still remain very costly, and remove partially the interest of using spectral methods. The problem is less important when considering Legendre or Chebyshev polynomials, since the error committed by the interpolation is $\|f - p_N\| = \mathcal{O}(K^{-N})$ where $K > 1$. In that case, even if f has strong discontinuities, the errors converge to zero.

Other kinds of errors that have to be considered in spectral methods are round-off errors. If the floating-point representation of real numbers is at a given relative precision 10^{-m}, one has to be careful when recurrence relations are used for computing derivatives, since they can be an important source of loss of accuracy by accumulation for a large number of arithmetic operations. This is the case when the number of operations is of order 10^m (a figure which can be easily reached when considering three-dimensional problems with a resolution larger than 200 in each direction os space) in single precision computations. If that is the case, a re-conditioning of the problem can be proposed by re-formulating the problem.

In general, transformation methods do not amplify truncation errors. It appears that, in terms of rounding errors, the evaluation of convolution integrals using Fourier transforms provides better results than the direct computation of these integrals.

The stability of spectral methods is not guaranteed for any values of the numerical parameters N and time-step Δt. As for finite differences schemes, conditions of stability have to be established. Since the discretization operators in spectral methods generally have a spectrum of eigenvalues wider than that of finite differences operators, it is a good idea to use higher order time-discretization, e.g. such as third-order Runge-Kutta schemes, or third-order Adams-Bashforth. Moreover, the accuracy of the time discretization has to be consistent with the accuracy of the spatial discretization, so as not to lose the gain of the high-order spatial scheme.

3.6.7   Wavenumber, vortex, wavelet

When dealing with spectral methods, it is difficult to assign a physical meaning to the spectral coefficients in general. However, Fourier coefficients for instance can *to some extent* be given a significance in terms of structures in the fluid flow.

Let us recall that one can define auto-correlation functions of a space-varying function, $< u_i(\boldsymbol{x})u_j(\boldsymbol{x} + \boldsymbol{r}) >$ (where $<>$ stands for statistical ensemble averaging), that establish the link between a velocity component u_i at point $\boldsymbol{x}$ and the velocity component u_j at point $\boldsymbol{x} + \boldsymbol{r}$. When the fluid flow exhibits a structural coherence, i.e. the presence of vortices, the value of the correlation becomes closer to unity, with respect to a zero value for completely uncorrelated fluid motion.

A vortex is a structure which has a motion similar to that of a rotating solid body, with a given relationship between the displacement of fluid particles with distinct distance to the center of the vortex, thus with a specific auto-correlation function. Unlike the correlation function that depends on the spatial separation, the spectrum is a function of the wavenumber, and may be computed by a suitable average of $\hat{u}_i(k)\hat{u}_j(-k)$. However, it is directly

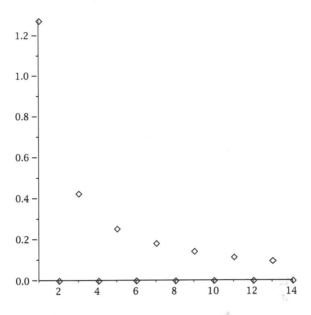

FIGURE 3.6: The spectrum of coefficients of the step function plotted on Figure 3.5.

linked to the correlation function by a Fourier transform. Using this link, a vortex of characteristic size ℓ may then be associated with a given wavenumber k with a scaling such that $\ell = 2\pi/k$ (the wavelength associated with the wavenumber k). A "vortex" of wavenumber k is therefore a structure in the fluid which possesses energy in spectral space around k, that is dominant spectral coefficients associated with this wavenumber. Such a point of view is of course a very sketchy description of the reality of fluid flows, since in reality several spectral coefficients are associated with a given vortex. For instance, if the variation of the velocity were similar to the step function shown in Figure 3.5—this would be a very clear-cut *vortex*—the distribution of spectral coefficients would be as shown on Figure 3.6.

However, one element of the success of spectral methods using trigonometric polynomials is the fact that Fourier coefficients, directly obtained by the method, allow a direct interpretation of the computed spectral coefficients distribution in terms of length scales in the fluid flow, thus in terms of a range of vortices with a wide range of length scales. This picture is better justified in homogeneous flows, with no significant variation in space of the statistical properties of the flow, which can be treated with periodic boundary conditions. In more complex flows, with strong inhomogeneities, adapted mathematical tools allow to extend the spectral description to be closer to the physical reality. Wavelets are a kind of decomposition functions that not only are associated to a given length scale, but have also a localization in physical space. Numerical methods have recently been developed for the resolution of fluid flow motion based on wavelets decomposition.

An illustration of the difference between the Fourier representation and wavelets distributions is given in Figures 3.7 and 3.8. On Figure 3.7, vortices in a periodic domain are plotted, corresponding to a stream function $\sin(x)\sin(y)$ whose spectrum in Fourier space correspond to a unique frequency. In that case, the wavelength associated with the vortices in a single value. The vortices are therefore perfectly characterized by one value of the frequency corresponding to the non zero Fourier coefficient and its associated wavelength.

Alternately, Figure 3.8 shows the physical distribution of a one-dimensional wavelet function, which can correspond to an isolated vortex structure, and its spectrum. Opposite to the previous example, the spectrum of Fourier coefficients is continuous and rather localized in Fourier space. Of course, in wavelets transform space, it corresponds to a single transform coefficient.

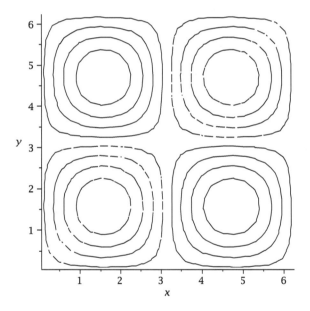

FIGURE 3.7: Taylor-Green vortices in a two-dimensional flow. Dashes indicate counter-rotating vortices.

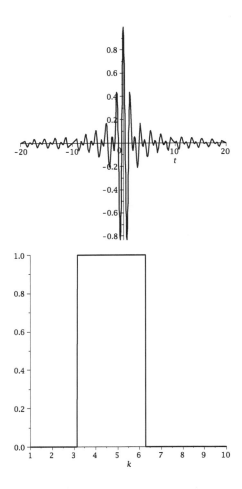

FIGURE 3.8: Top figure: one kind of base wavelet function showing the localization of its distribution, with respect to the sine dependence of the Fourier polynomials; bottom: the Fourier spectrum of the wavelet, showing that the frequency contents is not a single coefficient, as for a sine function, but a continuous distribution over an interval localized in spectral space.

Chapter 4

Smoothed-particle hydrodynamics (SPH) methods

Francis Leboeuf

Ecole Centrale de Lyon - LMFA
36 avenue Guy de Collongue
69134 Ecully Cedex, France

Jean-Christophe Marongiu

ANDRITZ-HYDRO
Rue des Deux-Gares 6
CH 1800 Vevey, Suisse

4.1 Introduction

Among the meshless numerical method, the SPH method (Smooth Particle Hydrodynamics) has been now theoretically studied and developed in order to allow numerous applications in fluid mechanics. It has been first introduced by Lucy (1977) and formalized by Gingold and Monaghan (1977, 1983) initially for astrophysical applications. Monaghan has proposed many important developments (see a list of references in this text).

The purpose of this text is to present the basic features of the SPH method from a practical point of view, avoiding then to refer to mathematical theorems. We will then introduce the use of SPH method for an ALE formalism (Arbitrary Lagrange Euler). The combination of this ALE description together with a conservative (or weak) form of the Euler equations allows the introduction of Godunov's and higher order possible schemes; most of the material in this part is issued from the work of Vila and his co-workers (1999).

A review of different techniques for accuracy improvement will be given, particularly close to the boundaries; this part is mainly based on the work of Marongiu (2007 to 2009). A particular analysis of the Finite Volume Particle Method (FVPM), a particular SPH method, will be also given as it has some nice features for conservation properties; it has been developed by Hietel (2005), Struckmeier (2002, 2008) and applied to some complex flow phenomena by Quinlan and his co-workers (2009). The influence of the boundaries $\partial D(x)$ of the kernel support will be underlined in the various expressions; these boundary terms are in general not considered in the various publication except in Marongiu (2007 to 2009).

REMARK 4.1   In this chapter, the following notations will be considered. The vector position are in general referred to as x. The position in the domain of the particle is also referred as x'. The unknown flow values are noted ϕ; this may be a scalar function or a vector, depending on the context. We have chosen not to overload the text with superscript, and the vectors or matrices will be specified when they appear the first time in the text.   ⬚

4.2   SPH approximation of a function

The attractiveness of SPH methods is related to its ability to generate numerical estimations for functions and their spatial derivatives without resorting to a mesh.

Consider in a domain Ω, limited by a boundary $\partial\Omega$, a set of disordered points (or particles) as in Figure 4.1, with Δx the distance between two neighbouring points, h the radius of the sphere of influence $D(x)$ of particle x, centred at x. Let a field ϕ sufficiently regular, defined in the domain Ω limited by a boundary $\partial\Omega$. We may define the value $\phi(x)$ as the following convolution product with δ the Dirac function:

$$\phi(x,t) = \int_{D(x,t)} \phi(x',t)\,\delta(x - x')\,dx'^d \tag{4.1}$$

with d the number of dimension. The numerical estimation of this integral requires that the δ function is approached by a regular function W, also referred to as the smoothing kernel function or kernel. We then get an approximation of the function $\phi(x)$ as:

$$\phi^h(x,t) = \int_{D(x,t)} \phi(x',t)\,W(x - x',h)\,dx'^d \tag{4.2}$$

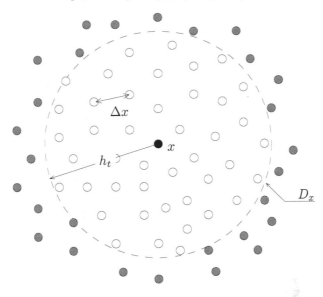

FIGURE 4.1: Interpolation domain in 2D and neighboring points.

As W is scaled to one (see (4.7) below), (4.2) can be interpreted as a smoothing process of ϕ over the domain $D(x)$. The numerical estimation of the integral (4.2) requires an integration formula.

$$\phi^h\left(x_i,t\right) \simeq \sum_{j\in D(x_i)} \phi\left(x_j,t\right)W\left(x_i - x_j, h\right)\omega_j\left(x_j,t\right) \tag{4.3}$$

where ω_j is a weight of the particle; ω_j can be seen as the volume of the particle or the jacobian of the transformation between the coordinate (ξ, t) and the Lagrangian coordinate (x, t). In (4.3), the summation is performed on all the neighbouring particles included in the domain $D(x)$. Let v_0 be a velocity field, used to transport the particles i along the trajectories; it is important to notice that v_0 may be different from the velocity of the flow. We have for a particle i and according to the Lagrangian transformation:

$$\frac{1}{\omega_i}\frac{d\omega_i}{dt} = div\left(v_o\right) \tag{4.4}$$

$$\frac{dx_i}{dt} = \dot{x}_i = v_o\left(x_i,t\right) \tag{4.5}$$

4.3   Properties of the kernel function W

The kernel function W should be a good approximation of the Dirac function:

$$\lim_{h \to 0} W(x - x', h) = \delta(x - x') \qquad (4.6)$$

In practice, W is usually defined on a compact support, which implies that $W = 0$ for $x \geq c\,h$ where c is of the order of 2 to 3 in general. W is continuous and derivable. It should decrease monotonously away from $x' = x$; and it is always non-negative; if this last condition is not fulfilled then the integral ϕ^h of a positive function (say the density for instance) could be negative according to (4.2). Moreover, if we perform a Taylor series expansion around $(x' = x)$, we get that (4.2) is a second order accurate approximation if:

$$\int_{D(x)} W(x - x', h)\, dx'^d = 1 \qquad (4.7)$$

$$\int_{D(x)} (x - x')\, W(x - x', h)\, dx'^d = 0 \qquad (4.8)$$

This last relation is satisfied if the kernel W is symmetric around $x' = x$, $W(x - x') = W(x' - x)$ and $\nabla_x W(x - x') = -\nabla_x W(x' - x)$. The discrete equivalent equations deduced from (4.7) and (4.8) are:

$$\sum_{j \in D(x)} W(x - x_j, h)\, \omega_j = 1 \qquad (4.9)$$

$$\sum_{j \in D(x)} (x - x_j)\, W(x - x_j, h)\, \omega_j = 0 \qquad (4.10)$$

REMARK 4.2   In order to get an order of approximation greater than two, this would require the zeroing of all the higher moments of W (Belytschko, 1998, Liu, 1995, Dilts, 1999, 2000)

$$M_k = \int_{D(x)} (x - x')^k\, W(x - x', h)\, dx' = 0, \forall k > 1.$$

This is not possible with a non-negative W function for all even k values. Therefore an SPH kernel approximation with a non-negative smoothing function is of second order accuracy at most. $\quad\square$

REMARK 4.3   If the particle is close to the boundary $\partial \Omega$ of the domain, then the conditions (4.9) and (4.10) may be difficult to fulfil as the integral could be truncated. This is a severe limitation of the method. We will see in a following chapter that FVPM method easily overcome this limitation, see Chapter 6. □

REMARK 4.4   Numerically, none of the conditions (4.9) and (4.10) can be exactly fulfilled because of particle disorder; it is then very difficult to guarantee the second order accuracy of the numerical approximation of the basic scheme without using other techniques. □

REMARK 4.5   According to Vila (1999, 2005), (4.3) is a valid approximation of (4.2) only if $\frac{\Delta x}{h} \to 0$, which requires that the number of particles in $D(x_i)$ tends towards infinity as h tends to zero. □

REMARK 4.6   Similarly if the derivatives of function $\phi(x)$ (up to kth order derivative) are to be exactly reproduced, another group of expressions about the smoothing function W can be obtained (Liu et al. 2003):

$$W^{k-1}\left(x - x', h\right)_{\partial D} = 0$$

where W^{k-1} is the $k-1$ derivative of the kernel function along $(x - x')$. It expresses that the derivatives of W should also vanish on the boundary $\partial D(x)$. □

4.4   Barycenter of $D(x_i)$

Let b_i be the barycenter of the domain $D(x_i)$. Its position is defined as:

$$b_i = \int_{D_i} x' W_i\left(x_i - x', h\right) dx'^d \tag{4.11}$$

or in a discrete form: $b_i = \sum x_j W\left(x_i - x_j, h\right) \omega_j$ Consider a Taylor series' expansion around $x' = b_i$:

$$\phi\left(x', t\right) \approx \phi\left(b_i\right) + \left(x' - b_i\right) \bullet \left. \nabla \phi \right|_{x = b_i} + O\left(\left|x' - b_i\right|^2\right)$$

Compute the average in the sense of (4.2):

$$\phi^h\left(x', t\right) = \int_{D(x)} \left\{\phi\left(b_i\right) + \left(x' - b_i\right) \bullet \left. \nabla \phi \right|_{x = b_i}\right\} W\left(x - x', h\right) dx'^d$$

Using the condition (4.7), far from the boundary $\partial\Omega$:

$$\int_{D(x)} W\left(x - x', h\right) dx'^d = 1$$

$$\Rightarrow \phi_i^h\left(x, t\right) = \phi\left(b_i\right) + \nabla\phi|_{x_i=b_i} \int_{D(x)} \left(x' - b_i\right) W\left(x - x', h\right) dx'^d + O\left(h^2\right)$$

With the definition of b_i, we get finally $\phi_i^h = \phi\left(b_i\right) + O\left(h^2\right)$, which shows that the average of the function f is in fact computed at the barycentre of the domain $D(x_i)$ with a second order of accuracy, provided that the condition (4.7) is fulfilled in practice. For any points x_i, and for any other points y in $D(x_i)$ such that $y \neq b_i$, we get only $\Phi\left(x_i\right) = \Phi\left(y\right) + O\left(h\right)$ because $y \neq \int_{D_i} x' W_i\left(x_i - x', h\right) dx'^d$

4.5   Choices of the kernel function W

Many possibilities exist for the kernel function W, but B-splines are often used, with d the number of dimension.

$$W\left(x, h\right) = \frac{C}{h^d} f\left(\frac{x}{h}\right) \tag{4.12}$$

Let $q = \frac{x}{h}$

Order 3   Monaghan and Lattanzio (1985) use a B-spline of order 3. Note that $W = 0$ only at $\frac{x}{h} \geq 2$ for this case

$$f\left(q\right) = \begin{cases} 1 - \frac{3}{2}q^2 + \frac{3}{4}q^3 & \text{if } 0 \leq q \leq 1 \\ \frac{1}{4}\left(2 - q\right)^3 & \text{if } 1 \leq q \leq 2 \\ 0 & \text{elsewhere} \end{cases} \tag{4.13}$$

$C = 2/3$, $10/7\pi$ and $1/\pi$ for $d = 1$, 2 and 3.

Order 4   For a B-spline of order 4, we get:

$$f\left(q\right) = \begin{cases} \left(\frac{5}{2} - q\right)^4 - 5\left(\frac{3}{2} - q\right)^4 + 10\left(\frac{1}{2} - q\right)^4 & \text{if } 0 \leq q \leq 0.5 \\ \left(\frac{5}{2} - q\right)^4 - 5\left(\frac{3}{2} - q\right)^4 & \text{if } 0.5 \leq q \leq 1.5 \\ \left(\frac{5}{2} - q\right)^4 & \text{if } 1.5 \leq q \leq 2.5 \\ 0 & \text{elsewhere} \end{cases} \tag{4.14}$$

$C = 96/1199\pi$, $10/7\pi$ and $1/20\pi$ for $d = 2$ and 3.

Order 5 Morris (1997) uses a B-spline of order 5.

$$f(q) = \begin{cases} (3-q)^5 - 6(2-q)^5 + 15(1-q)^5 & \text{if } 0 \leq q \leq 1 \\ (3-q)^5 - 6(2-q)^5 & \text{if } 1 \leq q \leq 2 \\ (3-q)^5 & \text{if } 2 \leq q \leq 3 \\ 0 & \text{elsewhere} \end{cases} \tag{4.15}$$

$C = 7/478\pi$, $10/7\pi$ and $1/120\pi$ for $d = 1, 2$ and 3.

Kernels of Wendland The kernels of Wendland (Wendland 1999) are also often used (see Robinson and Monaghan 2008). Let $q = \frac{r}{2h}$, some examples of Wendland's kernels are written here with $f = 0$ for $q > 1$:
Wendland C2: with $\sigma = \frac{7}{4\pi}$ for $d = 2$ and $\sigma = \frac{21}{16\pi}$ for $d = 3$

$$f(q) = \frac{\sigma}{h^d}(1-q)^4(4q+1) \tag{4.16}$$

Wendland C4: with $\sigma = \frac{3}{4\pi}$ for $d = 2$ and $\sigma = \frac{165}{256\pi}$ for $d = 3$

$$f(q) = \frac{\sigma}{h^d}(1-q)^6(35q^2 + 18q + 3) \tag{4.17}$$

Wendland C6: with $\sigma = \frac{39}{14\pi}$ for $d = 2$ and $\sigma = \frac{1365}{512\pi}$ for $d = 3$

$$f(q) = \frac{\sigma}{h^d}(1-q)^8(32q^3 + 25q^2 + 8q + 1) \tag{4.18}$$

4.6 SPH approximation of differential operators applied on a function ϕ

4.6.1 Basic formulation

Consider the average (4.2) of a differential operator ∇ (gradient or divergence for instance):

$$\nabla\phi(x) = \int_{D(x)} W(x - x')\nabla_{x'}\phi(x')\, dx'^d$$

Using integration by parts, we get:

$$\nabla\phi(x) = -\int_{D(x)} \phi(x')\nabla_{x'}W(x - x')\, dx'^d + \int_{\partial D(x)} \phi(x')W(x - x')\,n(x')\, dx'^{d-1}$$

where $n(x')$ stands for the unit vector normal to the boundary. And finally

$$\nabla\phi(x) = \int\limits_{D(x)} \phi(x') \nabla_x W(x - x') dx'^d + \int\limits_{\partial D(x)} \phi(x') W(x - x') n(x') dx'^{d-1}$$

$$(4.19)$$

The last term in the second member stands for the effect of the boundary $\partial D(x)$. It is zero in general for the previous choices of W, except if $D(x)$ intercepts the boundary ∂W of the computational domain. Provided that ϕ is known on ∂W, this term can be easily computed; in other cases, specific treatments on ∂W are required.

The nice feature of SPH method is that the computation of the average of differential operator is then no more complicated than the average of the function itself, according to (4.19) because ∇W can be analytically computed.

Performing a Taylor series expansion of $\phi(x')$ around x, and replacing in (4.19), it can be easily shown that this expression is second order accurate with the same condition (4.7) provided that the boundary $\partial D(x)$ does not intercept $\partial \Omega$.

The discrete equivalent of (4.19) is then:

$$(\nabla\phi)_i^h = \sum_{j \in D_i} \phi_j \nabla W_i(x_i - x_j, h) \omega_j + \sum_{j \in \partial D_i} \phi_j W_i(x_i - x_j, h) \omega_j^\partial n_j \quad (4.20)$$

where ω_j^∂ stands for a weight of the surface element dx'^{d-1}, not to be confused with the volume weight as in (4.3) for instance. If ϕ is a vector as the velocity v for instance, ∇v is a tensor of order d, we use a tensorial product $\otimes$ in the previous expression in order to be clearer.

$$(\nabla v)_i^h = \sum_{j \in D_i} v_j \otimes \nabla W_i(x_i - x_j, h) \omega_j + \sum_{j \in \partial D_i} v_j \otimes n_j W_i(x_i - x_j, h) \omega_j^\partial$$

4.6.2 Consistent formulation for a constant function or global conservation

Applying this equation (4.20) far from a boundary $\partial\Omega$ on a constant function, we should get $(\nabla\phi)_i^h = 0$ if $\sum\limits_{j \in D_i} \nabla_{x_i} W(x_i - x_j, h) \omega_j = 0$, which is the consequence of the relation (4.9).

This is in general a difficult result to get for a general set of disordered particles; for this reason, the differential operator is practically computed as:

$$(\nabla\phi)_i^h = (\nabla\phi)_i^h - \phi_i \nabla(1)^h = \sum_{j \in D_i} (\phi_j - \phi_i) \nabla W_i(x_i - x_j, h) \omega_j \quad (4.21)$$

$$+ \sum_{j \in \partial D_i} (\phi_j - \phi_i) W_i(x_i - x_j, h) \omega_j^\partial n_j$$

Note that this expression does not enforce $\sum\limits_{j\in D_i} \nabla W_i \left(x_i - x_j, h\right) \omega_j = 0$, but it only diminishes its effect in the computation of the differential operator $(\nabla\phi)_i^h$.

Although the previous relation enables the zeroing of the gradient of a constant function for instance, it has a severe drawback. Assume for instance that x is sufficiently far from $\partial\Omega$ such that the last term in (4.19) is zero because $W = 0$ on ∂D. Consider now for instance the case for which $\phi \equiv p$ the pressure. For two neighbouring particles i and j, we have:

$$\left(\phi_j - \phi_i\right) \nabla_{x_i} W \left(x_i - x_j, h\right) = \left(\phi_i - \phi_j\right) \nabla_{x_j} W \left(x_j - x_i, h\right)$$

because of the symmetry of the function W. The consequence is that (4.21) is not able to fulfil the action-reaction principle between pressure forces. For this reason, a different expression of the operator is often used as:

$$(\nabla\phi)_i^h = (\nabla\phi)_i^h + \phi_i \nabla (1)^h = \sum_{j\in D_i} \left(\phi_j + \phi_i\right) \nabla W_i \left(x_i - x_j, h\right) \omega_j \quad (4.22)$$
$$+ \sum_{j\in\partial D_i} \left(\phi_j + \phi_i\right) W_i \left(x_i - x_j, h\right) \omega_j^\partial n_j$$

We will see in a following paragraph 5.2.3 that global conservations of transported flow quantities can be also guarantee with (4.22).

However, it is clear that (4.22) cannot guarantee the zeroing of the gradient of a constant function. For this reason, (4.22) is often used with other techniques as the renormalization, which allows the proper computation of a linear function. Moreover in the literature, the choice of the sign in $(\nabla\phi)_i^h = (\nabla\phi)_i^h \pm \phi_i \nabla (1)^h$ is also a matter of the quantity whose gradient has to be computed. In the following, we will mainly use (4.22), except specifically. We will also call (4.21) the "constant zeroing" formula, and (4.22) the "global conservation" formula.

4.6.3 The use of an adjoint operator of $\nabla\phi$

Following Vila (1999), consider a discrete scalar product defined by:

$$\left(f, g\right)_\Delta = \sum_{i\in\Omega} \omega_i f_i g_i \quad (4.23)$$

We use the scalar product of $\nabla\phi$ defined by (4.21) with g, and we assume the particles to be located far from the boundary $\partial\Omega$:

$$\left((\nabla\phi)_i^h, g\right)_\Delta = \sum_{i\in\Omega} \omega_i \, (\nabla\phi)_i^h \, g_i$$

$$= \sum_{i\in\Omega} \omega_i \left(\sum_{j\in D_i} (\phi_j - \phi_i) \, \nabla W_i \, (x_i - x_j, h) \, \omega_j \right) g_i$$

$$= \sum_{i\in\Omega} \sum_{j\in D_i} \phi_j \nabla W_i \, (x_i - x_j, h) \, \omega_i \omega_j g_i$$

$$- \sum_{i\in\Omega} \sum_{j\in D_i} \phi_i \nabla W_i \, (x_i - x_j, h) \, \omega_i \omega_j g_i$$

In the first term, we interchange the role of the indices i and j:

$$\left((\nabla\phi)_i^h, g\right)_\Delta = \sum_{i\in\Omega} \omega_i \, (\nabla\phi)_i^h \, g_i$$

$$= \sum_{j\in\Omega} \sum_{i\in\Omega} \phi_i \nabla W_j \, (x_j - x_i, h) \, \omega_i \omega_j g_j$$

$$- \sum_{i\in\Omega} \sum_{j\in\Omega} \phi_i \nabla W_i \, (x_i - x_j, h) \, \omega_i \omega_j g_i$$

$$= \sum_{i\in\Omega} \phi_i \omega_i \sum_{j\in\Omega} (\nabla W_j \, (x_j - x_i, h) \, g_j - \nabla W_i \, (x_i - x_j, h) \, g_i) \, \omega_j$$

Using now the symmetry of the kernel function

$$\nabla W_j \, (x_j - x_i, h) = -\nabla W_i \, (x_i - x_j, h)$$

we get:

$$\left((\nabla\phi)_i^h, g\right)_\Delta = \sum_{i\in\Omega} \omega_i \, (\nabla\phi)_i^h \, g_i = -\sum_{i\in\Omega} \phi_i \omega_i \sum_{j\in\Omega} \nabla W_i \, (x_i - x_j, h) \, (g_j + g_i) \, \omega_j$$

$$(4.24)$$

Let us define the adjoint operator (∇^*) of $\nabla\phi$ by:

$$\left((\nabla\phi)_i^h, g\right)_\Delta = -\left(\phi, (\nabla^* g)_i^h\right)_\Delta \Rightarrow \sum_{i\in\Omega} \omega_i \, (\nabla\phi)_i^h \, g_i = -\sum_{i\in\Omega} \omega_i \, (\nabla^* g)_i^h \, \phi_i$$

$$(4.25)$$

By identification of (∇^*) in (4.25) with the previous expression (4.24), we get:

$$(\nabla^* g)_i^h = \sum_{j\in\Omega} (g_j + g_i) \, \nabla W_i \, (x_i - x_j, h) \, \omega_j \qquad (4.26)$$

We observe that the adjoint operator (∇^*) satisfies the global conservation condition.

An important result due to Vila (1999) is that a weak discrete formulation of the transport equation such as (5.1) should be based on the adjoint operator (4.26). This allows the use of the global conservation formulation.

4.6.4 Consistent formulation for a linear function – Renormalization

Johnson and Beisel (1996) and Randles and Liberski (1996) have introduced renormalization in order to improve the accuracy of SPH method.

4.6.4.1 The work of Vila

Vila (1999, 2005) has also shown that this renormalization allows the condition $\frac{\Delta x}{h} \to 0$ to be relaxed, and that it is now sufficient that $\frac{\Delta x}{h} = O\,(1)$ for a consistent approximation. This technique is basically a correction of the differential operators with the help of a weight matrix $B(x)$, also called the renormalization matrix.

The equation (4.21) (for instance) is then modified as:

$$(\nabla\phi)_i^h = B\,(x)\,(\nabla\phi)_i^h - \phi_i B\,(x)\,\nabla\,(1)^h \tag{4.27}$$

$B(x)$ is chosen such that (4.27) is consistent for linear function. Vila (2005) has shown that this is performed if and only if the matrix $B(x)$ is given by

$$
\begin{aligned}
B\,(x) &= E\,(x)^{-1} \\
E\,(x)^{\alpha\beta} &= \sum_{j\in D(x)} \left(x_j^\beta - x^\beta\right) \frac{\partial W(x-x_j,h)}{\partial x^\alpha}\,\omega_j
\end{aligned}
\tag{4.28}
$$

where x_j^β stands for the component β of the vector position x_j and $E\,(x)^{\alpha\beta}$ is the (α,β) component of the matrix $E(x)$, which appears as an SPH approximation of

$$E\,(x)^{\alpha\beta} = \frac{\partial x^\beta}{\partial x^\alpha} - x^\beta \frac{\partial\,(1)}{\partial x^\alpha} \approx \delta^{\alpha\beta}$$

For (4.21) and according to (4.27), we have then:

$$
\begin{aligned}
(\nabla\phi)_i^h &= B\,(x)\,(\nabla\phi)_i^h - B\,(x)\,\phi_i\nabla\,(1)^h = \\
&\sum_{j\in D_i} (\phi_j - \phi_i)\,B_i\nabla W_i\,(x_i - x_j,h)\,\omega_j + \sum_{j\in\partial D_i} (\phi_j - \phi_i)\,B_i W_i\,(x_i - x_j,h)\,\omega_j^\partial n_j
\end{aligned}
\tag{4.29}
$$

Vila (1999) has also introduced a slightly more compact formulation that comes from the idea that $B(x)$ is applied on the gradient of kernel ∇W and not on $(\nabla\phi)_i^h$:

$$
\begin{aligned}
(\nabla\phi)_i^h = &\sum_{j\in D_i} (\phi_j B_j - \phi_i B_i)\,\nabla W_i\,(x_i - x_j,h)\,\omega_j \\
&+ \sum_{j\in\partial D_i} (\phi_j - \phi_i)\,W_i\,(x_i - x_j,h)\,\omega_j^\partial n_j
\end{aligned}
$$

Note that the summation on the boundary ∂D_i does not include the renormalisation matrix as the gradient of kernel ∇W is now corrected and not $(\nabla\phi)_i^h$.

Considering also that $B_j \approx B_i$ because the two particles are very closed from each other, we can also write that:

$$\phi_j B_j - \phi_i B_i = (\phi_j - \phi_i) \frac{(B_i + B_j)}{2} + \frac{(\phi_i + \phi_j)}{2} (B_j - B_i)$$
$$\approx (\phi_j - \phi_i) \frac{(B_i + B_j)}{2}$$

Vila (2005) has also shown that all the results of convergence are also valid for a symmetric variant of the method. Stating $B_{ij} = \frac{1}{2} (B_i + B_j)$:

$$(\nabla \phi)_i^h = \sum_{j \in D_i} (\phi_j - \phi_i) B_{ij} \nabla W_i (x_i - x_j, h) \omega_j \qquad (4.30)$$
$$+ \sum_{j \in \partial D_i} (\phi_j - \phi_i) W_i (x_i - x_j, h) \omega_j^\partial n_j$$

and for this symmetric variant $B_{ij} \nabla W_i (x_i - x_j, h) = -B_{ji} \nabla W_j (x_j - x_i, h)$

REMARK 4.7   This renormalization method has some similarity with the correction of the kernel function, as introduced by the RKPM method (Reproducing Kernel Particle Method) proposed by Liu, Jun and Zhang (1995) or the MLS method (Moving Least Square) proposed by Dilts (1999, 2000). Although the renormalization is less general than RKPM or MLS, it provides a very useful mathematical framework for stability and accuracy analysis as shown by Vila (2005). □

4.6.4.2   The approach of Bonet and Lok (1999)

Bonet and Lok have introduced a specific method of correction of the gradient with the purpose of preserving the angular momentum for non-dissipative systems and in the absence of external forces. Note that rotational invariance requires the correct evaluation of linear velocity fields. In this paragraph, we restrict the field ϕ to the velocity field v.

Consider a solid body rotation defined by an angular velocity vector ω with components $(\omega_x, \omega_y, \omega_z)$. The matrix of rotation M_ω is also given by:

$$M_\omega = \begin{bmatrix} 0 & -\omega_z & \omega_y \\ \omega_z & 0 & -\omega_x \\ -\omega_y & \omega_x & 0 \end{bmatrix}$$

The velocity vector is given by $v = \omega \wedge x = M_\omega x$. The gradient of v is:

$$\nabla v = M_\omega \qquad (4.31)$$

The gradient of the velocity field can be computed with the SPH approximation such as (4.21) for instance.

$$(\nabla v)_i^h = \sum_{j \in D_i} (v_j - v_i) \otimes \nabla W_i (x_i - x_j, h) \, \omega_j$$

$$+ \sum_{j \in \partial D_i} (v_j - v_i) \otimes n_j W_i (x_i - x_j, h) \, \omega_j^\partial$$

Note that because ∇v is a tensor of order d, we have then introduced a tensorial product $\otimes$ in the previous expression in order to be clearer.

We use now the expression of $v = M_\omega x$ where M_ω is constant for a solid rotation, and is then assumed identical between particles i and j.

$$(\nabla v)_i^h = \sum_{j \in D_i} (M_\omega x_j - M_\omega x_i) \otimes \nabla W_i (x_i - x_j, h) \, \omega_j$$

$$+ \sum_{j \in \partial D_i} (M_\omega x_j - M_\omega x_i) \otimes W_i (x_i - x_j, h) \, \omega_j^\partial n_j$$

$$(\nabla v)_i^h = M_\omega [\sum_{j \in D_i} (x_j - x_i) \otimes \nabla W (x_i - x_j, h) \, \omega_j$$

$$+ \sum_{j \in \partial D_i} (x_j - x_i) \otimes W_i (x_i - x_j, h) \, \omega_j^\partial n_j]$$

The last term in the previous expression is zero except if ∂D_i intercepts the boundary $\partial \Omega$. The condition for preserving the solid body rotation is then according to (4.31):

$$\sum_{j \in D_i} (x_j - x_i) \otimes \nabla W_i (x_i - x_j, h) \, \omega_j + \sum_{j \in \partial D_i} (x_j - x_i) \otimes W_i (x_i - x_j, h) \, \omega_j^\partial n_j = I$$

(4.32)

Bonet and Lok have proposed three practical implementations of condition (4.32).

Gradient correction The first implementation of (4.32) is similar to the renormalization as described by Vila (2005).

Consider equation (4.29) with $\phi = v$ in the form:

$$(\nabla v)_i^h = (\nabla v)_i^h B_i (x) - v_i \nabla (1)^h B_i (x) \qquad (4.33)$$

$$= \sum_{j \in D_i} (v_j - v_i) \otimes \nabla W_i (x_i - x_j, h) \, \omega_j B_i$$

$$+ \sum_{j \in \partial D_i} (v_j - v_i) \otimes W_i (x_i - x_j, h) \, \omega_j^\partial B_i n_j$$

$$= M_\omega [\sum_{j \in D_i} (x_j - x_i) \otimes \nabla W_i (x_i - x_j, h) \, \omega_j$$

$$+ \sum_{j \in \partial D_i} (x_j - x_i) \otimes W_i (x_i - x_j, h) \, \omega_j^\partial n_j] B_i$$

The condition for preserving the solid body rotation is then

$$[\sum_{j \in D_i} (x_j - x_i) \otimes \nabla W_i (x_i - x_j, h) \, \omega_j$$
$$+ \sum_{j \in \partial D_i} (x_j - x_i) \otimes W_i (x_i - x_j, h) \, \omega_j^{\partial} n_j] B_i = I$$

or

$$B_i = [\sum_{j \in D_i} (x_j - x_i) \otimes \nabla W_i (x_i - x_j, h) \, \omega_j$$
$$+ \sum_{j \in \partial D_i} (x_j - x_i) \otimes W_i (x_i - x_j, h) \, \omega_j^{\partial} n_j]^{-1} \qquad (4.34)$$

Equation (4.34) is similar to (4.28), but note the position of the matrix $B(x)$ in (4.33) (because of the gradient of a vector) compared to (4.29) (which is correct for the gradient of a scalar ϕ).

These expressions (4.32) and (4.34) will ensure that the gradient of a linear vector field is exactly evaluated; moreover, if the vector field is a velocity field, then angular momentum is preserved provided that the internal forces are derived from variational principle (Bonet and Lok, 1999). Nestor and al. (2009) uses this method also in FVPM.

REMARK 4.8 If we use the (4.22), the "global conservation" formula for the gradient, then we get:

$$B_i = [\sum_{j \in D_i} (x_j + x_i) \otimes \nabla W_i (x_i - x_j, h) \, \omega_j$$
$$+ \sum_{j \in \partial D_i} (x_j + x_i) \otimes W_i (x_i - x_j, h) \, \omega_j^{\partial} n_j]^{-1} \qquad (4.35)$$

⬚

Kernel correction This technique is similar to the MLS method of Liu, already mentioned in the renormalization paragraph.

Consider a new corrected kernel ψ such that (4.3) is now written as:

$$\phi^h (x) = \sum_{j \in D(x)} \phi (x_j, t) \, \psi (x - x_j, h) \, \omega_j (x_j, t)$$

with

$$\psi (x - x_j, h) = W (x - x_j, h) \, \alpha (x) \, [1 + \beta (x) (x - x_j)] \qquad (4.36)$$

The scalar $\alpha(x)$ and vector $\beta(x)$ are computed by enforcing that any linear vector field is exactly evaluated, that is:

$$\phi^h(x) = \phi_0 + \phi_1 x = \sum_{j \in D(x)} (\phi_0 + \phi_1 x_j) \, \psi(x - x_j, h) \, \omega_j(x_j, t)$$

As ϕ_0 and ϕ_1 are arbitrary, we get:

$$\sum_{j \in D(x)} \psi(x - x_j, h) \, \omega_j(x_j, t) = 1 \tag{4.37}$$

$$\sum_{j \in D(x)} (x - x_j) \, \psi(x - x_j, h) \, \omega_j(x_j, t) = 0 \tag{4.38}$$

These two equations enable the evaluations of $\alpha(x)$ and $\beta(x)$ by using (4.36). After some computations we get finally:

$$\beta(x) = \left[\sum_{j \in D(x)} (x - x_j) \otimes (x - x_j) \, W(x - x_j, h) \, \omega_j \right]^{-1}$$
$$\sum_{j \in D(x)} (x_j - x) \, W(x - x_j, h) \, \omega_j \tag{4.39}$$

$$\alpha(x) = \frac{1}{\sum\limits_{j \in D(x)} [1 + \beta(x)(x - x_j)] \, W(x - x_j, h) \, \omega_j} \tag{4.40}$$

This will allow condition (4.32) expressed in term of the corrected kernel ψ to be satisfied thereby enforcing also the conservation of the solid body rotation.

REMARK 4.9 Although the evaluations of the scalar $\alpha(x)$ and vector $\beta(x)$ are easy, as it only requires the evaluation of a $d \times d$ matrix, the evaluation of the gradient of ψ may be very costly because the α and β coefficients depend on x through the local distribution of particles. It is then tempting to consider the case $\beta(x) = 0$, then

$$\alpha(x) = \frac{1}{\sum\limits_{j \in D(x)} W(x - x_j, h) \, \omega_j} \Rightarrow \tag{4.41}$$

$$\psi(x - x_j, h) = \frac{W(x - x_j, h)}{\sum\limits_{j \in D(x)} W(x - x_j, h) \, \omega_j} \tag{4.42}$$

and

$$\phi^h(x) = \frac{\sum\limits_{j \in D(x)} \phi(x_j, t) \, W(x - x_j, h) \, \omega_j(x_j, t)}{\sum\limits_{j \in D(x)} W(x - x_j, h) \, \omega_j} \tag{4.43}$$

This is just the Shepard's interpolation of ϕ (Shepard, 1968). Although (4.43) fails to preserve the rotational invariance, according to Bonet and Lok it provides an improved interpolation especially close to the boundary $\partial\Omega$. ☐

REMARK 4.10   Using the Shepard's kernel ψ (4.42), we fulfil automatically the condition (4.9) for at least a first order approximation. ☐

Mixed kernel correction and gradient correction   It is possible to combine the constant kernel correction proposed by (4.41) and a correction of the gradient by means of a renormalization matrix. This will enable also to recover the preservation of linear gradients that we lose with $\beta(x) = 0$. We use then (4.43):

$$\phi^h(x) = \sum_{j \in D(x_i)} \phi(x_j, t)\, \psi(x - x_j, h)\, \omega_j(x_j, t)$$

$$= \frac{\displaystyle\sum_{j \in D(x_i)} \phi(x_j, t)\, W(x - x_j, h)\, \omega_j(x_j, t)}{\displaystyle\sum_{j \in D_i} W(x - x_j, h)\, \omega_j}$$

with the gradient of ϕ given by $(\nabla\phi)_i^h = (\nabla\phi)_i^h B(x)$ in function of the corrected kernel ψ

$$(\nabla\phi)_i^h - (\nabla\phi)_i^h B(x)$$

$$= \sum_{j \in D_i} \phi_j \otimes \nabla_{x_i} \psi(x_i - x_j, h)\, \omega_j B_i$$

$$+ \sum_{j \in \partial D_i} \phi_j \otimes \psi(x_i - x_j, h)\, \omega_j^\partial n_j B_i$$

In the previous expression of $(\nabla\phi)_i^h$, we have considered ϕ as a vector. Note that the term $-\phi_i$ is no more necessary in this expression because the Shepard's approximation (4.43) allows the gradient of a constant field to vanish.

Applying this expression for a gradient of a velocity field created by a solid rotation, we get:

$$(\nabla v)_i^h = M_\omega \Big[\sum_{j \in D_i} x_j \otimes \nabla_{x_i} \psi(x_i - x_j, h)\, \omega_j \tag{4.44}$$

$$+ \sum_{j \in \partial D_i} x_j \otimes \psi(x_i - x_j, h)\, \omega_j^\partial n_j \Big] B_i$$

In order to preserve the solid body rotation, the matrix B_i is given by

$$B_i = \left[\sum_{j \in D_i} x_j \otimes \nabla_{x_i} \psi(x_i - x_j, h)\, \omega_j + \sum_{j \in \partial D_i} x_j \otimes \psi(x_i - x_j, h)\, \omega_j^\partial n_j \right]^{-1}$$

$$\tag{4.45}$$

The previous expression was given by Bonet and Lok without the boundary term.

We can also modify the Bonet and Lok's expression in order to fulfil the action-reaction principle:

$$(\nabla\phi)_i^h = \sum_{j\in D_i} (\phi_j + \phi_i) \otimes \nabla_{x_i}\psi\,(x_i - x_j, h)\,\omega_j B_i$$

$$+ \sum_{j\in \partial D_i} (\phi_j + \phi_i) \otimes \psi\,(x_i - x_j, h)\,\omega_j^\partial n_j B_i$$

$$B_i = [\sum_{j\in D_i} (x_j + x_i) \otimes \nabla_{x_i}\psi\,(x_i - x_j, h)\,\omega_j \tag{4.46}$$

$$+ \sum_{j\in \partial D_i} (x_j + x_i) \otimes \psi\,(x_i - x_j, h)\,\omega_j^\partial n_j]^{-1}$$

4.6.5 Derivatives with a Shepard's kernel ψ

4.6.5.1 About the Shepard's kernel

If W_i stands for the kernel function at x_i, the Shepard's kernel ψ_i is defined by

$$\psi_i\,(x_i, x_i - x_j, h) = \frac{W_i\,(x_i - x_j, h)}{\sum\limits_{j\in D(x_i)} W_i\,(x_i - x_j, h)\,\omega_j}$$

or

$$\psi_i\,(x_i, q_j, h) = \frac{W_i\,(q_j, h)}{\sum\limits_{j\in D(x_i)} W_i\,(q_j, h)\,\omega_j}, \quad q_j = x_i - x_j \tag{4.47}$$

Note that $W_i\,(q, h) = W_i\,(-q, h)$, $\nabla_q W_i\,(q, h) = -\nabla_q W_i\,(-q, h)$ and $\psi_i\,(x_i, q_j, h) = \psi_i\,(x_i, -q_j, h)$, $\nabla_q \psi_i\,(x_i, q_j, h) = -\nabla_q \psi_i\,(x_i, -q_j, h)$
If $h_i = h_j$ then for two different particles we get also

$$W_i\,(q, h) = W_j\,(-q, h), \quad \nabla_q W_i\,(q, h) = -\nabla_q W_j\,(-q, h)$$

We will note thereafter

$$\sigma\,(x_i) = \sum_{j\in D(x_i)} W_i\,(q_j, h)\,\omega_j \tag{4.48}$$

But for SPH approximation, in general $\sigma_i \neq \sigma_j \neq 1$ particularly close to a boundary $\partial\Omega$. Then even if $h_i = h_j$ we get also

$$\psi_i\,(x_i, q_j, h) \nsucc \psi_j\,(x_j, -q_j, h), \quad \nabla_q \psi_i\,(x_i, q_j, h) \neq \nabla_q \psi_i\,(x_j, -q_j, h)$$

The SPH average of a function ϕ is then with the Shepard's kernel:

$$\phi^h\left(x_i\right) = \sum_{j \in D(x_i)} \phi\left(x_j, t\right) \psi_i\left(x_i, q_j, h\right) \omega_j\left(x_j, t\right)$$

$$= \frac{\displaystyle\sum_{j \in D(x_i)} \phi\left(x_j, t\right) W_i\left(q_j, h\right) \omega_j\left(x_j, t\right)}{\displaystyle\sum_{j \in D(x_i)} W_i\left(q_j, h\right) \omega_j}$$

Note that the definition of the Shepard's kernel allows writing a condition that enables the preservation of constant functions:

$$\int_{D(x)} \psi\left(x, x - x', h\right) dx'^d = 1 \tag{4.49}$$

or in a discrete form:

$$\sum_{j \in D(x_i)} \psi_i\left(x_i, q_j, h\right) \omega_j\left(x_j, t\right) = 1 \tag{4.50}$$

We will note thereafter:

$$D_i = D\left(x_i\right)$$

$$W_{ij} = W_i\left(q_j, h\right)$$

$$\psi_{ij} = \psi_i\left(x_i, q_j, h\right) = \frac{W_i\left(q_j, h\right)}{\sigma\left(x_i\right)}$$

$$\sigma_i = \sigma\left(x_i\right) = \sum_{j \in D_i} W_{ij} \omega_j$$

$$\nabla \psi_{ij} = \nabla_q \psi_i\left(x_i, q_j, h\right)$$

$$\dot{x}_i = v_0$$

From (4.49) or (4.50) and using the SPH derivative of the constant function $\phi = 1$, we get also:

$$\int_{D(x)} \nabla_x \psi\left(x, x - x'\right) dx'^d + \int_{\partial D(x)} \psi\left(x, x - x'\right) n\left(x'\right) dx'^{d-1} = 0$$

$$\sum_{j \in D_i} \nabla \psi_i\left(x_i, q_j, h\right) \omega_j + \sum_{j \in \partial D_i} \psi_i\left(x_i, q_j, h\right) \omega_j^\partial n_j = 0$$

or

$$\sum_{j \in D_i} \nabla \psi_{ij} \omega_j + \sum_{j \in \partial D_i} \psi_{ij} \omega_j^\partial n_j = 0 \tag{4.51}$$

4.6.5.2 Derivatives of the Shepard's kernel

The gradient $\nabla \psi_i$ is easy to compute:

$$\nabla \psi_{ij} = \frac{\nabla W_{ij}}{\sigma_i} - \frac{\psi_{ij}}{\sigma_i} \nabla \sigma_i = \frac{\nabla W_{ij}}{\sigma_i} - \frac{\psi_{ij}}{\sigma_i} \sum_{j \in D_i} (\nabla W_{ij} \omega_j + W_{ij} \nabla \omega_j)$$

and particularly when the weight ω_j is constant in Ω:

$$\nabla \psi_{ij} = \frac{\nabla W_{ij}}{\sigma_i} - \frac{W_{ij}}{\sigma_i^2} \sum_{j \in D_i} (\nabla W_{ij} \omega_j) \tag{4.52}$$

We compute the time derivative of ψ in function of the gradient of W. Starting from $\psi_{ij} = \frac{W_{ij}}{\sigma_i}$ and according to (4.48):

$$\frac{\partial \psi_{ij}}{\partial t} = \frac{1}{\sigma_i} \frac{\partial W_{ij}}{\partial t} - \frac{\psi_{ij}}{\sigma_i} \frac{\partial \sigma_i}{\partial t} = \frac{1}{\sigma_i} \frac{\partial W_{ij}}{\partial t} - \frac{\psi_{ij}}{\sigma_i} \sum_{j \in D_i} \left(\frac{\partial W_{ij}}{\partial t} \omega_j + W_{ij} \frac{\partial \omega_j}{\partial t} \right)$$

In the framework of an ALE description, we consider a particular Lagrangian derivative with a transport field $v_0 = \dot{x}$ Assuming that h is constant, then

$$\frac{d}{dt} W_{ij} = 0 \Rightarrow \frac{\partial W_{ij}}{\partial t} + \dot{x}_i \nabla W_{ij} = \frac{\partial W_{ij}}{\partial t} - \dot{x}_j \nabla W_{ij} = 0$$

$$\frac{\partial \psi_{ij}}{\partial t} = -\frac{\dot{x}_i}{\sigma_i} \nabla W_{ij} - \frac{\psi_{ij}}{\sigma_i} \sum_{j \in D_i} \left(-\dot{x}_i \omega_j \nabla W_{ij} + W_{ij} \frac{\partial \omega_j}{\partial t} \right)$$

$$\frac{\partial \psi_{ij}}{\partial t} = -\frac{\dot{x}_i}{\sigma_i} \nabla W_{ij} + \frac{\dot{x}_i \psi_{ij}}{\sigma_i} \sum_{j \in D_i} (\omega_j \nabla W_{ij}) - \frac{\psi_{ij}}{\sigma_i} \sum_{j \in D_i} \left(W_{ij} \frac{\partial \omega_j}{\partial t} \right)$$

Using the transport of ω_i in equation (4.4):

$$\frac{1}{\omega_i} \frac{d\omega_i}{dt} = div\,(v_o) \Rightarrow \frac{\partial \omega_i}{\partial t} = -\dot{x}_i \nabla \omega_i + \omega_i div\,(v_o)$$

$$\frac{\partial \psi_{ij}}{\partial t} = -\frac{\dot{x}_i}{\sigma_i} \nabla W_{ij}$$

$$+ \dot{x}_i \frac{\psi_{ij}}{\sigma_i} \sum_{j \in D_i} (\omega_j \nabla W_{ij})$$

$$- \frac{\psi_{ij}}{\sigma_i} \sum_{j \in D_i} (W_{ij} \{ -\dot{x}_j \nabla \omega_j + \omega_j div\,(v_o) \})$$

or in function of the gradient of ψ only: $\frac{\nabla W_{ij}}{\sigma_i} = \nabla \psi_{ij} + \frac{\psi_{ij}}{\sigma_i} \nabla \sigma_i$

$$\frac{\partial \psi_{ij}}{\partial t} = -\dot{x}_i \left(\nabla \psi_{ij} + \frac{\psi_{ij}}{\sigma_i} \nabla \sigma_i \right) \qquad (4.53)$$

$$+ \dot{x}_i \psi_{ij} \sum_{j \in D_i} \left(\omega_j \left(\nabla \psi_{ij} + \frac{\psi_{ij}}{\sigma_i} \nabla \sigma_i \right) \right)$$

$$- \frac{\psi_{ij}}{\sigma_i} \sum_{j \in D_i} \left(W_{ij} \left\{ -\dot{x}_j \nabla \omega_j + \omega_j div\left(v_o \right) \right\} \right)$$

Using the normalisation of the kernel (4.50) and its gradient (4.51):

$$\frac{\partial \psi_{ij}}{\partial t} = -\dot{x}_i \nabla \psi_{ij} + \dot{x}_i \psi_{ij} \sum_{j \in D_i} \left(\omega_j \nabla \psi_{ij} \right)$$

$$- \frac{\psi_{ij}}{\sigma_i} \sum_{j \in D_i} \left(W_{ij} \left\{ -\dot{x}_j \nabla \omega_j + \omega_j div\left(v_o \right) \right\} \right)$$

$$\frac{\partial \psi_{ij}}{\partial t} = -\dot{x}_i \nabla \psi_{ij} - \dot{x}_i \psi_{ij} \sum_{j \in \partial D_i} \psi_{ij} \omega_j^\partial n_j \qquad (4.54)$$

$$- \psi_{ij} \sum_{j \in D_i} \left(\psi_{ij} \left\{ -\dot{x}_j \nabla \omega_j + \omega_j div\left(v_o \right) \right\} \right)$$

From (4.54) we get immediately the Lagrangian derivative of the Shepard's kernel:

$$\frac{d\psi_{ij}}{dt} = \frac{\partial \psi_{ij}}{\partial t} + \dot{x}_i \nabla \psi_{ij} \qquad (4.55)$$

$$= -\dot{x}_i \psi_{ij} \sum_{j \in \partial D_i} \psi_{ij} \omega_j^\partial n_j - \psi_{ij} \sum_{j \in D_i} \left(\psi_{ij} \left\{ -\dot{x}_j \nabla \omega_j + \omega_j div\left(v_o \right) \right\} \right)$$

We observe that if the transport field v_0 of the particles is divergence free, and if the initial set of particles has uniform weight ω_i over Ω, then we get

$$\frac{d\psi_{ij}}{dt} = \frac{\partial \psi_{ij}}{\partial t} + \dot{x}_i \nabla \psi_{ij} = -\dot{x}_i \psi_{ij} \sum_{j \in \partial D_i} \psi_{ij} \omega_j^\partial n_j \qquad (4.56)$$

The second member is also zero if ∂D_i does not intercept $\partial \Omega$. In this particular case, we recover the particular feature of the original kernel W whose Lagrangian derivative is zero everywhere.

4.7   Using a Taylor series expansion

Instead of applying sequentially the Shepard's kernel in order to get C^0 consistency, followed by the renormalisation in order to get a C^1 consistency as in Bonet and Lok approach, Liu and Liu (2006) have proposed to derive a SPH approximation that allows to get simultaneously C^1 consistency. Their method is to write a Taylor series expansion for a function $\phi(x)$ around a position x_i:

$$\phi\left(x'\right) = \phi_i + \sum_{\alpha}\left(x'_\alpha - x_{\alpha,i}\right)\left(\frac{\partial\phi}{\partial x_\alpha}\right)_i + \sum_\alpha\sum_\gamma\frac{\left(x'_\alpha - x_{\alpha,i}\right)\left(x'_\gamma - x_{\gamma,i}\right)}{2}\left(\frac{\partial^2\phi}{\partial x_\alpha\partial x_\gamma}\right)_i$$
$$+ o\left(x'_\alpha - x_{\alpha,i}\right)^3$$

$$(4.57)$$

In this expression, α and γ vary between 1 and d the number of dimensions. We neglect thereafter all terms of order $o\left(x'_\alpha - x_{\alpha,i}\right)^2$, then all second order derivatives $\left(\frac{\partial^2\phi}{\partial x_\alpha\partial x_\gamma}\right)_i$ disappear.

Multiply then (4.57) first by the kernel function W, and then by its derivatives $\frac{\partial W}{\partial x_\beta}$, and integrate over the support of the kernel $D(x_i)$, we get $d + 1$ equations:

$$\int_{D(x_i)}\phi\left(x'\right)W_i\left(x' - x_i\right)dx'^d = \phi_i\int_{D(x_i)}W_i\left(x' - x_i\right)dx'^d \qquad (4.58)$$

$$+ \sum_{\alpha=1,d}\left(\frac{\partial\phi}{\partial x_\alpha}\right)_i\int_{D(x_i)}\left(x'_\alpha - x_{\alpha,i}\right)W_i\left(x' - x_i\right)dx'^d$$

$$\int_{D(x_i)}\phi\left(x'\right)\frac{\partial W_i\left(x' - x_i\right)}{\partial x_\beta}dx'^d = \phi_i\int_{D(x_i)}\frac{\partial W_i\left(x' - x_i\right)}{\partial x_\beta}dx'^d$$

$$+ \sum_{\alpha=1,d}\left(\frac{\partial\phi}{\partial x_\alpha}\right)_i\int_{D(x_i)}\left(x'_\alpha - x_{\alpha,i}\right)\frac{\partial W_i\left(x' - x_i\right)}{\partial x_\beta}dx'^d$$

Or in a discrete SPH form:

$$\sum_{j\in D(x_i)}\phi_j W\left(x_i - x_j, h\right)\omega_j = \phi_i\sum_{j\in D(x_i)}W\left(x_i - x_j, h\right)\omega_j \qquad (4.59)$$

$$+ \sum_{\alpha=1,d}\left(\frac{\partial\phi}{\partial x_\alpha}\right)_i\sum_{j\in D(x_i)}\left(x_{\alpha,j} - x_{\alpha,i}\right)W\left(x_i - x_j, h\right)\omega_j$$

$$\sum_{j\in D(x_i)}\phi_j\frac{\partial W_i\left(x' - x_i, h\right)}{\partial x_\beta}\omega_j = \phi_i\sum_{j\in D(x_i)}\frac{\partial W_i\left(x' - x_i, h\right)}{\partial x_\beta}\omega_j$$

$$+ \sum_{\alpha=1,d}\left(\frac{\partial\phi}{\partial x_\alpha}\right)_i\sum_{j\in D(x_i)}\left(x_{\alpha,j} - x_{\alpha,i}\right)\frac{\partial W_i\left(x' - x_i, h\right)}{\partial x_\beta}\omega_j\left(x_j\right)$$

The system of equations (4.59) is then used in order to compute the $d + 1$ unknowns

$$\left(\phi_i, \left(\frac{\partial\phi}{\partial x_\alpha}\right)_i\right)_{\alpha=1,d}$$

Liu and Liu (2006) have obtained very good results for model problems with this method, even close to the boundary $\partial\Omega$ of the domain.

REMARK 4.11   We observe that the original kernel is not modified, thereby keeping the basic properties of W (positive, symmetric).   ⬚

REMARK 4.12   Compared to classical SPH method, we need to solve a $(d+1, d+1)$ system for each particle.   ⬚

REMARK 4.13   This method can be extended in principle to higher order by keeping for instance the second order derivatives in equation (4.57), with a price to be paid in the computation time.   ⬚

4.8 Concluding remarks

We have given the formulation of the SPH method in a general way because the particles are transported with an arbitrary velocity v_0, which is not necessarily equal to the flow velocity v. This is a framework of the SPH-ALE method.

The advantage of this SPH-ALE method is clearly to provide a tool for computing a quantity in a domain Ω where a disordered set of particles is allowed to move at a specific velocity v_0. In principle, the computation of a differential operator applied on flow quantities is no more difficult. Because the choice of the transport velocity v_0 is left to the user, v_0 appears as a free parameter that can be used for specific purpose as improving the accuracy of the numerical results for instance.

The disadvantage of SPH methods is first the great number of particles that is required to get sufficient accuracy; as shown by Vila (1999), the basic scheme requires $\frac{\Delta x}{h} \to 0$.

The second disadvantage is clearly the difficulty of getting sufficient accuracy for the computation of differential operators. Various techniques have been presented that allow at least treating linear fields. We recommend the use of the renormalization technique, that Vila (2005) has shown to allow also $\frac{\Delta x}{h} = O(1)$; this is an important evolution, because we are now justified to use a finite set of particles. This renormalization technique is very close to the method proposed by Bonet and Lok (1999), which has been shown to preserve also the angular momentum. Finally, one of the most promising methods seems to be the mixed kernel correction and gradient correction also proposed by Bonet and Lok: this is a combination of the Shepard's kernel, which allows the proper treatment of constant field, coupled to a renormalization matrix. This technique has some common features with the Finite Volume Particle Method (FVPM), which will be presented at the end of this text.

An interesting point has been underlined in the care to preserve global flow quantities simultaneously with the proper computation of a constant field; in general, these two purposes can be reached with quite different and sometimes incompatible schemes. We have chosen to underline the interaction with the boundary $\partial\Omega$ of the domain; the boundary treatment is a key parameter in engineering applications, but most of the references do not treat in general this important point.

Chapter 5

Application of SPH methods to conservation equations

Francis Leboeuf

Ecole Centrale de Lyon - LMFA
36 avenue Guy de Collongue
69134 Ecully Cedex, France

Jean-Christophe Marongiu

ANDRITZ-HYDRO
Rue des Deux-Gares 6
CH 1800 Vevey, Suisse

5.1   General form of conservation equations

Consider a conservation equation written in the following form:

$$\frac{\partial \phi}{\partial t} + \nabla . F\left(\phi\right) = S\left(\phi\right) \tag{5.1}$$

For instance for a fluid, ϕ stands here for the vector of unknowns $\phi\left(\rho, \rho v, \rho e\right)$ where ρ is the density, v the velocity vector and e is the specific internal energy. $F(\phi)$ is the matrix of flux and $S(\phi)$ is a source term.

$$F\left(\phi\right) = \begin{pmatrix} \rho v \\ \rho\left(v \otimes v + pI\right) \\ \left(\rho e + p\right) v \end{pmatrix} \tag{5.2}$$

We look for a weak SPH formulation for (5.1) written in an ALE form (Arbitrary Lagrange Euler).

5.2   Weak SPH-ALE formulation of the conservation equations

5.2.1   SPH approximation of conservation equations

Let us perform the SPH approximation of the conservation equations. Multiply (5.1) by a kernel function W and integrate over a domain $D(x)$ linked to a particle; note that W could be either one of the function defined in paragraph 4.5 or a corrected kernel such as the Shepard's function ψ in equation (4.42)

$$\int_{D(x)} \left(\frac{\partial \phi}{\partial t} + \nabla_{x'}.F(\phi) \right) W dx'^d = \int_{D(x)} S(\phi) W dx'^d$$

Integrate by parts to get:

$$\int_{D(x)} \left(\frac{\partial \phi W}{\partial t} + \nabla_{x'}.(F(\phi)W) \right) dx'^d - \int_{D(x)} \phi \frac{\partial W}{\partial t} dx'^d - \quad (5.3)$$

$$\int_{D(x)} F(\phi) \nabla_{x'} W dx'^d = \int_{D(x)} S(\phi) W dx'^d$$

In order to compute $\frac{\partial W}{\partial t}$, we will use the fact that h is constant both in space and time. According to (4.12), $W = W\left(h, q = \frac{x}{h}\right)$ and we can write with v_0 the transport velocity field of the particles: $\frac{dW}{dt} = \frac{\partial W}{\partial t} + v_0 \nabla W$ and under the previous hypothesis $\frac{dW}{dt} = 0$ so that:

$$\frac{\partial W}{\partial t} = -v_o \nabla W \quad (5.4)$$

Note that v_o can be different from the velocity of the fluid v. If $v_o = 0$, then we get a Eulerian description, while for $v_o = v$ we get a pure Lagrangian description.

In order to compute the Lagrangian derivative of the SPH approximation of ϕ: $\int_{D(x)} \phi W dx'^d$ we invoke the Reynolds theorem, which allows writing:

$$\frac{d_{v_o}}{dt} \int_{D(x)} \phi W dx'^d = \int_{D(x)} \frac{\partial \phi W}{\partial t} dx'^d + \int_{D(x)} \nabla_{x'}.(\phi W \otimes v_o) dx'^d \quad (5.5)$$

We then replace into (5.5), the $\int_{D(x)} \frac{\partial \phi W}{\partial t} dx'^d$ term deduced from (5.3), and use (5.4) to get:

$$\int_{D(x)} \frac{\partial \phi W}{\partial t} dx'^d = \int_{D(x)} \phi \frac{\partial W}{\partial t} dx'^d + \int_{D(x)} F(\phi) \nabla_{x'} W dx'^d$$

$$- \int_{D(x)} \nabla_{x'} \cdot (F(\phi) W) dx'^d + \int_{D(x)} S(\phi) W dx'^d$$

$$\frac{d_{v_o}}{dt} \int_{D(x)} \phi W dx'^d = \int_{D(x)} \phi \frac{\partial W}{\partial t} dx'^d + \int_{D(x)} F(\phi) \nabla_{x'} W dx'^d$$

$$- \int_{D(x)} \nabla_{x'} \cdot (F(\phi) W) dx'^d$$

$$+ \int_{D(x)} \nabla_{x'} \cdot (\phi W \otimes v_o) dx'^d + \int_{D(x)} S(\phi) W dx'^d$$

$$\frac{d_{v_o}}{dt} \int_{D(x)} \phi W dx'^d = - \int_{D(x)} \phi \otimes v_o \nabla_{x'} W dx'^d + \int_{D(x)} F(\phi) \nabla_{x'} W dx'^d$$

$$- \int_{D(x)} \nabla_{x'} \cdot (F(\phi) W) dx'^d$$

$$+ \int_{D(x)} \nabla_{x'} \cdot (\phi W \otimes v_o) dx'^d + \int_{D(x)} S(\phi) W dx'^d$$

$$\int_{D(x)} \frac{\partial \phi W}{\partial t} dx'^d = \int_{D(x)} \phi \frac{\partial W}{\partial t} dx'^d + \int_{D(x)} F(\phi) \nabla_{x'} W dx'^d$$

$$- \int_{D(x)} \nabla_{x'} \cdot (F(\phi) W) dx'^d$$

$$+ \int_{D(x)} S(\phi) W dx'^d$$

In the previous equation, we will use $\nabla_{x'} W = -\nabla_x W$; this seems to be more stable in practical situations. Collecting similar terms, invoking Green's theorem for the divergence term, we get finally:

$$\frac{d_{v_o}}{dt} \int_{D(x)} \phi W dx'^d = - \int_{D(x)} (F(\phi) - \phi \otimes v_o) \nabla_x W dx'^d \qquad (5.6)$$

$$- \int_{\partial D(x)} (F(\phi) - \phi \otimes v_o) W n dx'^{d-1} + \int_{D(x)} S(\phi) W dx'^d$$

Where n is the normal to $\partial D(x)$. This is a weak formulation of the SPH approximation of the conservation equations. It is written in an ALE form (Arbitrary Lagrange Euler).

Note that according to (4.19), the first two terms in second member of (5.6) is just the SPH approximation of minus the divergence of the matrix G defined as:

$$G = F(\phi) - \phi \otimes v_o \tag{5.7}$$

It remains to write the discrete form of (5.6). First note that $\frac{1}{\omega_i}\frac{d_{v_o}(\omega_i\phi)}{dt} = \frac{d_{v_o}\phi}{dt} + \frac{\phi}{\omega_i}\frac{d_{v_o}\omega_i}{dt} = \frac{\partial\phi}{\partial t} + \nabla.(\phi v_o)$ where the transport equation of ω_i in equation (4.4) has been used. According to the Reynolds theorem, the first member of (5.6) can then be written as:

$$\frac{d_{v_o}}{dt}\int_{D(x)}\phi W dx'^d = \frac{1}{\omega_i}\frac{d_{v_o}(\omega_i\phi_i)}{dt}$$

For the second member we get:

$$(\nabla.G)_i^h = \sum_{j\in D_i} G_j \nabla_{x_i} W(x_i - x_j, h)\,\omega_j + \sum_{j\in\partial D_i} G_j W(x_i - x_j, h)\,\omega_j^\partial n_j$$

So that we get finally:

$$\frac{1}{\omega_i}\frac{d_{v_o}(\omega_i\phi_i)}{dt} = -\sum_{j\in D_i} G_j \nabla_{x_i} W(x_i - x_j, h)\,\omega_j \tag{5.8}$$

$$- \sum_{j\in\partial D_i} G_j W(x_i - x_j, h)\,\omega_j^\partial n_j$$

$$+ \sum_{j\in D_i} S(\phi_j) W(x_i - x_j, h)\,\omega_j$$

5.2.2 Improved SPH approximation accurate to second order

A better SPH approximation can also be written using for instance the renormalization technique described by Vila (2005) also given in paragraph 4.6.3; using (4.29)

$$(\nabla.G)_i^h = B(x)(\nabla.G)_i^h - B(x)G_i\nabla(1)^h$$

$$= \sum_{j\in D_i}(G_j - G_i)B_i\nabla_{x_i} W(x_i - x_j, h)\,\omega_j$$

$$+ \sum_{j\in\partial D_i}(G_j - G_i)B_i W(x_i - x_j, h)\,\omega_j^\partial n_j$$

As shown before, this expression allows treating exactly the components of G, which depend linearly on x. We get for the renormalization form:

$$\frac{1}{\omega_i}\frac{d_{v_o}\left(\omega_i\phi_i\right)}{dt} = -\sum_{j\in D_i}\left(G_j - G_i\right)B_i\nabla_{x_i}W\left(x_i - x_j, h\right)\omega_j \qquad (5.9)$$

$$-\sum_{j\in\partial D_i}\left(G_j - G_i\right)B_iW\left(x_i - x_j, h\right)\omega_j^\partial n_j$$

$$+\sum_{j\in D_i}S\left(\phi_j\right)W\left(x_i - x_j, h\right)\omega_j$$

or the other form based on (4.22) and (4.30) which allows the global conservation of quantities ϕ.

Note that the matrix of renormalisation does not appear in the surface summation as the normalisation is now assumed to apply on the gradient of the kernel and not on the gradient of the function ϕ itself.

$$\frac{1}{\omega_i}\frac{d_{v_o}\left(\omega_i\phi_i\right)}{dt} = -\sum_{j\in D_i}\left(G_j + G_i\right)B_{ij}\nabla_{x_i}W\left(x_i - x_j, h\right)\omega_j \qquad (5.10)$$

$$-\sum_{j\in\partial D_i}\left(G_j + G_i\right)W\left(x_i - x_j, h\right)\omega_j^\partial n_j$$

$$+\sum_{j\in D_i}S\left(\phi_j\right)W\left(x_i - x_j, h\right)\omega_j$$

5.2.3 Global conservation of transported quantities ϕ

Consider the use of (5.10) far from the boundary $\partial\Omega$. Remember that if $B_{ij} = \frac{1}{2}\left(B_i + B_j\right)$, then $B_{ij}\nabla_{x_i}W\left(x_i - x_j, h\right) = -B_{ji}\nabla_{x_j}W\left(x_j - x_i, h\right)$; this allows the cancellation of the mutual interactions between particles, so that we get by summing on all particles i in Ω:

$$\frac{d_{v_o}\left(\sum_{i\in\Omega}\omega_i\phi_i\right)}{dt} = \sum_{i\in\Omega}\omega_i\sum_{j\in D_i}S\left(\phi_j\right)W\left(x - x_j, h\right)\omega_j$$

This is just the discrete form of the conservation equation of ϕ in Ω:

$$\frac{d_{v_o}\int_\Omega \phi dx'^3}{dt} = \int_\Omega S dx'^3$$

5.2.4 Numerical viscosity

Equation (5.10) looks like a central scheme, so that it is generally necessary to add some numerical dissipation in order to allow a stable integration in time with explicit schemes. We introduce then a Π_{ij} term, such that $\Pi_{ij} = \Pi_{ji}$ defined as:

$$\Pi_{ij} = \begin{cases} \frac{\beta\mu_{ij}^2 - \alpha\mu_{ij}c_{ij}}{\frac{(\rho_i + \rho_j)}{2}} & \text{if } (v_i - v_j).(x_i - x_j) < 0 \\ \\ 0 & \text{elsewhere} \end{cases} \tag{5.11}$$

$$\mu_{ij} = \frac{h(v_i - v_j).(x_i - x_j)}{(x_i - x_j)^2 + \epsilon h^2}$$

$c_{ij} = \frac{(c_i + c_j)}{2}$ is an averaged velocity of sound, $\epsilon \in \left[10^{-3},\ 10^{-6}\right] \ll 1$ allows to avoid a division by zero, and α, β are free parameters of order one that arc case dependant.

Monaghan has proposed these expressions. In practice, their main objective is to correct the pressure p by a viscous term Π_v, so that $p + \Pi_v$ will be used in the transport equation of momentum and energy. This is related to the work of Richtmeyer (1967) who defined the viscous pressure Π_v by:

$$\Pi_v = \begin{cases} \beta\rho l^2 \left(\nabla.v\right)^2 - \alpha\rho lc\nabla.v & \text{if } \nabla.v < 0 \\ 0 & \text{elsewhere} \end{cases} \tag{5.12}$$

with $l \sim h$ or Δx. From the minus sign in (5.11), and the original expression (5.12), it is clear that (5.11) is based on a "constant zeroing" formula (4.21) of the divergence. See also (Issa, 2005, Violeau 2002, 2004). Introducing the numerical viscosity in the conservation equations, we get for instance from (5.10):

$$\frac{1}{\omega_i}\frac{d_{v_o}(\omega_i\phi_i)}{dt} = -\sum_{j\in D_i}(G_j + G_i + \Pi_{ij})\,B_{ij}\nabla_{x_i}W\,(x_i - x_j, h)\,\omega_j \tag{5.13}$$

$$-\sum_{j\in\partial D_i}(G_j + G_i + \Pi_{ij})\,W\,(x_i - x_j, h)\,\omega_j^\partial n_j$$

$$+\sum_{j\in D_i}S\,(\phi_j)\,W\,(x_i - x_j, h)\,\omega_j$$

As mentioned by Gingold and Monaghan (1977) and Vila (2005), the nice feature of (5.11) is its compatibility with the second principle of thermodynamics. Considering the definition of specific entropy:

$$Tds = de - \frac{p}{\rho^2}d\rho$$

or its SPH discrete equivalent equation:

$$T_i\frac{ds_i}{dt} = -\frac{1}{2}\sum_{j\in D_i}\rho_j\omega_j\Pi_{ij}\,(v_j - v_i)\,\nabla_{x_i}W\,(q = x_i - x_j, h)$$

whose second member is non positive if $\frac{dW}{dq} < 0$, which is provided by a B-spline kernel for instance.

5.2.5   Godunov's scheme and Riemann solver

Riemann solver is a successful tool in order to increase the stability of a numerical scheme. In practice, this type of solver also includes implicitly a certain amount of numerical dissipation (see Vila (2005)). We will also show how the summation over the neighbouring particles has some connection with the summation over the faces of a control volume, as in finite volume method for instance.

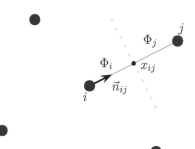

FIGURE 5.1: Flux balance at the interface x_{ij} between two particles i and j.

5.2.6   The analogy with finite volume method

Following the approach of Vila (2005), consider first the line connecting two particles i and j (Figure 5.1). Let n_{ij} the unit vector defined as:

$$n_{ij} = \frac{x_j - x_i}{|x_j - x_i|}$$

We have then, according to (4.12):

$$\begin{aligned}
\nabla_{x_i} W\left(x_i - x_j, h\right) &= -\nabla_{x_j} W\left(x_i - x_j, h\right) \\
&= -\frac{C n_{ij}}{h^{d+1}} \\
&= \frac{df\left(q_{ij} \frac{|x_i - x_j|}{h}\right)}{dq} \\
&= -n_{ij} \frac{C}{h^{d+1}}\left(\frac{df}{dq}\right)_{ij}
\end{aligned}$$

$$\nabla_{x_i} W\left(x_i - x_j, h\right) = n_{ij} df_q^* \tag{5.14}$$

In the equation of conservation (5.10) , we may also write:

$$B_{ij}\nabla_{x_i} W\left(x_i - x_j, h\right) = -B_{ij} n_{ij} df_q^* \approx -\delta_{ij} n_{ij} df_q^* = -n_{ij} df_q^*,$$

and $dA_{ij} = \omega_j df_q^*$ so that:

$$\frac{d_{v_o}(\omega_i \phi_i)}{dt} = \omega_i \sum_{j \in D_i} (G_j + G_i)\, dA_{ij} n_{ij} \tag{5.15}$$

$$-\omega_i \sum_{j \in \partial D_i} (G_j + G_i)\, W(x_i - x_j, h)\, \omega_j^\partial n_j$$

$$+\omega_i \sum_{j \in D_i} S(\phi_j)\, W(x_i - x_j, h)\, \omega_j$$

Note that on the boundary ∂D_i, ω_j^∂ stands for an element of surface. If we define $dA_{ij}^{\partial \Omega} = \omega_j^\partial W(x_i - x_j, h)$, we can then write also:

$$\frac{d_{v_o}(\omega_i \phi_i)}{dt} = \omega_i \sum_{j \in D_i} (G_j + G_i)\, dA_{ij} n_{ij} \tag{5.16}$$

$$-\omega_i \sum_{j \in \partial D_i} (G_j + G_i)\, dA_{ij}^{\partial \Omega}\, n_j$$

$$+\omega_i \sum_{j \in D_i} S(\phi_j)\, W(x - x_j, h)\, \omega_j$$

Define now $2g_{ij}(n_{ij}, \phi_i, \phi_j) = (G_j + G_i) n_{ij}$ where $g_{ij}(n_{ij}, \phi_i, \phi_j)$ stands for a numerical flux computed at x_{ij} midway between the two particles i and j, and submitted to the two following conditions:

$$\begin{cases} g(n, \phi, \phi) = G(\phi)\, n \\ g(n, \phi, \phi) = -g(-n, \phi, \phi) \end{cases}$$

We get finally:

$$\frac{d_{v_o}(\omega_i \phi_i)}{dt} = \sum_{j \in D_i} 2g_{ij} dA_{ij} - \sum_{j \in \partial D_i} 2g_{ij} dA_{ij}^{\partial \Omega} + \omega_i \sum_{j \in D_i} S(\phi_j)\, W(x_i - x_j, h)\, \omega_j$$

$$\tag{5.17}$$

This equation can be used in order to identify an analogy with the finite volume method, in the domain D_i. We observe that the point where the flux g_{ij} is computed could be considered as being located at $x_{ij} = \frac{x_i + x_j}{2}$. ω_i stands for a volume, $g_{ij}(n_{ij}, \phi_i, \phi_j)$ is a flux in the direction of n_{ij}, and dA_{ij} corresponds to a surface located at x_{ij}.

Similarly, $dA_{ij}^{\partial \Omega}$ stands for the surface element located on the boundary $\partial \Omega$ that intercept ∂D_i, while n_j is the local normal to $\partial \Omega$.

For a constant field ϕ_{ij}, we have for the first term of the second member:

$$\sum_{j \in D_i} 2g_{ij} dA_{ij} = 2G(\phi_i) \sum_{j \in D_i} n_{ij} dA_{ij}$$

This quantity is zero for a closed finite volume, while it is equivalent in SPH to $\sum_{j \in D(x_i)} W(x - x_j)\, \omega_j = 1$ in (4.9) or $\sum_{j \in D_i} \nabla_{x_i} W(x_i - x_j, h)\, \omega_j = 0$.

We have already discussed in chapter 9, various techniques in order to force this last quantity to zero. See for instance (4.21), or the renormalization technique or the use of Shepard's kernel ψ (4.42). Again, we will see that FVPM method theoretically fulfil the "closed volume" condition.

REMARK 5.1   If ϕ may be considered as constant over the intersection of $\partial \Omega$ and ∂Di, then

$$\sum_{j \in \partial D_i} 2g_{ij} dA_{ij}^{\partial D} = 2G\left(\phi_i\right) \sum_{j \in \partial D_i} n_j dA^{\partial \Omega}$$

$\square$

5.2.7   Riemann solver

Vila (1999) was the first to build a variant of the standard SPH method based on a weak form. He noticed that the discretization of this conservative formulation by the SPH method leads to the appearance of one-dimensional Riemann problems between pairs of neighbouring points. Between the two points, we thus consider a one-dimensional set of conservation laws, whose initial condition is discontinuous at the mid-point. Considering two interacting particles whose indexes are noted i and j, this can be expressed in the homogeneous case as:

$$\begin{cases} \dfrac{\partial \phi}{\partial t} + \dfrac{\partial}{\partial x^{(n_{ij})}} \left(G\left(\phi\right).n_{ij}\right) = 0 \\ \phi\left(x_{ij}, t = 0\right) = \begin{cases} \phi_i \text{ if } x^{(n_{ij})} < 0 \\ \phi_j \text{ if } x^{(n_{ij})} > 0 \end{cases} \end{cases} \qquad (5.18)$$

The exact solution to (5.18) is self-similar and is given by:

$$\begin{cases} \phi = \phi_E \left(\dfrac{x^{(n_{ij})} + x_o(t)}{t}, \phi_i, \phi_j\right) \\ x_o\left(t\right) = \int_O^t v_o\left(x_{ij}, \tau\right).n_{ij} d\tau \end{cases} \qquad (5.19)$$

Here $v_o\left(x_{ij}, t\right)$ is the velocity of the moving interface between points i and j.

With the previous considerations, it is then possible to discretize the conservation equation (5.10) as a true balance of numerical fluxes, these latter's being computed using Godunov's schemes. Using the results of Vila (2005), the system of discrete conservation laws can be written as:

$$\frac{1}{\omega_i} \frac{d_{v_o}(\omega_i \phi_i)}{dt} = -\sum_{j \in D_i} 2 G_E B_{ij} \nabla_{x_i} W\left(x_i - x_j, h\right) \omega_j \qquad (5.20)$$

$$-\sum_{j \in \partial D_i} 2 G_E W\left(x_i - x_j, h\right) \omega_j^\partial n_j$$

$$+\sum_{j \in D_i} S\left(\phi_j\right) W\left(x_i - x_j, h\right) \omega_j$$

G_E stands for Godunov's fluxes adapted to the ALE description and is given from (5.7) by:

$$G_E\left(\phi_i, \phi_j\right) = F_E\left(\phi_{ij}\left(\lambda_0^{ij}\right)\right) - \phi_{ij}\left(\lambda_0^{ij}\right) \otimes v_o\left(x_{ij}, t\right)$$
$$\phi_{ij}\left(\lambda_0^{ij}\right) = \phi_E\left(\lambda_0^{ij}, \phi_i, \phi_j\right) \qquad (5.21)$$
$$\lambda_0^{ij} = v_o\left(x_{ij}, t\right).n_{ij}$$

Equation (5.21) means that the solution of the moving Riemann problem is searched along the direction $x = \lambda_0^{ij} t$ that is to say along the move of the interface x_{ij} in the direction n_{ij}.

This gives a numerical method, which is quite different from the standard SPH one. First of all, the ALE description enables the use of an upwind solution for velocity thanks to the convective fluxes, which are missing in a pure Lagrangian description. Incidentally, one can notice that even if a Lagrangian transport field, $v_0 = v$, is chosen, mass fluxes are not necessarily reduced to zero because of the upwind component of $v_{E,ij}$, which means that in all cases the mass of particles is no longer a constant variable along time (even though total mass is globally conserved in Ω).

Another difference is that the geometric transformation of points distribution is explicitly taken into account through (4.4), which adds one equation compared to the standard SPH set of equations. Clearly calculation points should no longer be taken for particles but for moving control volumes, which exchange not only momentum but also mass between each other's. This seems to be more appropriated to describe fluid flows than material particles. It also connects this hybrid method to finite volumes formalism, at the difference that the flux balance is not computed on the boundary surface of the control volume but in a small surrounding volume.

5.2.8   Numerical viscosity and Riemann solver

Ben Moussa and Vila (2000) have indicated that the replacement in (5.10) of $2g_{ij}\left(n_{ij}, \phi_i, \phi_j\right) = \left(G_j + G_i\right) n_{ij}$, where $g_{ij}\left(n_{ij}, \phi_i, \phi_j\right)$ stands for the numerical flux, is equivalent to introducing a numerical viscosity in (5.10):

$$\frac{1}{\omega_i}\frac{d_{v_o}(\omega_i\phi_i)}{dt} = -\sum_{j\in D_i}(G_j+G_i)\,B_{ij}\nabla_{x_i}W\,(x_i-x_j,h)\,\omega_j \tag{5.22}$$

$$-\sum_{j\in D_i}Q\,(n_{ij},\phi_i,\phi_j)\,(\phi_i-\phi_j)\,\nabla_{x_i}W\,(x_i-x_j,h)$$

$$-\sum_{j\in\partial D_i}(G_j+G_i)\,W\,(x_i-x_j,h)\,\omega_j^{\partial}n_j$$

$$+\sum_{j\in D_i}S\,(\phi_j)\,W\,(x_i-x_j,h)\,\omega_j$$

where $Q\,(n_{ij},\phi_i,\phi_j)$ is a numerical viscosity defined from one-dimensional analysis and a scalar case by:

$$Q\,(n_{ij},\phi_i,\phi_j)\,(\phi_i-\phi_j) = G\,(\phi_i,x,t)\,n_{ij} - 2g_{ij}\,(n_{ij},\phi_i,\phi_j) + G\,(\phi_j,x,t)\,n_{ij} \tag{5.23}$$

5.3 Application to flow conservation equations

5.3.1 Euler equation for a non-viscous fluid

The full set of discrete equations describing this hybrid method is finally written here from (5.20) and (5.21) with $B_{ij}=\delta_{ij}$ and without the energy equation for Euler equations:

$$
\begin{cases}
(a) & \frac{dx_i}{dt} = v_0\,(x,t) \\
(b) & \frac{d\omega_i}{dt} = \omega_i\sum_{j\in D_i}\omega_j\,(v_0\,(x_j,t)-v_0\,(x_i,t))\,.\nabla_i W\,(x_i-x_j,h) \\
(c) & \frac{d(\omega_i\rho_i)}{dt} = -\omega_i\sum_{j\in D_i}\omega_j 2\rho_{E,ij}\,(v_{E,ij}-v_0\,(x_{ij},t))\,.\nabla_i W\,(x_i-x_j,h) \\
& -\omega_i\sum_{j\in\partial D_i}2\rho_{E,ij}\,(v_{E,ij}-v_0\,(x_{ij},t))\,W\,(x_i-x_j,h)\,\omega_j^{\partial}n_j \\
(d) & \frac{d(\omega_i\rho_i v_i)}{dt} = -\omega_i\sum_{j\in D_i}\omega_j 2\,[\rho_{E,ij}v_{E,ij}\otimes(v_{E,ij}-v_0\,(x_{ij},t))+p_{E,ij}\delta_{ij}] \\
& .\nabla_i W\,(x_i-x_j,h) \\
& -\omega_i\sum_{j\in\partial D_i}2\,[\rho_{E,ij}v_{E,ij}\otimes(v_{E,ij}-v_0\,(x_{ij},t))+p_{E,ij}\delta_{ij}] \\
& W\,(x_i-x_j,h)\,\omega_j^{\partial}n_j + \omega_i\rho_i g
\end{cases}
\tag{5.24}
$$

In the system of equations (5.24), (a) gives the trajectory of the particle, (b) gives the evolution of the weight ω_i according to (4.4), (c) is the conservation of mass and (d) is the conservation of momentum in which only gravity forces are considered as external source term.

5.3.2   Practical implementation of Riemann solver in an SPH method

The equation (5.24) requires the solution of the non-linear Riemann problems of the form (5.18), which appear between each pair of neighbouring points. The form of this solution depends also on the equation of state. We will use the system of equations (5.24) for water in Chapter 5.5. The equation of Tait (1888) will be used in order to link the density ρ and the pressure p for a barotropic fluid:

$$p = \frac{\rho_0 c_0^2}{\gamma} \left(\left(\frac{\rho}{\rho_0} \right)^{\gamma} - 1 \right) \tag{5.25}$$

where the density ρ_0 and the velocity of sound c_0 are references quantities. We use $\gamma = 7$ for water.

Only the principle of an exact resolution will be presented here. More details can be found in Marongiu (dec. 2007); the practical use of Riemann solver for the treatment of boundary conditions is described in a following paragraph as for instance on solid boundaries.

The general form of the one-dimensional Riemann problem is rewritten as:

$$\begin{cases} \frac{\partial \phi}{\partial t} + \frac{\partial}{\partial x^{(n_{ij})}} \left(F\left(\phi\right) . n_{ij} \right) = 0 \\ \phi\left(x_{ij}, t = 0\right) = \begin{cases} \phi_L & \text{if } x < 0 \\ \phi_R & \text{if } x > 0 \end{cases} \end{cases} \tag{5.26}$$

where subscripts L and R denote the so called left and right states on both sides of the initial discontinuity situated on $x = 0$. The dimension of this problem should only be two (density and velocity) but as this problem has been introduced in the frame of multidimensional method, we add a second component to the velocity in the tangential direction. The vectors of conservative variables and fluxes are so given by:

$$\phi = \left(\rho, \rho v^{(1)}, \rho v^{(2)} \right)$$
$$F_E\left(\phi\right) = \left(\rho v^{(1)}, \rho v^{(1)^2} + \kappa \rho^{\gamma}, \rho v^{(1)} v^{(2)} \right)$$

with $\kappa = \frac{B}{\rho_0^{\gamma}}$. Superscript (1) denotes the direction normal to the discontinuity (the direction linking the two particles) while (2) denotes the direction tangential to the discontinuity.

The system (5.26) has three real eigenvalues:

$$\begin{cases} \lambda_1 = v^{(1)} \\ \lambda_2 = v^{(1)} + c \\ \lambda_3 = v^{(1)} - c \end{cases}$$

As we have low Mach numbers, $\lambda_2 > 0$ represents a non-linear wave moving to the right and $\lambda_3 < 0$ represents a non-linear wave moving to the left, each of these can be either a rarefaction wave, or a shock wave. Figure 5.2 shows that these waves delineate three regions in which the states Φ are constant: the left and right states and an intermediate state Φ_* called the star region. The eigenvalue λ_1 corresponds to a shear wave across which the tangential component $v^{(2)}$ of the velocity varies discontinuously.

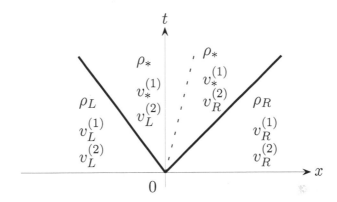

FIGURE 5.2: Structure of a typical solution to the one-dimensional Riemann problem.

In the frame of an ALE numerical method, the solution along $\frac{x}{t} = v_0^{(1)}$ is needed. In practice, although the transport field can be chosen arbitrarily, we are principally interested in Eulerian, Lagrangian and intermediate descriptions; this means that we can roughly consider that $0 \leq \|v_0\| \leq \|v\|$. This leads to $\|v_0\| \ll c$ and consequently the solution has to be searched in the star region.

Solving for the Φ_* state can be achieved by using the Rankine-Hugoniot jump relations across shock waves and the Riemann invariants across isentropic rarefaction waves. States on both sides of a non-linear wave can hence be linked, leading to a system of non linear equations relating Φ_L, Φ_R and Φ_*. The whole procedure is very well described in Ivings et al. (1998). The principle of this methodology is illustrated in Figure 5.3. Two lines in the ρ-v diagram can represent the set of states; they can be reached from Φ_L or Φ_R through either a rarefaction or a shock wave. The solution is therefore the intersection point of these two lines. In practice, because of the non-linearity, an iterative method is required. We can use a Newton-Raphson procedure and numerical experiments show that a maximum of four iterations are usually necessary to reach convergence in our applications. Nevertheless the evaluation of each intermediate solution is costly and a simpler linear solver is useful.

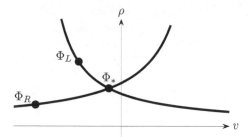

FIGURE 5.3: Exact solution to the Riemann problem.

5.4 Boundary conditions

Boundary conditions are certainly the major weakness for a practical use of SPH in real cases. The free surface condition for a liquid is rather easily handled, thanks to the Lagrangian description and provided that the influence of gas and surface tension can be neglected. But other kinds of condition do not benefit from a proper and general modeling. Inlet/outlet type boundary conditions for instance are most of the time simply treated as supersonic ones, meaning that the downward flow has no influence on the imposed physical boundary condition, which restricts the field of applications.

We shall first describe three methods used in classical SPH methods, the boundary repulsive force, the mirror particles and the ghost particles; one other method uses a normalization of the density close to the boundary. We then introduce the semi-analytical method of Vila (1999) that allows to avoid the difficult task to determine the normal on a boundary. We shall then present how SPH-ALE method can easily handle complex geometries and flows with high dynamics.

5.4.1 Boundary repulsive forces

The case of wall boundary condition has been more widely considered. Sirovich (1968) formulated the boundary condition as a boundary force.

Peskin (1977) developed a similar technique for immersed boundary method. Monaghan (1994, 1995) proposed then to ensure non-penetration of fluid particles across walls using repulsive forces in SPH, first similar to Lennard-Jones forces of molecular dynamics and then considered as normal boundary forces. The solid geometry is hence discretized with points, and these wall points apply centred or normal forces on fluid particles, as shown in Figure 5.4. The strength of these forces depends on the relative distance between fluid particles and wall points and increases quickly enough to push fluid particles back and prevent them from leaking. This model is simple, cheap, and can easily be applied to walls of any shape, allowing the use of SPH for complicated real cases. Monaghan and Kajtar (2009) have given a good description of the practical use of this model; they consider radial and sufficiently smooth forces to ensure that, when the total force on a fluid particle is obtained by summing over the boundary particle forces, the final result is independent of the discrete nature of the boundary to a high degree of accuracy. They mention that the boundary particles must have a spacing relative to the fluid particles of 1/3 to guarantee that the magnitude of the tangential force relative to the normal force, and the relative variation in the normal force for a fixed distance above the boundary, are both $< 10^{-5}$. The expression of the boundary force on a fluid particle i due to a boundary particle j is:

$$F_{j \to i} = \frac{1}{\beta} \left(\frac{V_{max}^2}{r_{ij} - d} \right) \frac{r_{ij}}{|r_{ij}|} W(r_{ij}) \frac{2m_j}{m_j + mi} \tag{5.27}$$

where β is the ratio of the fluid particle spacing to the boundary points spacing. $d \approx 0$.

Apart from avoiding the penetration of fluid particles through the wall, this model can trigger numerical instabilities in the vicinity of the boundary. In fact this model doesn't compensate for the truncated kernel support and hence doesn't restore mathematical consistency in the near wall region, making gradient approximations very inaccurate. Moreover, wall pressure is not explicitly involved in this model.

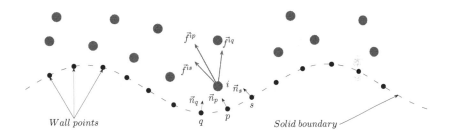

FIGURE 5.4: Repulsive wall forces directed along the normal vector to the boundary.

5.4.2   Mirror particles

A more complex model is the mirror (or fictitious) particles developed by Morris (1997). Mathematical consistency is first restored by extending the computational domain outside the wall using a regular and pre-established fictitious points distribution, so that the volume interpolation scheme has not to be modified for near wall fluid particles (see Figure 5.5).

Physical properties of these mirror particles are then set so as to reproduce the desired boundary condition. In particular the non-penetration condition $v.n = 0$ is obtained through assigning to mirror particles the symmetric pressure field of fluid particles. Issa (2005) noticed that numerical results could be quite sensitive to the fictitious point's distribution. Besides, the imposed symmetric pressure field between fluid and mirror particles turns out to be equivalent to $\frac{\partial p}{\partial n} = 0$, which takes into account neither hydrostatic nor hydrodynamic effects. Finally, the fictitious point distribution has to be defined initially, which is not always straightforward for complicated shaped solid walls. For more than one fluid, the nature of the mirror particles has to be chosen in advance, and the mirror should also change their state if an interface is close to the boundary.

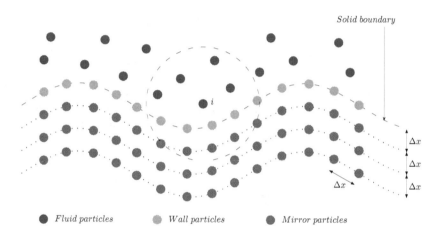

FIGURE 5.5: Regular distribution of wall and mirror particles.

5.4.3 Ghost particles

A more accurate model available to treat solid walls in SPH is the ghost particles model (Colagrossi, Landrini, 2003). It retains the previous idea of extending the computational domain so as to restore mathematical consistency but this is achieved using the images of the fluid particles themselves (Figure 5.6). Thus the position of fluid particles in the vicinity of a boundary are symmetrized across the boundary; this creates the image particles, called ghost particles, which are also involved in the discrete equations of motion. A real particle produces an image, which has the same state quantities (density, pressure and temperature) as their real particle. In order to fulfil the non-penetration condition, the normal component of ghost particles' velocity

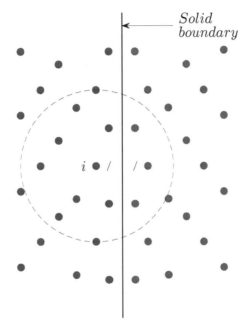

• *Fluid particles*

• *Ghost particles*

FIGURE 5.6: Ghost particles distribution for a planar boundary.

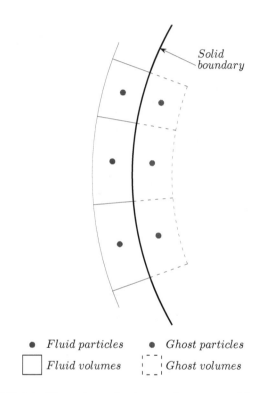

FIGURE 5.7: Variation of volumes for a curved boundary.

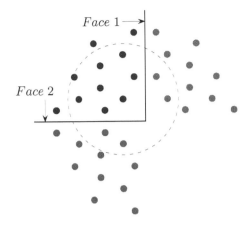

● *Fluid particles*

● *Ghost particles, face* 1

● *Ghost particles, face* 2

FIGURE 5.8: Separated treatment of faces for a corner.

has an opposite sign to normal component of the real particle. The tangential velocity component has the same sign for free slip condition or opposite for no slip condition.

Some modifications are required to treat moving walls (Doring, 2006). A pressure correction term can be added to take static effects (gravity) into account (Oger, 2006). The main difficulty using this model is to obtain satisfactory symmetry transformations in the case of complicated shaped walls. Weights (volumes) of ghost particles should hence depend on the local radius of curvature of the wall, as mentioned in Vila (1999) (Figure 5.7). The case of corners is very tricky and corner sides are generally treated separately (Figure 5.8). The treatment of more than one fluid requires the interface to appear among the ghost particles.

Unfortunately, this is still not sufficient to handle real and complicated solid geometries particularly for flows with very high dynamics.

The standard SPH method is apparently very simple to implement and comes along with a lot of boundary treatments that enables very easily to conduct free surface simulations which would require more sophisticated numerical techniques otherwise. But from the accuracy point of view, it seems that a shift is needed to bring SPH to the same quality standards as mesh-based methods. It will be shown in the sequel that an SPH-ALE description of the flow enables the adoption of more sophisticated numerical schemes directly inspired from existing finite volumes schemes.

5.4.4 Normalizing conditions

This method is due to Feldman and Bonet (2007). They obtain the SPH density by using a Shepard's kernel (4.42), thereby summing the density over the particles and correcting the summation by a normalizing function.

$$\rho^h(x) = \alpha(x) \sum_{j \in D(x_i)} \rho(x_j, t) \, W(x - x_j, h) \, \omega_j(x_j, t) \tag{5.28}$$

$$= \frac{\displaystyle\sum_{j \in D(x_i)} \rho(x_j, t) \, W(x - x_j, h) \, \omega_j(x_j, t)}{\displaystyle\sum_{j \in D_i} W(x - x_j, h) \, \omega_j} \tag{5.29}$$

In general, $\gamma(x) = \sum_{j \in D_i} W(x - x_j, h) \, \omega_j$ is close to one inside the flow domain Ω. However, close to the boundary $\partial\Omega$, its value may be very different from one because of the truncation of the particle domain D_i; this term may then correct for the absence of particles beyond the boundary $\partial\Omega$. Feldman and Bonet have then shown that using a Shepard's kernel for the density (then the pressure) allows extra forces to appear in the equations of motion, which depends on $\nabla\gamma(x)$. These forces act normal to the surfaces, and tend towards zero far from the boundary; their net effect is then to prevent the particles to cross a solid boundary.

In order to avoid costly computation of the volume integral $\gamma(x)$, the authors introduce a vector function w such that:

$$\begin{aligned} \nabla.w &= W(x - x_j, h) \\ w &= f(q)\,x, \quad q = \frac{|x|}{h} \end{aligned} \tag{5.30}$$

In this way, the volume integral γ is obtained from the divergence theorem and its computation involves only the boundary of the compact support of W:

$$\gamma(x) = \int_{D(x)} W(x - x', h) \, dx'^d = \int_{D(x)} \nabla.w \, dx'^d = \int_{\partial D(x)} w.n \, dx'^{d-1} \tag{5.31}$$

Feldman and Bonet have given explicit expression for $f(q)$, $\gamma(x)$ and $\nabla\gamma(x)$ for a two-dimensional case including the treatment of sharp corners:

$$f(q) = \frac{1}{q^2} \int W q' dq'$$

Monaghan and Kajtar (2009) report that this method works well in the case of fixed surfaces, but in more complicated situations it can be cumbersome. For example in a flow containing several bodies which can interact, where any two of the bodies are in rolling and sliding contact, the normalizing function is different for every configuration of the bodies, and its evaluation is costly, especially in three dimensions.

5.4.5   The semi-analytical method

Vila (1999) has proposed a method that treats the boundary term in the conservation equation as a term distributed in the volume.

Consider the conservation equation in the following form:

$$\frac{1}{\omega_i}\frac{d_{v_o}(\omega_i\phi_i)}{dt} = -\sum_{j\in D_i} 2G_E B_{ij}\nabla_{x_i}W\left(x_i - x_j, h\right)\omega_j$$

$$-\sum_{j\in\partial D_i} 2G_E W\left(x_i - x_j, h\right)\omega_j^\partial n_j$$

$$+\sum_{j\in D_i} S\left(\phi_j\right)W\left(x_i - x_j, h\right)\omega_j$$

The Figure 5.9 shows the interaction of a flow domain D_i with the boundary

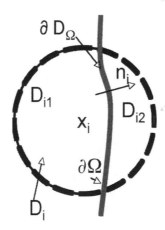

FIGURE 5.9: Interaction of the particle domain D_i with the flow boundary $\partial\Omega$.

$\partial\Omega$ of the domain Ω. The boundary $\partial\Omega$ separates the volume of the support of the kernel into two parts D_{i1} and D_{i2} where D_{i1} is assumed to be located in the flow domain Ω while D_{i2} is outside Ω. Let assume also that the normal to the boundary $\partial\Omega$ is directed from D_{i1} towards D_{i2}. Because $W\left(x_j \in \partial D_i, h\right) = 0$ and as the boundary term $-\sum\limits_{j\in\partial D_i} 2G_E W\left(x_i - x_j, h\right)\omega_j^\partial n_j$ must be under-
stood as computed on the intersection ∂D_Ω of $\partial\Omega$ and of the volume of the

support of the kernel D_i ($\partial D_\Omega = D_i \cap \partial\Omega$), we can then write on the boundary of D_{i2}:

$$- \sum_{j \in \partial D_i} 2 G_E W \left(x_i - x_j, h\right) \omega_j^\partial n_j = - \sum_{j \in \partial D_\Omega} 2 G_E W \left(x_i - x_j, h\right) \omega_j^\partial n_j$$

$$= \sum_{j \in \partial D_{i2}} 2 G_E W \left(x_i - x_j, h\right) \omega_j^\partial n_j$$

Using then Green's theorem, we get for the last integral:

$$\sum_{j \in \partial D_{i2}} 2 G_E W \left(x_i - x_j, h\right) \omega_j^\partial n_j = - \sum_{j \in D_{i2}} \nabla_{x_j} \left(2 G_E W \left(x_i - x_j, h\right)\right) \omega_j$$

$$= -2 \bar{G}_{E(\partial D_\Omega)} \sum_{j \in D_{i2}} \nabla_{x_j} W \left(x_i - x_j, h\right) \omega_j$$

where the last result is obtained by assuming that the numerical flux G_E is constant over D_{i2} and defined as its average $\bar{G}_{E(D_i \cap \partial\Omega)}$ over $\partial D_\Omega = D_i \cap \partial\Omega$. Owing to the fact that over D_i we have also as a consequence of (4.9)

$$\sum_{j \in D_i} \nabla_{x_i} W \left(x_i - x_j, h\right) \omega_j = 0 \tag{5.32}$$

so that $\sum\limits_{j \in D_{i2}} \nabla_{x_i} W \left(x_i - x_j, h\right) \omega_j = - \sum\limits_{j \in D_{i1}} \nabla_{x_i} W \left(x_i - x_j, h\right) \omega_j$, we get finally:

$$- \sum_{j \in \partial D_i} 2 G_E W \left(x_i - x_j, h\right) \omega_j^\partial n_j = 2 \bar{G}_{E(\partial D_\Omega)} \sum_{j \in D_{i1}} \nabla_{x_j} W \left(x_i - x_j, h\right) \omega_j$$

Note that in practice (5.32) is never numerically fulfilled exactly in the flow domain Ω, so that using (5.32) explicitly is a way to force this condition for the particles in contact with the boundary.

For the particles i whose domain D_i intercepts the flow boundary $\partial\Omega$ (then if $\partial D_\Omega \neq 0$), we get then the following conservation equation:

$$\frac{1}{\omega_i} \frac{d_{v_o} \left(\omega_i \phi_i\right)}{dt} = - \sum_{j \in D_i, \partial D_\Omega \neq 0} 2 \left(G_E - \bar{G}_{E(\partial D_\Omega)}\right) B_{ij} \nabla_{x_i} W \left(x_i - x_j, h\right) \omega_j$$

$$+ \sum_{j \in D_i} S \left(\phi_j\right) W \left(x_i - x_j, h\right) \omega_j \tag{5.33}$$

Consider for instance the interface between two flows (say a liquid and a gas). Let P_{atm} be the pressure in the gas. The particle close to the interface will then feel a relative pressure $P - P_{atm}$. This method is then particularly well adapted to the treatment of the pressure boundary condition for a free surface in a liquid. It avoids also the determination of the normal to the free interface, which is particularly difficult in this case.

A similar treatment can be also used for a solid wall; however the wall pressure cannot be estimated in advance as for the free surface from the gas. The SPH-ALE boundary treatment will then propose a specific treatment in order to determine the pressure from the flow behavior.

5.4.6   SPH-ALE boundary treatment

In the SPH-ALE framework, the boundary conditions can be much more adequately set than in the standard SPH method. This is mainly due to the choice of an ALE description, which enables the treatment of a boundary surface travelling with its own velocity, independently of the fluid velocity. It will be shown that upwinding is also of great interest to transmit effects of the fluid onto the boundary, and reciprocally. The SPH-ALE boundary treatment can be split into two steps. In the first, a mathematically consistent approximation of the boundary term is deduced from the kernel approximation itself. Its discrete counterpart, similar to a particle approximation, leads to the setup of boundary fluxes. These boundary fluxes are, in the second step, computed in an upwind method, after an interpretation of the mutual influence of the fluid and the boundary condition as a partial Riemann problem.

5.4.6.1   Surface integral boundary term

Ghost or mirror particles boundary treatments are facing the difficult task to add extra calculation points outside the computation domain. This is mandatory so as to cope with truncated kernel supports in the vicinity of a boundary. In practice the spatial distribution of these points is not easy to set because of geometric issues and field values attributed to these points can be obtained only through a long-range (the size of the kernel support) extrapolation of the field from inside the fluid domain. Besides, for reason of accuracy, this extrapolation should be non-linear, as the Euler equations are.

If we go back to the kernel approximation of gradients (4.19), we observe that the surface integral term is non-zero when the kernel support intersects the boundary of the flow domain. We shall then compute directly this surface term without resorting to added particles. In practice, the particle approximation of this surface term is obtained using the same quadrature formula, as the one used for the volume term, provided a satisfactory discretization of the boundary surface is available. Boundary fluxes are finally expressed in (5.20) as:

$$2 \int_{\partial D(x)} \left(F\left(\phi\right) - \phi \otimes v_o \right) W n dx'^{d-1} = 2 \sum_{j \in \partial D_i} G_E\left(\phi_i, \phi_j\right) W\left(x_i - x_j, h\right) \omega_j^{\partial} n_j$$

(5.34)

where G_E stands for boundary Godunov's fluxes adapted to the ALE description and is given from (5.21); it is given explicitly in the next step. The quantity ω_j^{∂} is the weight of the boundary element j (area of the surface element in 3D), which samples the boundary, and n_j is the unit normal vector to the boundary at location x_j.

5.4.6.2   Partial Riemann problem

The Riemann problem described in 5.3.2 is used to determine the mutual influence of two fluid states, namely the left state and the right state. Considering now the interaction of a fluid state, let's say a left state, and a boundary condition; it is clear that no right state can be defined. However, the influence of the boundary on the numerical solution is computed through boundary fluxes arising at the boundary surface. The boundary surface thus plays a similar role as the one played by an interface between two fluid states. Following Dubois (2001), the boundary surface can hence be taken for an interface of a partial Riemann problem, for which one state is missing. Nevertheless, the solution to this partial Riemann problem is partially defined by the physical boundary conditions imposed on the boundary surface. The whole solution consequently results from the selection, among all the states compatible with the imposed boundary condition, of the one, which can be reached from the left state through a shock or a rarefaction wave. Figure 5.10 helps understanding this concept.

We consider a one-dimensional partial Riemann problem along the normal direction to the boundary, between a given left (fluid) state Φ_L and a boundary surface. Primitive variables are noted ρ and v where v stands for the component of the velocity field normal to the boundary. The imposed boundary condition turns out to be a given relation (implicit or explicit) between ρ and v and is stated in a general form $B(\rho, v) = 0$. If two conditions are imposed, the solution is fully determined by the boundary condition and the fluid state plays no role. This configuration corresponds to a supersonic inlet for instance. On the contrary, if no boundary condition is imposed, the solution is fully determined by the fluid state, as it is the case for a supersonic outlet. The general case where only one condition is imposed is now considered. In the ρ, v diagram, the relation $B(\rho, v) = 0$ can be represented by a continuous curve noted C_B. This curve describes the set of states that are compatible with the boundary condition. We can also define the curve, noted C^-, which describes the set of states that can be reached from the left state through either a shock or a rarefaction wave. The solution Φ_* to the partial Riemann problem is finally the intersection point between these two curves. Some usual configurations are given in Figure 5.10, for an imposed velocity (a), an imposed pressure (b) and an imposed mass flow (c). It is worth noticing that this boundary modeling can be used for any kind of boundary condition, and that ensures a full compatibility of the boundary treatments with fluid interactions in the volume.

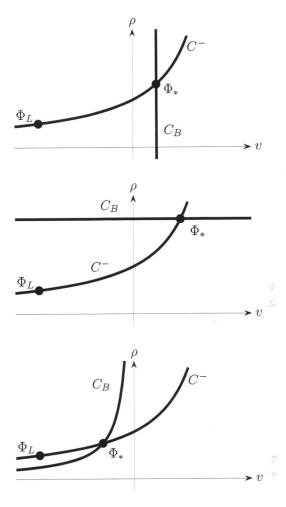

FIGURE 5.10: Various configurations corresponding to usual boundary conditions: (a) Case of a solid wall or a subsonic inlet: velocity imposed; (b) Case of a subsonic outlet: pressure imposed; (c) Case of a subsonic inlet: mass flow imposed.

5.4.6.3 Global state

For a specific particle, the partial Riemann problem only assigns the contribution of the particle to the boundary element. In order to get a solution that accounts for the influence on the boundary surface of all the particles in the field, for instance the pressure map at a point on a solid wall, we have to sum up the interactions of that boundary element with all the fluid particles in the neighbourhood.

In practice, the global pressure flux exchanged between the fluid particles indexed by j in the fluid domain and the boundary element indexed by i with the weight ω_i^∂ can be expressed from the discrete momentum equation in (5.24) (d) with the boundary terms as in (5.20) as:

$$F_{i \to \Omega}^p = -\omega_i^\partial \sum_{j \in D_i} \omega_j 2 p_{E,ij} W (x_i - x_j, h) \, n_i \qquad (5.35)$$

On the other hand if p_i stands for the pressure value assigned to the boundary element i, this total pressure flux can be written:

$$F_{i \to \Omega}^p = -\omega_i^\partial p_i n_i \qquad (5.36)$$

Equating the two above relations, we get the relationship between the pressure p_i on the boundary element n°i and the partial pressure $p_{E,ij}$ induced by each local j particle of the fluid domain:

$$p_i = \sum_{j \in D_i} \omega_j 2 p_{E,ij} W (x_i - x_j, h) \qquad (5.37)$$

This is simply an averaged sum of the partial pressures resulting from the partial Riemann problems. This definition is conservative, which means that the total numerical force exerted by the fluid on the boundary is equal, in the continuous limit, to the integral of the pressure field on the boundary surface:

$$F_{\Omega \to \partial\Omega} = \sum_{j \in \partial\Omega} \omega_j^\partial p_j n_j \approx \int_{\partial\Omega} p n dx'^{d-1} \qquad (5.38)$$

5.5   Applications of SPH and SPH-ALE methods

In this part the SPH-ALE method will be used to simulate unsteady complex free-surface flows. The examples are taken from hydraulic turbines and in particular from the Pelton turbine, which is a hydraulic impulse machine well adapted for high head (from 200 to 2 000 meters) and low discharge installations. It is composed of several hydraulic components, among which the runner plays the most important role. It is composed of 18 to 26 buckets that are impinged by high velocity water jets (from 1 to 6). It will be shown that the SPH-ALE method is particularly suited to study the interaction of the jets and the buckets.

Static and rotating configurations are presented. Numerical simulations are done at model scale, which is a usual setup for mesh-based simulations. In order to save some computational cost, a symmetry condition is used. Indeed the main components of a Pelton turbine are usually symmetric against the middle plane of the runner. It is thus possible to consider the flows we are interested in as symmetric, and to divide the computational domain size by a factor of two. But for the correctness of simulations, the removed part must be replaced by a symmetry condition. This is obtained by mirroring the calculation points that are in the vicinity of the symmetry plane, i.e. points whose kernel support intersects the plane. These duplicated points are used only to compute fictitious fluxes that are then involved in the flux balance of real computational points, in a way similar to the well-known technique of ghost particles (Colagrossi and Landrini, 2003).

Water is modeled as a weakly compressible fluid whose behavior is governed by the Tait's equation of state (5.25):

$$p = \frac{\rho_0 c_0^2}{\gamma} \left[\left(\frac{\rho}{\rho_0} \right)^\gamma - 1 \right]$$

where ρ_0 and c_0 are the reference density and speed of sound, respectively. The parameter γ is set to 7, which is a classical practice among the SPH practitioners community. The value of the speed of sound is not taken equal to the real one for water, as it would lead to very small time increments. A common practice is to take this parameter equal to ten times a velocity scale of the case considered.

The first case presented is a single steady Pelton bucket impinged by a water jet; the second involves a complete rotating Pelton runner impinged by a water jet. On both cases a constant velocity profile is imposed at the inlet boundary condition, where new particles are periodically injected in the computational domain. Vortices are then not considered at inlet section.

5.5.1 Flow in a single steady Pelton bucket

The first application considered is a steady Pelton bucket impinged by a water jet. The jet velocity is $C_{jet} = 19.61$ m/s and its diameter is $d_0 = 0.03$ m. The discretization size is $\Delta x = 1$ mm. The simulation represents a physical time of approximately 22 ms, so as to obtain a converged state in time. The number of particles involved is 93 000 at the end of the simulation. Only the inner surface of the bucket is considered, its discretization uses a surface triangulation achieved with the commercial mesh tool ICEMR. This triangulation is refined near sharp geometrical details like the leading edge. The bucket is thus represented by a set of 48 738 surface elements.

The artificial speed of sound used in the Tait's equation of state is set to $c_0 = 200$ m/s, and the reference density of water is $\rho = 1000$ kg/m^3. Figure 5.11 shows a view of the case. It can be seen that the escaping water sheet can be properly represented far from the trailing edge of the bucket, showing that the Lagrangian description allows a proper tracking of interfaces on long distances.

Pressure distribution on the bucket surface is closely examined (see Figure 5.12). Results are compared with measurements and CFX$_R$ results at

FIGURE 5.11: Global view of the flow in a steady Pelton bucket.

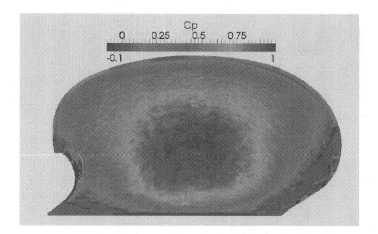

FIGURE 5.12: Pressure coefficient map on the bucket surface.

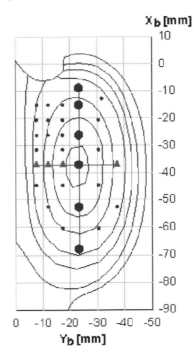

FIGURE 5.13: Position of the pressure sensors and mark of the plotted lines.

some given locations corresponding to pressure sensors locations (see Figure 5.13). Figure 5.14 shows the pressure profiles along two lines of sensors. The hybrid method tends to underestimate the value of the pressure coefficient. Results are here presented using one sensor as the reference, which enables to compare the evolution only.

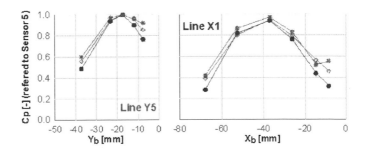

FIGURE 5.14: Pressure profiles along the X1 and Y5 lines boundary conditions: (stars) experiments; (diamonds) VOF method; (dots) SPH-ALE method.

Computational Fluid Dynamics

FIGURE 5.15: Rotating Pelton runner impinged by a single jet of water.

5.5.2 Flows in a rotating Pelton runner

The runner is made of 21 buckets and at model scale its diameter is $D_1 = 327.71$ mm and the buckets width is $B_2 = 80$ mm. A single water jet impinges it. The chosen operating point corresponds to a water head of $H = 60$ m and a discharge of $Q = 0.20205$ m^3/s and is situated in the high load part of the operating range.

With these settings, the jet diameter is $d_0 = 0.02766$ m and its velocity is $C_{jet} = 34.3$ m/s. The typical discretization size is also $\Delta x = 1$ mm, while the artificial speed of sound is set to $c_0 = 430$ m/s. Duplicating and rotating 21 times the surface of a single bucket, so that all the buckets are equivalent, make the surface representation of the runner. Each half-bucket is hence represented by around 91 000 surface elements. The time integration scheme used is the 3rd order Runge-Kutta with a CFL number of 0.8. Linearized Riemann solver and 2nd order MUSCL reconstruction are chosen in the flow solver.

Figure 5.15 shows the flow configuration at several instants along the simulation. The last picture corresponds to almost one complete rotation of the runner. Two view angles are provided, one along the rotation axis, one along the symmetry plane. Fluid particles are coloured by their velocity in the absolute frame of reference. It can be observed that the escaping water sheets have a very small residual velocity, which is the result of the exchange of energy between the jet and the buckets along the rotation. The SPH-ALE method is also able to track the water sheets on a long distance after the trailing edges of the buckets, which is the clear advantage of the Lagrangian description. In particular the water sheets remain separated as it can be observed on the model in the hydraulic laboratory (see Figure 5.16). These clean water sheets are more difficult to obtain with a classical mesh based technique like the VOF method (Volume Of Fluid), which tends to diffuse the free surface interface and to merge water sheets. This ability of the SPH-ALE method to capture water sheets can then be used to study the flow when the runner is located in the casing, which is very difficult to observe experimentally (see Figure 5.17). One can also see that some isolated particles escape the buckets later, which could be the result of the emptying process.

The evolution of the hydraulic torque on one bucket during the rotation is presented in Figure 5.18 and compared to the torque obtained with a flow solver based on the VOF method. For the SPH-ALE method, the torque is obtained by integrating the torque contributions coming from the pressure field on the bucket surface. For the VOF method, viscous contributions are also taken into account. In Figure 5.18, torque values have been divided by the maximum values, so that superposition of the curves is possible. It can be seen that some differences exist at the beginning of the jet-bucket interaction and also after the peak. But the agreement between the two curves is globally good. Without the scaling of curves, the SPH-ALE method tends to over-predict the torque values compared to the VOF method.

FIGURE 5.16: Water sheets observed at model scale.

FIGURE 5.17: Impact of water sheets on the casing.

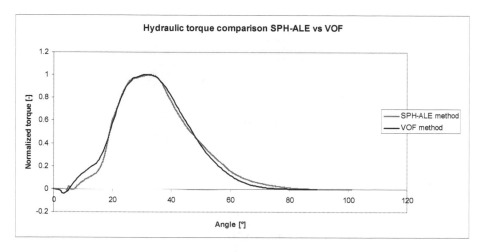

FIGURE 5.18: Hydraulic torque predicted by the SPH-ALE and VOF methods.

Chapter 6

Finite volume particle methods (FVPM)

Francis Leboeuf

Ecole Centrale de Lyon - LMFA
36 avenue Guy de Collongue
69134 Ecully Cedex, France

Jean-Christophe Marongiu

ANDRITZ-HYDRO
Rue des Deux-Gares 6
CH 1800 Vevey, Suisse

6.1 Introduction

The Finite Volume Particle Method (FVPM) is a particular SPH method that has some nice features for conservation properties; it has been developed by Hietel (2005), Struckmeier (2002, 2008) and applied to some complex flow phenomena by Quinlan and his co-workers (2009).

FVPM is based on the use of a sort of Shepard's kernel ψ (4.42) but without a weighting of the kernel W by the ω_j.

$$\psi\left(x\left(t\right), x_j, h\right) = \frac{W\left(x\left(t\right) - x_j, h\right)}{\sum\limits_{j \in N_x} W\left(x_j - x\left(t\right), h\right)} = \frac{W\left(x\left(t\right) - x_j, h\right)}{\sum\limits_{j \in N_x} W\left(x\left(t\right) - x_j, h\right)}$$

where N_x is the set of particles j which contain the point x in their own domain of influence D_j. Let

$$\sigma\left(x\left(t\right), h\right) = \sum\limits_{i \in N_x} W\left(x_j - x\left(t\right), h\right) \Rightarrow \psi\left(x\left(t\right), x_j, h\right) = \frac{W\left(x\left(t\right) - x_j, h\right)}{\sigma\left(x\left(t\right), h\right)}$$

Note that σ is different from one. In general (see Figure 6.1), this is because the summation is done on the values at x of the kernel function of the particles j that belongs to N_x; it does not represent an integral of the kernel function associated to the particle at x as in (4.9). As a consequence, this is quite different from the normalization used by Feldman and Bonet (2007) in their CSPH method (see paragraph 5.4.4).

The disadvantage of using ψ instead of W is that the norm σ depends on (x, t) and so ψ also, while W depends only on $(x - x_j, t)$; as a consequence, the general shape of ψ as to be recomputed for each position (x, t).

6.2 Partition of unity

The main advantage of the kernel function ψ_i is that they form a partition of unity:

$$\sum_{j \in N_x} \psi\left(x(t), x_i, h\right) = 1 \tag{6.1}$$

because

$$\sum_{i \in N_x} \psi\left(x(t), x_i, h\right) = \sum_{i \in N_x} \frac{W\left(x(t) - x_i, h\right)}{\sigma\left(x(t), h\right)} = \frac{\sum_{i \in N_x} W\left(x(t) - x_i, h\right)}{\sum_{j \in N_x} W\left(x(t) - x_j, h\right)} = 1$$

As a consequence, the analogy with Finite Volume is closer because the SPH "box" is closed (see paragraph 5.2.5).

Note that condition (6.1) is always fulfilled (at least theoretically) even close to the boundary $\partial\Omega$, where the domain $D(x_i)$ of the particles are truncated. To be more clear, consider the Figures 6.1, 6.2 and 6.3 similar to those found in Junk and Struckmeier (2002). In Figure 6.1, we get the particle positions

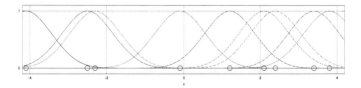

FIGURE 6.1: Particles position x_i and function W_i.

x_i and the associated function W_i. Note the symmetry of each function W_i around each position x_i. The sum σ of function W_i is given in Figure 6.2 for

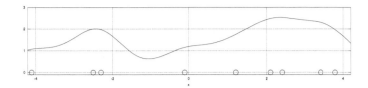

FIGURE 6.2: The function σ corresponding to Figure 6.1.

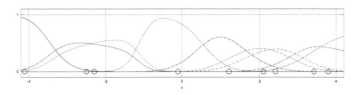

FIGURE 6.3: The kernel ψ_i.

each position x. The kernel ψ_i is given in Figure 6.3. Note that this function is non-symmetric around x_i because σ depends on (x, t) and not on $(x - x_j, t)$ and that the maxima of ψ_i vary along x.

6.3 Average of a function ϕ

A difference with the classical SPH method is related to the expression of the average of a function over the domain $D(x)$:

$$\phi(x, t) = \frac{1}{V(x, t)} \int_{D(x)} \phi(x', t) \, \psi(x(t), x', h) \, dx'^d \tag{6.2}$$

This discrete value ϕ_i is associated with each particle, as well as the volume V:

$$V(x, t) = \int_{D_i} \psi(x(t), x', h) \, dx'^d \tag{6.3}$$

Their discrete analogues are:

$$\tilde{\phi}(x_i, t) \simeq \frac{1}{V(x_i, t)} \sum_{j \in N_i} \phi(x_j, t) \, \psi(x_i(t), x_j) \, \omega_j \tag{6.4}$$

and

$$V(x_i, t) \simeq \sum_{j \in N_i} \psi(x_i(t), x_j) \, \omega_j$$

Note also that

$$V\left(x_i, t\right) = \int_{D_i} \psi\left(x_i, x', t\right) dx'^d = \int_{D_i} \frac{W\left(x_i - x', h\right)}{\sigma\left(x_i, t\right)} dx'^d \approx \frac{1}{\sigma\left(x_i, t\right)}$$

This means also that the density can be computed from $\rho_i = m_i \sigma_i = m_i \sum_{j \in D_i} W\left(x_i - x_j, h\right)$. This expression is particularly suited to the computation of the density closed to a boundary $\partial\Omega$ on a liquid-gas interface, provided that the summation is performed on the particles located on one side of the interface only, thereby allowing for a density gradient almost "naturally". This is the basis of the method developed by Hu and Adams (2006).

6.4   Derivatives of ψ

6.4.1   Lagrangian derivative of ψ

As we have a partition of unity $\sum_{i \in N_x} \psi\left(x\left(t\right), x_i\right) = 1$, then

$$\nabla\left(\sum_{i \in N_x} \psi\left(x\left(t\right), x_i\right)\right) = \sum_{i \in N_x} \nabla\psi\left(x\left(t\right), x_i\right) = 0$$

We will note thereafter:

$$W_i = W\left(x_i\left(t\right) - x_j\right)$$

and

$$\psi_i = \psi\left(x_i\left(t\right), x_j\right) = \frac{W\left(x_i\left(t\right) - x_j\right)}{\sigma\left(x_i\left(t\right)\right)}$$

Assuming that h is constant, then $\frac{d}{dt}W_i = 0 \Rightarrow \frac{\partial W_i}{\partial t} + \dot{x}_i\nabla W_i = 0$. We compute the gradient of ψ from the gradient of W. We start from

$$\psi_i = \frac{W_i}{\sigma_i} \Rightarrow \nabla\psi_i = \frac{\nabla W_i}{\sigma_i} - \frac{\psi_i}{\sigma_i}\nabla\sigma_i = \frac{\nabla W_i}{\sigma_i} - \frac{\psi_i}{\sigma_i}\sum_{j \in N_x} \nabla W_j$$

and

$$\frac{\partial\psi_i}{\partial t} = \frac{1}{\sigma_i}\frac{\partial W_i}{\partial t} - \frac{\psi_i}{\sigma_i}\frac{\partial\sigma_i}{\partial t}$$

$$= \frac{1}{\sigma_i}\frac{\partial W_i}{\partial t} - \frac{\psi_i}{\sigma_i}\sum_{j \in N_x}\frac{\partial W_j}{\partial t}$$

$$= -\frac{\dot{x}_i}{\sigma_i}\nabla W_i + \frac{\psi_i}{\sigma_i}\sum_{j \in N_x}\dot{x}_j\nabla W_j$$

In order to compute the Lagrangian derivative of ψ_i, we need:

$$\frac{\partial \psi_i}{\partial t} = -\dot{x}_i \frac{\nabla W_i}{\sigma_i} + \frac{\psi_i}{\sigma_i} \sum_{j \in N_{x_i}} \dot{x}_j \nabla W_j = \sum_{j \in N_{x_i}} \left(\dot{x}_j \psi_i \frac{\nabla W_j}{\sigma_i} - \dot{x}_i \psi_j \frac{\nabla W_i}{\sigma_i} \right)$$

$$\frac{\partial \psi_i}{\partial t} = - \sum_{j \in N_x} \left(\dot{x}_i \psi_j \nabla \psi_i - \dot{x}_j \psi_i \nabla \psi_j + \psi_i \psi_j \frac{\nabla \sigma_i}{\sigma_i} (\dot{x}_i - \dot{x}_j) \right) \qquad (6.5)$$

$$\nabla \psi_i = \sum_{j \in N_{x_i}} \left(\psi_j \frac{\nabla W_i}{\sigma_i} - \psi_i \frac{\nabla W_j}{\sigma_i} \right)$$

$$= \sum_{j \in N_x} \left(\psi_j \left(\nabla \psi_i + \psi_i \frac{\nabla \sigma_i}{\sigma_i} \right) - \psi_i \left(\nabla \psi_j + \psi_j \frac{\nabla \sigma_i}{\sigma_i} \right) \right)$$

$$\nabla \psi_i = \sum_{j \in N_x} (\psi_j \nabla \psi_i - \psi_i \nabla \psi_j) \qquad (6.6)$$

Assume that

$$\Gamma_{ij} = \psi_i \frac{\nabla W_j}{\sigma_i} = \psi_i \left(\nabla \psi_j + \psi_j \frac{\nabla \sigma_i}{\sigma_i} \right) \qquad (6.7)$$

We can write:

$$\frac{\partial \psi_i (x,t)}{\partial t} = - \sum_{j \in N_x} (\dot{x}_i \Gamma_{ji} - \dot{x}_j \Gamma_{ij}) \qquad (6.8)$$

$$\nabla \psi_i (x,t) = \sum_{j \in N_x} (\Gamma_{ji} - \Gamma_{ij}) \qquad (6.9)$$

The Lagrangian derivative of ψ_i is then

$$\frac{d\psi_i}{dt} = \frac{\partial \psi_i}{\partial t} + \dot{x}_i \nabla \psi_i = - \sum_{j \in N_x} (\dot{x}_i \Gamma_{ji} - \dot{x}_j \Gamma_{ij}) + \dot{x}_i \sum_{j \in N_x} (\Gamma_{ji} - \Gamma_{ij})$$

$$\frac{d\psi_i}{dt} = \sum_{j \in N_x} (\dot{x}_j - \dot{x}_i) \Gamma_{ij} = -\psi_i \sum_{j \in N_x} (\dot{x}_i - \dot{x}_j) \left(\nabla \psi_j + \psi_j \frac{\nabla \sigma_i}{\sigma_i} \right) \qquad (6.10)$$

Note that the summation in all these equations (6.5) to (6.10) is not related to integration as in (6.4) but to the normalisation in the kernel ψ_i.

6.4.2   Other useful coefficients and "closed box" condition

Define the following integrated coefficients by

$$\gamma_{ij} = \int_{\Delta \Omega} \Gamma_{ij} dx^d$$

$$\beta_{ij} = \gamma_{ij} - \gamma_{ji}$$

$$\beta_{ij} = \int_{\Delta\Omega} (\Gamma_{ij} - \Gamma_{ji})\, dx^d$$

$$= \int_{\Delta\Omega} \left(\psi_i \nabla \psi_j + \psi_i \psi_j \frac{\nabla \sigma_i}{\sigma_i} - \psi_j \nabla \psi_i - \psi_j \psi_i \frac{\nabla \sigma_i}{\sigma_i} \right) dx^d$$

$$\beta_{ij} = \int_{\Delta\Omega} (\psi_i \nabla \psi_j - \psi_j \nabla \psi_i)\, dx^d$$

Note that the integration has to be done on a domain $\Delta\Omega$ that is the intersection of the spheres of influence of the particles i and j. This is certainly a key point because these coefficients β play an important role in the transport equations. This seems also to be a very time consuming process. For instance, Nestor (2009) reports 6^d Gauss points to get an acceptable accuracy.

Integrating by parts, we get:

$$\beta_{ij} = 2 \int_{\Delta\Omega} \psi_i \nabla \psi_j dx^d - \int_{\partial\Delta\Omega} \psi_i \psi_j n dx^{d-1}$$

$$\sum_j \beta_{ij} = \sum_j \left(2 \int_{\Delta\Omega} \psi_i \nabla \psi_j dx^d - \int_{\partial\Delta\Omega} \psi_i \psi_j n dx^{d-1} \right)$$

$$= 2 \int_{\Delta\Omega} \psi_i \sum_j \nabla \psi_j dx^d - \int_{\partial\Delta\Omega} \psi_i n \sum_j \psi_j dx^{d-1}$$

$$\sum_j \beta_{ij} = - \int_{\partial\Delta\Omega} \psi_i n dx^{d-1} \tag{6.11}$$

Setting $n_{ij} = \frac{\beta_{ij}}{|\beta_{ij}|}$, β_{ij} is the analogue in FVPM of the product of a surface element $|\beta_{ij}|$ and a normal n_{ij} joining two cell centres. So that (6.11) expresses the closed "box" condition if the FVPM cell is far from the boundary $\partial\Omega$: $\sum_j \beta_{ij} = 0$. We note also that $\beta_{ij} = -\beta_{ji}$, $\beta_{ii} = 0$

REMARK 6.1 It seems that (6.11) is not numerically fulfilled despite of the normalisation of the kernel function through the use of the FVPM kernel ψ_i probably because of the numerical errors on the computation of β_{ij}. For instance Teleaga (2008) reports how to correct the fluxes terms in the conservation equation through an artificial term $\beta_{ii}^n = -\sum_{j \neq i} \beta_{ij}^n$. Nothing is said on the impact of this correction for the global conservation of the flow quantities. A different technique of correction of β_{ij} is also described by Keck (2002). □

6.4.3 Transport of the volume V_i

We have

$$\frac{d}{dt}\left(V_i\left(t\right)\right) = \frac{d}{dt}\int_{D_i} \psi_i\left(t, x\right) dx^d = \int_{D_i} \frac{\partial \psi_i}{\partial t} dx^d + \int_{\partial D_i} \psi_i \dot{x} n dx^{d-1}$$

$$\frac{d}{dt}\left(V_i\left(t\right)\right) = -\sum_{j\in N_i}\left(\dot{x}_i \gamma_{ji} - \dot{x}_j \gamma_{ij}\right) + \int_{\partial D_i} \psi_i \dot{x} n dx^{d-1} \qquad (6.12)$$

This is the equivalent for FVPM of (4.4) used for classical SPH method.

6.4.4 On the computation of the gradient $\nabla\phi_i$

Three formulations may be used:

- the mixed method of Bonet and Lok (1999) (4.44) that is based on Shepard's kernel.

- the method of Nestor (2009)

- the method of Keck (2005).

We shall detail the last two methods as the first one has already been given.

6.4.5 Method of Nestor

Nestor proposes to use a renormalized kernel $\tilde{\nabla}S_j\left(b_i\right)$ based on the barycentre b_i of each particles, where $b_i = \frac{1}{V_i}\int_{D_i} x\psi_i dx$. We have then

$$\nabla\phi_i = \sum_{j\in N_i} V_j\left(\phi_j - \phi_i\right)\tilde{\nabla}S_j\left(b_i\right)$$

$$\tilde{\nabla}S_j\left(b_i\right) = \left[\sum_{j\in N_i} V_j \nabla S_j\left(b_i\right) \otimes \left(b_j - b_i\right)\right]^{-1} \nabla S_j\left(b_i\right)$$

S_j is a cubic-spline used by Monaghan and Lattanzio (1985).

6.4.6   Method of Keck

The method of Keck is based on consistency points x_{ij}, which may be interpreted as midpoints of general cell faces. The consistency points are used to modify the discrete gradient operator so that we obtain a consistent approximation. We have then

$$V_i \overline{\overline{I}} = \frac{1}{2} \sum_{j \in N_i} (b_j \otimes \beta_{ij}) = \frac{1}{2} \sum_{j \in N_i} \left(\frac{1}{V_i} \int_{D_i} x \psi_j dx \otimes \beta_{ij} \right)$$

In practice, x_{ij} are explicitly given by:

$$x_{ij} = b_i + \left(\sum_{k \in N_i} \beta_{ik} \otimes \beta_{ik} \right)^{-1} \beta_{ij} V_i$$

Note that in general $x_{ij} \neq x_{ji}$. The consistent form of the gradient is then given by:

$$(\nabla \phi)_i = \frac{1}{V_i} \sum_{i \in N_i} \beta_{ij} \phi^{rec}(x_{ij})$$
$$\phi^{rec}(x_{ij}) = \phi(b_i) + (x_{ij} - b_i) \nabla^T \phi(b_i) + o(h^2)$$
$$(\nabla \phi)_i = \nabla \phi(b_i) + o(h)$$

where ϕ^{rec} is the reconstructed function at the consistency point x_{ij}. Linear functions are then properly reproduced.

6.5   Conservation equation and FVPM

We introduce again a weak formulation of the conservation equation (5.1)

$$\int_{D(x)} \left(\frac{\partial \phi}{\partial t} + \nabla . F(\phi) \right) \psi_i dx^d = \int_{D(x)} S(\phi) \psi_i dx^d$$

Integrating the first member by parts, we get

$$\int_{D(x)} \left(\frac{\partial \phi \psi_i}{\partial t} + \nabla . (F(\phi) \psi_i) \right) dx^d = \int_{D(x)} \phi \frac{\partial \psi_i}{\partial t} dx^d + \int_{D(x)} F(\phi) \nabla \psi_i dx^d$$

We use also the Reynolds theorem (5.5) in the form

$$\frac{d_W}{dt} \int_{D(x)} \phi \psi_i dx^d = \int_{D(x)} \frac{\partial \phi \psi_i}{\partial t} dx^d + \int_{D(x)} \nabla . (\phi \psi_i W) dx^d$$

Combining the two equations, we get

$$\frac{d_W}{dt} \int_{D(x)} \phi\psi_i dx^d + \int_{D(x)} \nabla.\left(\left[F\left(\phi\right) - \phi W\right]\psi_i\right) dx^d$$

$$= \int_{D(x)} \phi\frac{\partial\psi_i}{\partial t} dx^d + \int_{D(x)} F\left(\phi\right)\nabla\psi_i dx^d + \int_{D(x)} S\left(\phi\right)\psi_i dx^d$$

The derivatives of ψ_i are expressed with (6.8) and (6.9) which gives:

$$\frac{d}{dt} \int_{D(x)} \phi\psi_i dx^d = - \int_{D(x)} \phi\sum_{j\in N_x}\left(\dot{x}_i\Gamma_{ji} - \dot{x}_j\Gamma_{ij}\right) dx^d$$

$$+ \int_{D(x)} F\left(\phi\right)\sum_{j\in N_x}\left(\Gamma_{ji} - \Gamma_{ij}\right) dx^d$$

$$- \int_{\partial D(x)} \left(F\left(\phi\right) - \phi\dot{x}_i\right)\psi_i n dx^{d-1}$$

$$+ \int_{D(x)} S\left(\phi\right)\psi_i dx^d$$

$$\frac{d}{dt} \int_{D(x)} \phi\psi_i dx^d = - \sum_{j\in N_x}\left(\int_{D(x)}\left(F\left(\phi\right) - \phi\dot{x}_j\right)\Gamma_{ij} dx^d\right.$$

$$- \int_{D(x)}\left(F\left(\phi\right) - \phi\dot{x}_i\right)\Gamma_{ji} dx^d\right)$$

$$- \int_{\partial D(x)}\left(F\left(\phi\right) - \phi\dot{x}_i\right)\psi_i n dx^{d-1} + \int_{D(x)} S\left(\phi\right)\psi_i dx^d$$

This equation appears in an ALE form with the transport of the particle at the velocity $v_0 = \dot{x}_i$

We state $G_i = F\left(\phi\right) - \dot{x}_i\phi$ so that we get:

$$\frac{d}{dt} \int_{D(x)} \phi\psi_i dx^d = - \sum_{j\in N_x}\left(\int_{D(x)} G_j\Gamma_{ij} dx^3 - \int_{D(x)} G_i\Gamma_{ji} dx^d\right) \quad (6.13)$$

$$- \int_{\partial D(x)} G_i\psi_i n dx^{d-1} + \int_{D(x)} S\left(\phi\right)\psi_i dx^d$$

Two important hypotheses are now used in order to reduce the cost of the integrations:

1. The variation of ϕ are small over the domain Ω of intersection of the two spheres i and j.

2. $\int_{\Omega} G_j \Gamma_{ji} dx \approx G_j \gamma_{ji}$

Using these hypotheses, the conservation equation can be written:

$$\frac{d}{dt} \int_{D(x)} \phi \psi_i dx^d \approx \sum_{j \in N_x} \{G_i \gamma_{ji} - G_j \gamma_{ij}\} - \int_{\partial D(x)} G_i \psi_i n dx^{d-1} + \int_{D(x)} S(\phi) \psi_i dx^d$$

(6.14)

We can also write:

$$\sum_{j \in N_x} \{G_i \gamma_{ji} - G_j \gamma_{ij}\}$$

$$= \sum_{j \in N_x} (G_i - G_j) \frac{(\gamma_{ij} + \gamma_{ji})}{2} - \sum_{j \in N_x} \frac{(G_i + G_j)}{2} (\gamma_{ij} - \gamma_{ji})$$

Using the first hypothesis, we then have:

$$\sum_{j \in N_x} \{G_i \gamma_{ji} - G_j \gamma_{ij}\}$$

$$\approx - \sum_{j \in N_x} \frac{(G_i + G_j)}{2} (\gamma_{ij} - \gamma_{ji})$$

$$= - \sum_{j \in N_x} \frac{(G_i + G_j)}{2} \beta_{ij}$$

So that:

$$\frac{d}{dt} \int_{D(x)} \phi \psi_i dx^d \quad (6.15)$$

$$= - \sum_{j \in N_x} |\beta_{ij}| \frac{(G_i + G_j)}{2} n_{ij} - \int_{\partial D(x)} G_i \psi_i n dx^{d-1} + \int_{D(x)} S(\phi) \psi_i dx^d$$

$\frac{(G_i + G_j)}{2} n_{ij}$ is just a numerical flux computed in the direction n_{ij}, at the point x_{ij}. We shall write it in a more general form $G_{ij}^m (t, \phi_i, x_i, \phi_j, x_j, n_{ij})$. Note that ϕ_i is influenced by the particles j whose domain has a non-zero intersection with the domain D_i.

Finally, equation (6.15) is then:

$$\frac{d}{dt} (\phi_i V_i) = - \sum_{j \in N_i} |\beta_{ij}| G_{ij}^m - \int_{\partial D(x)} G_i \psi_i n dx^{d-1} + \int_{D(x)} S(\phi) \psi_i dx^d \quad (6.16)$$

This form looks very similar to the weak form of Vila (2005) (5.17) for instance. It is adapted to the use of Riemann solvers.

6.6 Concluding remarks

FVPM is in practice very close to SPH methods. It is possible to apply the same techniques of improvement of the computation of the gradient, and an ALE formulation can also be derived, allowing the use of various Riemann solvers. The theoretical advantage of local conservation associated with the partition of unity linked to the FVPM kernel is limited to the accuracy of the numerical integration of coefficient β_{ij}. The FVPM method will then have an interest only if a quick algorithm can be used for the computation of the integrals in the intersection of two spheres.

Chapter 7

Numerical algorithms for unstructured meshes

Bruno Koobus
Université Montpellier 2, Département de Mathématiques
Place Eugène Bataillon
34095 Montpellier cedex, France

Frédéric Alauzet
Institut National de Recherche en Informatique et Automatique
Rocquencourt, Domaine de Voluceau
78153 Le Chesnay Cedex, France

Alain Dervieux
Institut national de recherche en informatique et automatique
2004 Route des Lucioles
06902 Sophia-Antipolis, France

7.1 Introduction

The simulation of compressible flows experienced in the '90s a small revolution with the arising of new algorithms able to compute flows through (or around) any kind of shape. This was due to new numerical algorithms and to new mesh generation algorithms. For both type, the main innovation was related to unstructured meshes, and the way to do it first was to rely on tetrahedrisations. The "any kind of shape" slogan has been progressively completed by the adaptation to any kind of flow, for flows in moving meshes and for automatic mesh adaptation, appearing as an important issue to address in order to improve the expected benefits from a numerical simulation.

Several numerical methods have been developed for computing compressible flows on unstructured meshes. First, low-order methods (typically second-order) were developed. Let us mention central-differenced cell-centered and vertex-centered finite-volume methods [Jameson, 1993], [Mavriplis, 1997], Taylor-Galerkin methods [Donea et al., 1987], [Löhner, 2001], Galerkin Least-Square methods [Hughes and Mallet, 1986], and the distributive schemes [Deconinck et al., 1993] for second order accurate methods. Higher accurate schemes have then been developed, let us mention unstructured ENO methods [Abgrall, 1994] and Discontinuous Galerkin methods [Cockburn, 2003], [Cockburn and Shu, 1989].

The purpose of this chapter is to describe and discuss a Mixed finite-Element/finite-Volume (MEV) low order discretization for the Euler models of aerodynamics applicable to a very general class of tetrahedrisations, and to consider, a few crucial numerical issues for the application of an Euler scheme:

- mastering numerical dissipation,

- mastering positiveness,

- evaluating the synergy between such kind of numerics and high performance mesh adaptation methods.

The application of the MEV method to Navier-Stokes is considered in the companion chapter dealing with Large Eddy Simulation of turbulent flows. The MEV discretisation method is a combination of a Finite-Element method (FEM) with a vertex-centered Finite-Volume method (FVM). Like any vertex-centered approximation, it enjoys the property of handling the smallest number of unknowns for a given mesh and the possibility to assemble the fluxes on an edge-based mode. The underlying FEM is the standard Galerkin method with continuous piecewise linear approximation on triangles or tetrahedra. The FEM is applied directly for discretising second-order derivatives (diffusion or viscosity terms). For hyperbolic terms, the FEM needs extra stabilization terms which are derived from an upwind FVM. The underlying FVM is a vertex centered edge-based method. The finite-volume cell is built around each vertex, generally by using medians (2D) or median planes (3D), advection terms are stabilized with upwinding or artificial dissipation, and second order "viscous" terms are discretized with finite-elements. Among the different ways of constructing second-order accurate upwind schemes, the MUSCL formulation introduced by Van Leer in [Van Leer, 1979] for finite-volume methods is particularly attractive and has been generally chosen.

This family of schemes was initiated by Baba and Tabata [Baba and Tabata, 1981] for first-order upwind diffusion-convection models and Fezoui, Dervieux and coworkers for Euler flows (see [Fezoui, 1985], [Dervieux, 1987], [Fezoui and Stoufflet, 1989], [Fezoui and Dervieux, 1989], [Stoufflet et al., 1996]). It has been studied by many CFD teams (see in particular [Whitaker et al., 1989], [Anderson and Bonhaus, 1994], [Venkatakrishnan, 1996], [Barth, 1994], [Catalano, 2002]). The framework proposed in [Selmin and Formaggia, 1998] can also be considered as an extension of MEV. Current developments and results relying on this family of schemes are regularly reported by Farhat and co-workers (see [Farhat and Lesoinne, 2000]). A particular advantage of MEV is its ability to perform well in combination with very rather irregular meshes. As a consequence, this scheme was identified as particularly convenient for developing methods for shape design [Farhat, 1995], [Nielsen and Anderson, 2002], [Vàzquez et al., 2004], for fluid-structure interaction with moving meshes [Farhat and Lesoinne, 2000], and for anisotropic mesh adaptation [Loseille et al., 2007].

Several theoretical or methodological questions concerning MEV are addressed in this chapter:

Accuracy The basic scheme is introduced in section 1. In case of meshes with a bounded aspect ratio, the second order accuracy of the underlying Galerkin method holds for steady-state problems, even for rather irregular meshes. For the unsteady case, since the mass matrix diagonalisation is applied, the constraint on mesh regularity is somewhat stronger.

Highly stretched meshes The behavior of the upwind versions of the MEV for highly stretched structured meshes is the main drawback of this class of schemes. In [Barth, 1994], Barth suggests to modify the shape of finite-volume cells. This idea is examined and developed in section 2.

Low Mach number A common property of Godunovmethods is to loose accuracy when low Mach number flows are computed. We recall how to cure this problem for steady and unsteady problems in section 3.

Superconvergent low dissipation versions In the case where the flow field under study is smooth, the numerical dissipation can be importantly reduced while not allowing Gibbs oscillations. In the linear theory, oscillations arise when high frequency components of the solution are dispersed, *i.e.* propagated with large phase velocity error, without enough dissipation to damp them. For reducing overall dissipation while avoiding oscillation, we follow the lines of higher order upwinding. This is obtained by introducing a new type of MUSCL reconstruction. Dissipation appears as relying on higher order even derivatives. Some versions of the new family show higher order convergence on regular or very smooth meshes. We call this property superconvergence. This method is presented in section 4.

Robustness and positivity   In the '80s, robustness of numerical schemes for hyperbolics was put in relation with positiveness, monotony and Total Variation Diminishing properties. In section 5, we state several positivity results for MEV schemes.

Mesh adaptation   Mesh adaption is addressed in section 8. An interest of the tetrahedra-based approximation which we use is its ability to combine well with anisotropic metric-based mesh adaptation methods. We introduce and discuss a Hessian-based and a goal-oriented method.

7.2   Spatial representation

7.2.1   A particular P1 finite-element Galerkin formulation

7.2.1.1   Mathematical model

We write the unsteady Euler equations as follows in the computational domain $\Omega \subset \mathbb{R}^3$:

$$\Psi(W) = \frac{\partial W}{\partial t} + \nabla.\mathcal{F}(W) = 0 \quad \text{in} \quad \Omega, \tag{7.1}$$

where $W = {}^t(\rho, \rho u, \rho v, \rho w, \rho E)$ is the vector of conservative variables. $\mathcal{F}$ is the convection operator $\mathcal{F}(W) = (\mathcal{F}_1(W), \mathcal{F}_2(W), \mathcal{F}_3(W))$ with:

$$\mathcal{F}_1(W) = \begin{pmatrix} \rho u \\ \rho u^2 + p \\ \rho u v \\ \rho u w \\ (\rho E + p)u \end{pmatrix}, \mathcal{F}_2(W) = \begin{pmatrix} \rho v \\ \rho u v \\ \rho v^2 + p \\ \rho v w \\ (\rho E + p)v \end{pmatrix}, \mathcal{F}_3(W) = \begin{pmatrix} \rho w \\ \rho u w \\ \rho v w \\ \rho w^2 + p \\ (\rho E + p)w \end{pmatrix},$$
$$\tag{7.2}$$

so that the state equation becomes:

$$\frac{\partial W}{\partial t} + \frac{\partial \mathcal{F}_1(W)}{\partial x} + \frac{\partial \mathcal{F}_2(W)}{\partial y} + \frac{\partial \mathcal{F}_3(W)}{\partial z} = 0.$$

ρ, p and E hold respectively for the density, the thermodynamical pressure and the total energy per mass unit. Symbols u, v and w stand for the Cartesian components of velocity vector $\mathbf{u} = (u, v, w)$. For a calorically perfect gas, we have

$$p = (\gamma - 1)\left(\rho E - \frac{1}{2}\rho|\mathbf{u}|^2\right),$$

where γ is constant. A weak formulation of this system writes for $W \in V = \left[H^1(\Omega)\right]^5$ as follows:

$$\forall \phi \in V, \quad (\Psi(W), \phi) = \int_\Omega \left(\phi \frac{\partial W}{\partial t} + \phi \nabla.\mathcal{F}(W)\right) \, d\Omega -$$
$$\int_\Gamma \phi \hat{\mathcal{F}}(W).\mathbf{n} \, d\Gamma = 0, \qquad (7.3)$$

where Γ is the boundary of the computational domain Ω, $\mathbf{n}$ the outward normal to Γ and the boundary flux $\hat{\mathcal{F}}$ contains the boundary conditions. We are interested by this unsteady formulation together with the steady one, in which the time derivative is not introduced.

7.2.1.2 Discrete variational representation

We consider here the *steady* case, written:

$$\nabla.\mathcal{F}(W) = 0$$

also written in variational formulation:

$$\forall \phi \in V, \quad (\Psi(W), \phi) = \int_\Omega (\phi \nabla.\mathcal{F}(W)) \, d\Omega -$$
$$\int_\Gamma \phi \hat{\mathcal{F}}(W).\mathbf{n} \, d\Gamma = 0 \qquad (7.4)$$

The discretization chosen relies on two main focus. First, we consider a *tetrahedrisation* as the discretization of the computational domain. This choice is made in connection with the progresses made for automatically generating and adapting meshes of this kind. Once the mesh is chosen, we have to put on it a set of nodes, that are the geometrical supports of the degrees of freedom. This option has the smallest number of nodes, *viz.* the vertices.

Let $\mathcal{T}_h$ be a tetrahedrization of Ω which is admissible for Finite-Elements *i.e.*, Ω is partitioned in tetrahedra, and the intersection of two different tetrahedra is either empty, or a vertex, or an edge, or a face. The test functions are taken into the approximation space V_h made of continuous piecewise linear functions included in $V = [H^1(\Omega)]^5$:

$$V_h = \left\{\phi_h \, \middle| \, \phi_h \text{ is continuous and } \phi_{h|T} \text{ is linear } \forall T \in \mathcal{T}_h\right\}.$$

In order to avoid the management of projectors applicable in the whole H^1 space, we shall work inside the following spaces:

$$\bar{V} = ([H^2(\Omega)]^5) \text{ and } \bar{V}_h = \bar{V} \sqcup V_h.$$

It is useful to introduce Π_h, the corresponding P^1 interpolation operator:

$$\Pi_h : \bar{V}_h \longrightarrow V_h$$
$$\phi \longmapsto \Pi_h\phi \quad \text{with } \Pi_h\phi(i) = \phi(i) \;\; \forall i \text{ vertex of } \mathcal{T}_h \,.$$

Then from problem (7.3), the discrete steady formulation writes:

$$\forall \phi_h \in V_h, \quad \int_\Omega \phi_h \nabla.\mathcal{F}_h(W_h) \, d\Omega - \int_\Gamma \phi_h \hat{\mathcal{F}}_h(W_h).\mathbf{n} \, d\Gamma \;\; = \;\; 0 \,, \quad (7.5)$$

where $\mathcal{F}_h$ is by definition the P^1 interpolate of $\mathcal{F}$, in the sense that:

$$\mathcal{F}_h(W) = \Pi_h \mathcal{F}(W) \;\; \text{and} \;\; \mathcal{F}_h(W_h) = \Pi_h \mathcal{F}(W_h) \,, \quad (7.6)$$

and, as the operator $\mathcal{F}_h$ applies to the values of W at the mesh vertices, we have:

$$\mathcal{F}_h(W) = \mathcal{F}_h(\Pi_h W) = \Pi_h \mathcal{F}(\Pi_h W) \,. \quad (7.7)$$

We get the same relations for $\hat{\mathcal{F}}_h(W)$:

$$\hat{\mathcal{F}}_h(W) = \Pi_h \hat{\mathcal{F}}(\Pi_h W) \;\; \text{and} \;\; \hat{\mathcal{F}}_h(W_h) = \Pi_h \hat{\mathcal{F}}(W_h) \,. \quad (7.8)$$

Practically, this definition of $\mathcal{F}_h$ means that nodal fluxes values are evaluated at the mesh vertices. Consequently, discrete fluxes are derived from the nodal values by P^1 extrapolation inside every element. In contrast to the standard Galerkin approach, this definition emphasizes that the fluxes are projected in V_h.

7.2.2 Mixed-element-volume basic equivalence

The discrete formulation (7.5) can be transformed into a vertex-centered finite-volume scheme applied to tetrahedral unstructured meshes. This assumes a particular partition in control cells C_i of the discretized domain $\Omega_{\rm h}$:

$$\Omega_{\rm h} = \bigcup_{i=1}^{n_c} C_i \,, \quad (7.9)$$

each control cell being associated with a vertex i of the mesh. The corresponding test functions are the piecewise constant characteristic functions of cells:

$$\chi^i(\boldsymbol{x}) = \begin{cases} 1 \text{ if } \boldsymbol{x} \in C_i, \\ 0 \text{ otherwise.} \end{cases}$$

Then the weak form of (7.1) is integrated in this new formulation by writing, for each vertex i, *i.e.* for each cell C^i:

$$\int_{C^i} \frac{\partial W}{\partial t} \;\; + \;\; \int_{\partial C^i} \mathbf{n}_i \mathcal{F}(W) \, d\sigma = 0 \quad (7.10)$$

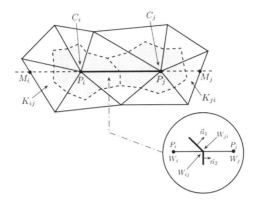

FIGURE 7.1: Illustration of finite-volume cells construction in two dimensions with two neighboring cells, C_i and C_j around i (P_i on the figure) and j (P_j on the figure) respectively, and of the upwind triangles K_{ij} and K_{ji} associated with the edge ij. Definition of the common boundary ∂C_{ij} with the representation of the solution extrapolated values for the MUSCL type approach.

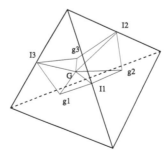

FIGURE 7.2: The planes which delimit finite-volume cells inside a tetrahedron (3D case).

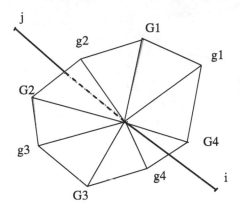

FIGURE 7.3: Illustration for 3D case of finite-volume cell interface ∂C_{ij} between two neighboring cells C_i and C_j.

The *dual finite-volume cell* is built by the rule of medians. In 2D, the median cell is limited by segments of medians between centroids and mid-edge (Figure 7.1). In 3D, each tetrahedron T of the mesh is split into four hexahedra constructed around each of its four vertices. For a vertex i, the hexahedron $C_i \cap T$ is defined by the following points (Figure 7.2):

  (i) the three middle points of the edges issued from i,

  (ii) the three gravity centers of the faces containing i,

 (iii) the center of gravity of the tetrahedron and

 (iv) the vertex i.

The cell C_i of vertex i is the collection of all hexahedra linked to i. The common boundary $\partial C_{ij} = \partial C_i \cap \partial C_j$ between two neighboring cell C_i and C_j is decomposed in several triangular interface facets. An illustration of this construction is shown in Figure 7.3 for the 3D case.

   The finite-volume fluxes between cells around vertices i and j are integrated through the common boundary ∂C_{ij} with a value of $\mathcal{F}_h$ equal to the half-sum of $\mathcal{F}_h(W_i)$ and $\mathcal{F}_h(W_j)$:

$$\Phi_{ij}^{MEV} = \frac{\mathcal{F}_h(W_i) + \mathcal{F}_h(W_j)}{2} \cdot \nu_{ij} , \qquad (7.11)$$

where ν_{ij} denotes the integral of the normal $\mathbf{n}_i$ to common boundary between cells C_i and C_j,

$$\nu_{ij} = \int_{\partial C_{ij}} \mathbf{n}_i \, d\sigma$$

and $W_i = W(i)$. The finite-volume formulation for an internal vertex i writes as the sum of all the fluxes evaluated from the vertices j belonging to $V(i)$

where $V(i)$ is the set of all neighboring vertices of i. Taking into account the boundary fluxes, the discrete Scheme (7.5) then writes:

$$\sum_{j \in V(i)} \Phi_{ij}^{MEV} - \int_{\Gamma \cap \partial C_i} \bar{\mathcal{F}}_h(W_h).\mathbf{n} \, d\Gamma = 0. \tag{7.12}$$

We obtain a vertex-centered finite-volume approximation which is P^1-exact with respect to the flux function $\mathcal{F}_h$. This scheme enjoys most of the accuracy properties of the Galerkin method [Mer, 1998b], such as the second-order accuracy on any mesh for diffusion-convection models. However, it lacks stability and cannot be applied to purely hyperbolic models such as the Euler equations. Before explaining how to stabilise this family of method, let us introduce a second way to build dual cells.

7.2.3 Circumcenter cells

Median-based cells are the exact counterbalance of the P^1 finite element formulation. They are well adapted to non-stretched unstructured meshes. In the case of highly stretched meshes, upwind finite volume methods show a truncation error which grows with aspect ratio. In some other case, cells with shapes closer to rectangles may allow a higher accuracy. We describe now a second option for defining the dual cells.

Two-dimensional circumcenter cells According to an idea of Barth [Barth, 1994], a cell is built around each vertex by joining the center of edges with the center of the smallest circle containing the considered triangle. We observe that this center is located at middle of largest edge in a triangle involving an angle larger or equal to 90°. As a result, a mesh made of rectangular triangles has cells which are rectangles, see Figure 7.7.

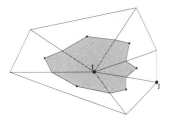 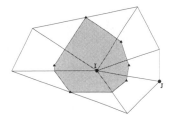

FIGURE 7.4: Median cell construction.

FIGURE 7.5: Circumcenter cell construction.

Three-dimensional circumcenter cells A 3D circumcenter cell can be defined as the union of polyhedra such that, for a given element, the common

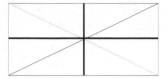

FIGURE 7.6: Trace of median division on two Friedrichs-Keller elements.

FIGURE 7.7: Trace of circumcenter division on two Friedrichs-Keller elements.

surface between two neighboring cells joins:

- the middle of the edge connecting these two vertices,

- the "surface center" of the faces of the element having this edge in common,

- the "volume center" of the element,

where

- the **"surface center"** of a given face is its center of circumscript circle if it comprises only acute angles, otherwise it is the middle of its longest edge,

- the **"volume center"** of an element is its center of circumscript sphere if this center is located inside the element; otherwise, it is the **"surface center"** of the largest surface.

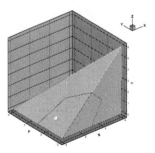

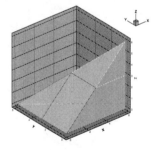

FIGURE 7.8: Trace of median cells on a tetrahedron in a particular Cartesian mesh.

FIGURE 7.9: Trace of a circumcenter cell on a tetrahedron in a particular Cartesian mesh.

In practical applications, we use the circumcenter cells for stretched meshes with aspect ratio larger than 10 and for Cartesian meshes.

REMARK 7.1 It is proposed in [Gourvitch et al., 2004] to combine the Barth cell construction with a particular splitting of cubes into six identical tetrahedra in a Cartesian mesh. With this combination, the cells are exactly cubes and the scheme is equivalent to the usual vertex-centered finite volume scheme. ▯

7.2.4 Flux integration

Once the cells are defined, the spatial divergence $div\mathcal{F}$ is transformed via the Stokes formula into integrals of normal fluxes $\mathcal{F}.\mathbf{n}$ at cell boundaries. In the proposed family of schemes, the accuracy of the integral geometrical quadrature is not as crucial. We define a very simple option, the *edge-based integration*. At the contrary, flux integration sets the important problem of scheme stabilization. The variables are assumed to be constant by cell, and therefore, they are discontinuous from a cell to its neighbor. Upwind integration will rely on the Godunov method based on the two different values at each side of the discontinuity.

7.2.4.1 Central differencing

Let us write a vertex-centered central differenced finite-volume scheme for the Euler equations applied to an unstructured mesh as follows:

$$\boldsymbol{\Psi}_h(\gamma, W)_j = 0, \quad \text{with}$$

$$\boldsymbol{\Psi}_h(\gamma, W)_j = \sum_{k \in V(j)} \Phi^{central}(W_j, W_k, \nu_{jk}) + \mathbf{B}_h(\gamma, W)_j \qquad (7.13)$$

where $V(j)$ is the set of vertices that are neighbors of j, ν_{jk} is the integral on interface between j and k of the normal vector. Symbol $\mathbf{B}_h(\gamma, W)_j$ holds for boundary fluxes. The centered integration for elementary flux Φ is written as follows:

$$\Phi^{central}(W_j, W_k, \nu_{jk}) = 0.5(\mathcal{F}_j + \mathcal{F}_k).\nu_{jk} \qquad (7.14)$$

where $\mathcal{F}_j = \mathcal{F}(W_j)$ are the Euler fluxes computed at W_j. This is equivalent to introducing the following discrete space operator ∇_h^*:

$$\nabla_h^*(f)_j = \sum_{k \in V(j)} (f_j + f_k)/2 \ \nu_{jk} \ / \ a(j) \qquad (7.15)$$

where $a(j)$ is the measure of cell C_j.

7.2.4.2 Godunov differencing

Godunov-type methods rely on discontinous representations of the unknown and computation of fluxes at discontinuities in function of both "left" and "right" values by applying an approximate or an exact Riemann solver. This process introduces numerical viscosity terms that are very useful for stabilizing transonic flows.

We write a vertex-centered first-order Godunov scheme for the Euler equations applied to an unstructured mesh as follows:

$$\boldsymbol{\Psi}_h(\gamma, W)_j = \sum_{k \in V(j)} \Phi(W_j, W_k, \nu_{jk}) + \mathbf{B}_h(\gamma, W)_j. \qquad (7.16)$$

The upwinding in elementary flux Φ is the Roe flux splitting. In the case of the standard Roe splitting, we have:

$$\Phi(W_j, W_k, \nu_{jk}) = 0.5(\mathcal{F}_j + \mathcal{F}_k).\nu_{jk} + 0.5|\mathcal{A}|(W_j - W_k) \qquad (7.17)$$

where $|\mathcal{A}|$ is the absolute value of the Jacobian flux along ν_{jk}:

$$\mathcal{A} = (\tfrac{\partial \mathcal{F}}{\partial W})_1(\nu_{jk})_1 + (\tfrac{\partial \mathcal{F}}{\partial W})_2(\nu_{jk})_2$$

$$\mathcal{A} = T\Lambda T^{-1}, \quad \Lambda \quad diagonal, \qquad (7.18)$$

$$|\mathcal{A}| = T|\Lambda|T^{-1}.$$

These matrices are computed at an intermediate value $\overline{W}_{jk}$ of W: of W_j and W_k, in short:

$$\overline{W}_{jk} = (\rho_j^{\frac{1}{2}} W_j + \rho_k^{\frac{1}{2}} W_k / (\rho_j^{\frac{1}{2}} \rho_k^{\frac{1}{2}})$$

which enjoys the following property:

$$\mathcal{F}(W_j) - \mathcal{F}(W_k) - \mathcal{A}(\overline{W}_{jk})(W_j - W_k).$$

In (high enough) supersonic case, $\mathcal{A}(\overline{W}_{jk}) = |\mathcal{A}(\overline{W}_{jk})|$ or $\mathcal{A}(\overline{W}_{jk}) = -|\mathcal{A}(\overline{W}_{jk})|$ and Roe's splitting is fully upwind. By the hyperbolicity assumption, matrix A can be diagonalized. the absolute value $|\mathcal{A}|$ writes:

$$|\mathcal{A}(\overline{W}_{jk})| = T^{-1}Diag(|\lambda_1|, |\lambda_2|, |\lambda_3|, |\lambda_4|)T = sign(A)\mathcal{A},$$

where

$$sign(\mathcal{A}) = T^{-1}Diag(sign(\lambda_1), sign(\lambda_2), sign(\lambda_3))T. \qquad (7.19)$$

thus this averaging also permits the following equivalent formulation:

$$(\mathcal{F}(W_j) + \mathcal{F}(W_k))/2 - sign(\mathcal{A}(\overline{W}_{jk}))(\mathcal{F}(W_R) - \mathcal{F}(W_L))/2. \qquad (7.20)$$

These schemes are spatially first-order accurate.

7.3 Towards higher spatial order

First-order upwind schemes of Godunov type enjoy a lot of interesting qualities and in particular monotonicity or, more or less equivalently, positivity. They can be extended to second order by applying the MUSCL method. We give in the next subsection a descrition of how this can be done. Unfortunately, even for the the second-order version, the amount of dissipation which is introduces seems larger than that needed in many applications. In particular, the dominant term of the numerical error is carried by the dissipation.

We get inspired by Direct Simulation techniques in which non-dissipative high-order approximations are stabilised in good accuracy conditions thanks to filters which rely on very-high even order derivatives. In order to do this, we have to further extend the discretization stencil. Then it can be also interesting to choose a stencil extension which also improve dispersion properties, since a less dispersive scheme needs less dissipation for avoiding Gibbs-like oscillations. This leads to the idea of superconvergent advection schemes which are more accurate for a subclass of applications, typically, for a nonlinear hyperbolic system solved Cartesian meshes (NLV6 version), or for a linear hyperbolic system solved on a Cartesian mesh (LV6 version).

7.3.1 The MUSCL method

The Godunov method builds fluxes between cells in which unknown variables are considered as constant. This results in a first-order accurate scheme, not enough accurate for most applications. Van Leer has proposed ([Van Leer, 1979], [Van Leer, 1977a]) to reconstruct a linear interpolation of the variables inside each cell and then to introduce in the Riemann solver the boundary values of these interpolations. Further, the slopes used for linear reconstruction can be limited in order to represent the variable without introducing new extremas. The resulting MUSCL method produces positive second-order schemes. We describe now an extension of MUSCL to unstructured triangulations with dual cells. The MUSCL ideas also applies to reconstructions which are different on each interface between cells, or equivalently on each edge. Several slopes of a dependant variable F are defined on the two vertices i and j of an edge ij as follows:

First, the *centered gradient* $(\nabla F)^c_{ij}$ is defined as

$$(\nabla F)^c_{ij} \cdot \vec{ij} = F_j - F_i.$$

We consider a couple of two triangles, one having i as a vertex, and the second having j as a vertex. With reference to Figure 7.10, we define ϵ_{ni}, ϵ_{mi}, ϵ_{jr} and ϵ_{js} as the components of vector $\vec{ji}$ (resp. $\vec{ij}$) in the oblique system of axes

$(\vec{in}, \vec{im})$ (resp. $(\vec{jr}, \vec{js})$):

$$\vec{ji} = \epsilon_{ni}\,\vec{in} + \epsilon_{mi}\,\vec{im}\,,$$

$$\vec{ij} = \epsilon_{jr}\,\vec{jr} + \epsilon_{js}\,\vec{js}\,.$$

We shall say that T_{ij} and T_{ji} are upstream and downstream elements with respect to edge ij if the components $\epsilon_{ni}, \epsilon_{mi}, \epsilon_{jr}, \epsilon_{js}$ are all nonnegative:

$$T_{ij} \text{ upstream and } T_{ji} \text{ downstream} \Leftrightarrow Min(\epsilon_{ni}, \epsilon_{mi}, \epsilon_{jr}, \epsilon_{js}) \geq 0.$$

The *upwind gradient* $(\nabla \Gamma)^u_{ij}$ is computed as the usual finite-element gradient on T_{ij} and the *downwind gradient* $(\nabla F)^d_{ij}$ on T_{ji}. This writes:

$(\nabla F)^u_{ij} = \nabla F\,|_{T_{ij}}$ and $(\nabla F)^d_{ij} = \nabla F\,|_{T_{ji}}$ where $\nabla F\,|_T = \displaystyle\sum_{k \in T} F_k \nabla \Phi_k|_T$ are

the P1-Galerkin gradients on triangle T.

We now specify our method for computing the *extrapolation slopes* $(\nabla F)_{ij}$ and $(\nabla F)_{ji}$:

$$(\nabla F)_{ij}.\vec{ij} = (1 - \beta)(\nabla F)^c_{ij}.\vec{ij} + \beta(\nabla F)^u_{ij}.\vec{ij}\,. \qquad (7.21)$$

The computation of F_{ji} is analogous:

$$(\nabla F)_{ji}.\vec{ij} = (1 - \beta)(\nabla F)^c_{ij}.\vec{ij} + \beta(\nabla F)^d_{ij}.\vec{ij}\,. \qquad (7.22)$$

The coefficient β is an upwinding parameter that controls the combination of fully upwind and centered slopes and that is generally taken equal to $1/3$, according to the error analysis below.

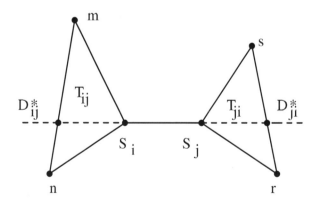

FIGURE 7.10: Localization of the extra interpolation points D^*_{ij} and D^*_{ji} of nodal gradients.

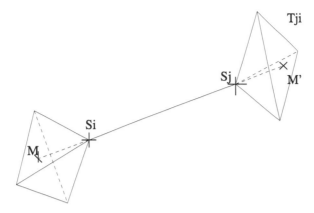

FIGURE 7.11: Downstream and upstream tetrahedra are tetrahedra having resp. S_i and S_j as a vertex and such that line S_iS_j intersects the opposite face.

The scheme description is completed by replacing the first-order formulation (7.16) by the following flux balance:

$$\Psi_h(\gamma, W)_j = \sum_{k \in V(j)} \Phi(W_{jk}, W_{kj}, \nu_{jk}) + \mathbf{B}_h(\gamma, W)_j. \qquad (7.23)$$

with

$$W_{jk} = W_j + \frac{1}{2}(\nabla W)_{ij}.\vec{ij} \quad , \quad W_{kj} = W_k + \frac{1}{2}(\nabla W)_{ji}.\vec{ji} \; .$$

This approximation is spatially second-order accurate. For a nonlinear flux function the accuracy of MUSCL schemes is limited to second order, as remarked by Wu and Wang [Wu and Wang, 1995]. Further, the method combines finite differences in the local reconstruction and finite volume for fluxes. As a consequence, improving the reconstruction to higher order interpolation does not carry a higher accuracy. We now examine how to change the reconstruction in order to improve the scheme.

7.3.2 Low dissipation advection schemes: 1D

7.3.2.1 Spatial 1D MUSCL formulation

Let us first consider the one-dimensional scalar conservation law

$$u_t + f(u)_x = 0 \; . \qquad (7.24)$$

As in a MUSCL approximation, a mixed finite-difference/finite-volume method is used for the discretization in space. Let x_j , $1 \leq j \leq N$ denote the discretization points of the mesh. For each discretization point, we state:

$u_j \approx u(x_j)$ and we define the control volume C_j as the interval $[x_{j-\frac{1}{2}}, x_{j+\frac{1}{2}}]$ where $x_{j+\frac{1}{2}} = \dfrac{x_j + x_{j+1}}{2}$.

As in a finite-difference method, we define the unknown vector $U = \{u_j\}$ as point approximation values of the function $u(x)$ in each node j of the mesh. The time advancing is written:

$$U_{j,t} + \Psi_j(U) = 0 \qquad (7.25)$$

where, similarly to finite-volumes, the vector $\Psi(U)$ is built according to approximations of $f(u)$ defined at cell boundaries:

$$\Psi_j(U) = \frac{1}{\Delta x}(\Phi_{j+\frac{1}{2}} - \Phi_{j-\frac{1}{2}}) \ ; \ \Phi_{j+\frac{1}{2}} = \Phi(u_j, u_{j+1}, f_{j+\frac{1}{2}-}, f_{j+\frac{1}{2}+}) \quad (7.26)$$

where

$$\Phi(u, v, f, g) = \frac{1}{2}\left[(1 + \delta sign(c))f + (1 - \delta sign(c))g\right] \ , \ c = f'(u/2 + v/2).$$

Let us write down a particular flux:

$$\Phi_{j+\frac{1}{2}} = \Phi(u_j, u_{j+1}, f_{j+\frac{1}{2}-}, f_{j+\frac{1}{2}+}) =$$

$$\frac{f_{j+\frac{1}{2}-} + f_{j+\frac{1}{2}+}}{2} + \delta sign(c)\frac{f_{j+\frac{1}{2}-} - f_{j+\frac{1}{2}+}}{2} \ . \qquad (7.27)$$

The coefficient δ controls the spatial dissipation. For defining the integration values $f_{j\pm\frac{1}{2}\pm}$ of f at boundaries of control volume C_j, we apply the MUSCL methodology [Van Leer, 1977a], to the left and right control volume boundary fluxes; $f_{j\pm\frac{1}{2}\pm}$ is built using linear interpolation formulas:

$$f_{j+\frac{1}{2}-} = f_j + \frac{1}{2}\Delta f_{j+\frac{1}{2}-} \quad ; f_{j+\frac{1}{2}+} = f_{j+1} - \frac{1}{2}\Delta f_{j+\frac{1}{2}+}$$
$$f_{j-\frac{1}{2}-} = f_{j-1} + \frac{1}{2}\Delta f_{j-\frac{1}{2}-} \ ; f_{j-\frac{1}{2}+} = f_j - \frac{1}{2}\Delta f_{j-\frac{1}{2}+}$$

where $\Delta f_{j\pm\frac{1}{2}\pm}$ are slopes, *i.e.* approximations of difference term $\dfrac{\partial f}{\partial x}\Delta x$:

$$\Delta f_{j+\frac{1}{2}-} = (1 - \beta)(f_{j+1} - f_j) + \beta(f_j - f_{j-1})$$
$$\Delta f_{j+\frac{1}{2}+} = (1 - \beta)(f_{j+1} - f_j) + \beta(f_{j+2} - f_{j+1}) \ . \qquad (7.28)$$

We observe that if the fluxes are based on polynomial reconstruction from the *average* values of the unknowns, as in ENO finite-volumes, choosing a higher-order reconstruction will produces a higher-order finite-volume scheme. In the present *vertex-centered* context, *high-order accuracy is not obtained by a*

Schemes	δ	β	θ^c	θ^d	Order
1	1	1/3	0	0	3
2	1	1/3	- 1/6	0	4
3	1	1/3	0	- 1/6	4
4	1	1/3	- 1/10	- 1/15	5
4'	0	1/3	- 1/10	- 1/15	6

Table 7.1: Accuracy of different versions of the V6 schemes in 1D case.

higher-order interpolation but by an interpolation that compensates the error
coming from the final central differencing in (7.26). This writes as follows:

$$\Delta f_{j+\frac{1}{2}-} = (1 - \beta)\,(f_{j+1} - f_j) + \beta\,(f_j - f_{j-1})$$
$$+ \theta^c\,(- f_{j-1} + 3f_j - 3f_{j+1} + f_{j+2})$$
$$+ \theta^d\,(- f_{j-2} + 3f_{j-1} - 3f_j + f_{j+1})$$
$$\Delta f_{j+\frac{1}{2}+} = (1 - \beta)\,(f_{j+1} - f_j) + \beta\,(f_{j+2} - f_{j+1})$$
$$+ \theta^c\,(- f_{j-1} + 3f_j - 3f_{j+1} + f_{j+2})$$
$$+ \theta^d\,(- f_j + 3f_{j+1} - 3f_{j+2} + f_{j+3}) \tag{7.29}$$

where θ^c and θ^d are parameters that control the combination of fully upwind
and centered corrections. In order to analyze more simply this scheme, we
assume that

$$c = f'(u) \text{ is constant and equal to 1.}$$

Then:

$$
\Psi_j(U) = \frac{1}{4\Delta x}(\qquad\qquad (1+\delta)\theta^d \qquad\qquad f_{j-3}
$$
$$
+ \qquad [(1+\delta)\beta + 2\delta\theta^c - 4\delta\theta^d - 4\theta^d]\qquad f_{j-2}
$$
$$
+ \quad [-2(\beta + 2\delta\beta + 1) - 8\delta\theta^c + 5\theta^d + 7\delta\theta^d]\; f_{j-1}
$$
$$
+ \qquad\qquad [6\delta\beta + 12\delta\theta^c - 8\delta\theta^d] \qquad\qquad f_j
$$
$$
+ \quad [2(\beta - 2\delta\beta + 1) - 8\delta\theta^c - 5\theta^d + 7\delta\theta^d]\; f_{j+1}
$$
$$
+ \qquad [-(1-\delta)\beta + 2\delta\theta^c - 4\delta\theta^d + 4\theta^d]\qquad f_{j+2}
$$
$$
-(1-\delta)\theta^d \qquad\qquad f_{j+3})
$$
$$\tag{7.30}$$

which gives:

$$\Psi_j(U) = \frac{\partial f}{\partial x} + C_3 \frac{\Delta x^2}{6}\frac{\partial^3 f}{\partial x^3} + C_4 \frac{\Delta x^3}{4}\frac{\partial^4 f}{\partial x^4} \tag{7.31}$$
$$+ C_5 \frac{\Delta x^4}{120}\frac{\partial^5 f}{\partial x^5} + C_6 \frac{\Delta x^5}{24}\frac{\partial^6 f}{\partial x^6} + O(\Delta x^6)$$

where:

$$C_3 = 1 - 3\beta$$
$$C_4 = \delta(\beta + 2\theta^c + 2\theta^d)$$
$$C_5 = 1 - 15\beta - 60\theta^d$$
$$C_6 = \delta(\beta + 2\theta^c + 8\theta^d) \ .$$

We observe that schemes described in (7.26)-(7.29) are in general second-order accurate but they become high-order accurate for some values of the parameters β, δ, θ^c and θ^d. Fifth-order accuracy is obtained with an adequate choice of the three coefficients, viz. $\beta = 1/3, \theta_c = -1/10, \theta^d = -1/15$. In that case, the numerical dissipation takes the form:

$$D_j(U) \ = \ \frac{\delta}{60\Delta x}(-f_{j-3} + 6f_{j-2} - 15f_{j-1} + 20f_j - 15f_{j+1} + 6f_{j+2} + f_{j+3}).$$
$$(7.32)$$

The corresponding dissipative flux writes:

$$\bar{D}_{j+\frac{1}{2}}(U) \ = \ \frac{\delta}{60}(-f_{j-2} + 5f_{j-1} - 10f_j + 10f_{j+1} - 5f_{j+2} + f_{j+3}). \ (7.33)$$

Since this dissipation term is based on an upwinding paradigm, it has good explicit linear stability (analyzed in the sequel), together with a balanced spatial stabilization property. Putting $\delta = 0$, leads to a central-differenced (non-dissipative) sixth-order accurate scheme.

7.3.2.2 Time advancing stability

We can combine the above scheme with the standard Runge-Kutta time advancing.

$$U^{(0)} = U^n$$
$$V_1 = \Delta t \ \Psi(U^n)$$
$$V_2 = \Delta t \ \Psi(U^n + V_1/2)$$
$$V_3 = \Delta t \ \Psi(U^n + V_2/2) \qquad (7.34)$$
$$V_4 = \Delta t \ \Psi(U^n + V_3)$$
$$U^{n+1} \ = \ U^n + V_1/6 + V_2/3 + V_3/3 + V_4/6$$

In many case a linearized version can be used. Let us recall the Jameson variant [Jameson, 1993] which writes as follows (N-stage version):

$$U^{(0)} = U^n$$
$$U^{(k)} = U^{(0)} + \frac{\Delta t}{N - k + 1}\Psi\left(U^{(k-1)}\right), \quad k = 1 \ldots N \qquad (7.35)$$
$$U^{n+1} = U^{(N)}.$$

An A-stability analysis as in [Hirsch, 1991] can be applied. We give in Table 7.2 some typical maximal CFL numbers for the six-stage RK scheme,

β	θ^c	θ^d	δ	CFL_{max}
			1	1
1/3	0	0	1	2.310
1/3	- 1/6	0	1	0.263
1/3	0	- 1/6	1	1.332
1/3	- 1/10	- 1/15	1	1.867
0	0	0	1	.303

Table 7.2: Maximal Courant numbers (explicit RK6 scheme) for the different LV6 spatial schemes (1D analysis).

which ensure a global accuracy order of five for the two best schemes of the proposed family. This table illustrates that the above schemes can be used with CFL number of the order of the unity.

All of these schemes can be advanced in time with implicit schemes such as BDF1 and BDF2, see [Debiez, 1996]. Combination with unsteady Defect Correction [Martin and Guillard, 1996] is also possible. Linear stability is unconditional in all cases.

7.3.3   Unstructured two-dimensional case

In order to increase the accuracy of the second-order MUSCL construction, we introduce an enriched method for computing the extrapolation slopes $(\nabla f)_{ij}$ and $(\nabla f)_{ji}$:

$$
\begin{aligned}
(\nabla f)_{ij}.\vec{ij} = &(1 - \beta)(\nabla f)^c_{ij}.\vec{ij} + \beta(\nabla f)^u_{ij}.\vec{ij} \\
&+\xi_c \left[(\nabla f)^u_{ij}.\vec{ij} - 2(\nabla f)^c_{ij}.\vec{ij} + (\nabla f)^d_{ij}.\vec{ij}\right] \\
&+\xi_d \left[(\nabla f)_{D^*_{ij}}.\vec{ij} - 2(\nabla f)_i.\vec{ij} + (\nabla f)_j.\vec{ij}\right] \quad,
\end{aligned}
\tag{7.36}
$$

The computation of f_{ji} is analogous:

$$
\begin{aligned}
(\nabla f)_{ji}.\vec{ij} = &(1 - \beta)(\nabla f)^c_{ij}.\vec{ij} + \beta(\nabla f)^d_{ij}.\vec{ij} \\
&+\xi_c \left[(\nabla f)^u_{ij}.\vec{ij} - 2(\nabla f)^c_{ij}.\vec{ij} + (\nabla f)^d_{ij}.\vec{ij}\right] \\
&+\xi_d \left[(\nabla f)_{D^*_{ji}}.\vec{ij} - 2(\nabla f)_j.\vec{ij} + (\nabla f)_i.\vec{ij}\right] \quad,
\end{aligned}
\tag{7.37}
$$

The term $(\nabla f)_{D^*_{ij}}$ is the gradient at the point D^*_{ij}. This last gradient is computed by interpolation of the nodal gradient values at the nodes contained in the face opposite to i in the upwind triangle T_{ij}.

$$
(\nabla f)_{D^*_{ij}} = \alpha\,(\nabla f)_m + (1 - \alpha)\,(\nabla f)_n \quad \text{with} \quad D^*_{ij} = \alpha m + (1 - \alpha)n.
$$

The coefficients ξ^c and ξ^d are upwinding parameters that control the combination of fully upwind and centered slopes.

	β	ξ^c	ξ^d	δ	Order
β-scheme	1/3	0	0	1	3
β-scheme	1/3	0	0	0	4
NLV6 Method	1/3	-1/30	-2/15	1	5

Table 7.3: Accuracy of different versions of the presented scheme in 2D Cartesian case.

Analysis of Cartesian case (2D/median/2D/circumcenter) We restrict to an advection model:

$$U_t + aU_x + bU_y = 0 . \tag{7.38}$$

The proposed schemes may have only sixth-order dissipation and are in general second-order accurate but they become higher-order accurate for some values of the parameters β (see [Desideri et al., 1987]), ξ^c and ξ^d, see [Carpentier, 1995] and Table 7.3. For the case of unstructured meshes we can show only first-order accuracy in general, and second-order for smooth variation of mesh size. Better convergence can be observed in practice, we refer to [Abalakin et al., 2002a]. Also, the level of dissipation of this family of schemes is much smaller than for usual MUSCL schemes, see [Debiez and Dervieux, 1999].

7.3.4   Extension to Euler: NLV6

The algorithm for assembling the new scheme, **NLV6**, can be summed up as follows.

0. A background flow $W = (\rho, \rho u, \rho v, E)$ on each vertex of the mesh are given.

1. Compute the fluxes $\bar F = F(W)$, $\bar G = G(W)$ on each vertex (vertexwise loop).

2. Compute the nodal gradients $\nabla \bar F$, $\nabla \bar G$ of the fluxes on each vertex (elementwise loop). This is done by applying the nodal gradient formula:

$$(\nabla \bar F)_i = \frac{1}{meas(C_i)} \sum_{T \in C_i} \frac{meas(T)}{3} \sum_{k \in T} (\bar F)_k \, \nabla \, \Phi_k^T. \tag{7.39}$$

3. Start *edgewise assembly loop*:
   Compute the extrapolated slopes:

$$\begin{aligned}
(\nabla \bar F)_{ij} \cdot \vec{ij} = {} & (1 - \beta)(\nabla \bar F))_{ij}^c \cdot \vec{ij} + \beta(\nabla \bar F))_{ij}^u \cdot \vec{ij} \\
& + \xi_c \left[(\nabla \bar F))_{ij}^u \cdot \vec{ij} - 2(\nabla \bar F))_{ij}^c \cdot \vec{ij} + (\nabla \bar F))_{ij}^d \cdot \vec{ij} \right] \\
& + \xi_d \left[(\nabla \bar F))_{D_{ij}^*} \cdot \vec{ij} - 2(\nabla \bar F))_i \cdot \vec{ij} + (\nabla \bar F))_j \cdot \vec{ij} \right] ,
\end{aligned} \tag{7.40}$$

and analog for $(\nabla \bar{F})_{ji}$.

Define flux interpolations:

$$\bar{\mathcal{F}} = (\bar{F}, \bar{G})$$

$$\bar{\mathcal{F}}_{ij} = \bar{\mathcal{F}}_i + \frac{1}{2} \nabla \bar{\mathcal{F}}_{ij}$$

$$\bar{\mathcal{F}}_{ji} = \bar{\mathcal{F}}_j - \frac{1}{2} \nabla \bar{\mathcal{F}}_{ji}$$

$$(7.41)$$

The central differenced flux then writes:

$$\Phi_{ij} = 0.5 \, (\bar{\mathcal{F}}_{ij} + \bar{\mathcal{F}}_{ji}) \cdot n_{ij} \qquad (7.42)$$

4. Evaluate the stabilisation term:

$$\mathcal{D}_{ij} = 0.5 \, \delta \, sign(\mathcal{A}_{ij})(\bar{\mathcal{F}}_{ji} \cdot n_{ij} - \bar{\mathcal{F}}_{ij} \cdot n_{ij}) \qquad (7.43)$$

where $\mathcal{A}_{ij}$ is defined by:

$$\mathcal{A}_{ij} = (F', G')((W_i + W_j)/2) \cdot n_{ij}. \qquad (7.44)$$

5. Compute the final edge flux as:

$$\Phi_{ij}^{upwind} = \Phi_{ij} - \mathcal{D}_{ij} \qquad (7.45)$$

and add (resp. substract) it to flux assembly at vertex i (resp. j).

REMARK 7.2   The combination of median cells and upwinding produces inconsistent error terms under the form of quotients $\frac{\Delta x}{\Delta y}$ or $\frac{\Delta y}{\Delta x}$ which can produce large errors when the previous quotients are large, that is when mesh is stretched. One way to escape this inconsistency without losing the low dispersion properties (typically, fifth-order accuracy) consists of using the Barth construction of cells. ▯

3D case: For the 3D case, high order is obtained only with the combination of circumcenter cell construction with the cube splitting proposed in [Gourvitch et al., 2004]. In that case a dissipative fifth-order accurate scheme is obtained, together with its non-dissipative sixth-order accurate variant.

7.3.5   High-order LV6 spatial scheme

Although very accurate on Cartesian meshes, NLV6 schemes may involve too much complexity for many flow problems. We present now a more simple family of scheme, the linear V6 schemes, **LV6**, introduced in [Debiez, 1996], [Debiez and Dervieux, 1999] in which *interpolation is applied to the primitive variables*. These schemes are built as follows:

0. A background flow $U = (\rho, \rho u, \rho v, E)$ on each vertex of the mesh is given.

1. Compute the primitive variable $\tilde{U} = (\rho, u, v, p)$ on each vertex (vertex-wise loop).

2. Compute the nodal gradients $\nabla \tilde{U}$.

$$(\nabla \tilde{U})_i = \frac{1}{meas(C_i)} \sum_{T \in C_i} \frac{meas(T)}{3} \sum_{k \in T} (\tilde{U})_k \nabla \Phi_k^T. \tag{7.46}$$

3. Start *edgewise assembly loop*:
   Compute the extrapolated slopes:

$$\begin{aligned}
(\nabla \tilde{U})_{ij} \cdot \vec{ij} = {}& (1 - \beta)(\nabla \tilde{U})_{ij}^c \cdot \vec{ij} + \beta(\nabla \tilde{U})_{ij}^u \cdot \vec{ij} \\
& + \xi_c \left[(\nabla \tilde{U})_{ij}^u \cdot \vec{ij} - 2(\nabla \tilde{U})_{ij}^c \cdot \vec{ij} + (\nabla \tilde{U})_{ij}^d \cdot \vec{ij} \right] \\
& + \xi_d \left[(\nabla \tilde{U})_{D_{ij}^*} \cdot \vec{ij} - 2(\nabla \tilde{U})_i \cdot \vec{ij} + (\nabla \tilde{U})_j \cdot \vec{ij} \right],
\end{aligned} \tag{7.47}$$

and analog for $\nabla(\tilde{U})_{ji}$.
Define left and right variable interpolations:

$$\begin{aligned}
\tilde{U}_{ij} &= \tilde{U}_i + \nabla \tilde{U}_{ij} \\
\tilde{U}_{ji} &= \tilde{U}_j - \nabla \tilde{U}_{ji}
\end{aligned} \tag{7.48}$$

and recover the left and right values of conservative variables U_{ij}, U_{ji}. The upwind differenced flux then writes:

$$\Phi_{ij} = \Phi^{Riemann}(U_{ij}, U_{ji}) \tag{7.49}$$

and add (substract) it to flux assembly at vertex i (j) and multiply flux assembly by the inverse mass matrix in order to obtain the update of the variable.

In order to recover consistency for stretched meshes, a modification similar to the one applied to NLV6 is necessary. Boundary conditions can also be addressed in the same way as for NLV6.

A crucial difference between both schemes is that positive extensions of the LV6 versions are easily derived, as will be explained in next section.

7.3.6 Time advancing

For explicit time advancing, we can use, as in the 1D case, a standart Runge-Kutta scheme or the linearized version defined above.

An implicit Backward-Differencing formula can also be applied with a spatially-first-order accurate simplified Jacobian. An interesting option is the second-order Backward-Differencing formula.

In case where we seek a steady solution or a slowly transient solution during a long time, the efficiency of an explicit scheme applying on unstructured mesh is severely limited by the Courant condition on the time step. It can be crucial to apply a multigrid iteration in combination with pseudo time-advancing (steady case) or (for both cases) an efficient implicit time advancing. Designing a multigrid scheme for unstructured meshes rises the problem of defining a series of coarser grids. In other words, we have to define several new meshes or to find an alternative strategy. In [Lallemand et al., 1992], [Francescatto and Dervieux, 1998], this is done in a transparent manner from the fine mesh by using the so-called cell agglomeration. Parallel multigrid extensions are proposed in [Mavriplis, 1997], [Fournier et al., 1998]. Concerning the implicit time-stepping, it needs a solution algorithms for at least a linearised problem. This also can be done with a multigrid algorithm. Another option well adapted to message passing parallelism is the Krylov-Newton-Schwarz (KNS) algorithm, as in [Knoll and Keyes, 2004]. A first version of KNS, under the form of the Restrictive Additive Schwarz **RAS** was developed in [Cai and Sarkis, 1999], [Sarkis and Koobus, 2000]. This method can produce second order convergence in space and time although using a spatially-first-order accurate simplified Jacobian. This is obtained by means of the two-step-Newton Defect Correction proposed by [Martin and Guillard, 1996].

7.3.7 Conclusion on superconvergent schemes

We have described a family of schemes for the Euler equations involving low numerical dissipation. Interested readers are refeered to [Debiez, 1996], [Debiez and Dervieux, 1999], [Debiez et al., 1998], [Abalakin et al., 2002a], [Abalakin et al., 2002b], [Abalakin et al., 2002c], [Abalakin et al., 2001], [Abalakin et al., 2004], [Camarri et al., 2001a], [Camarri et al., 2002b], [Camarri et al., 2004], [Gourvitch et al., 2004].

Based on MUSCL schemes, the LV6 schemes involve sophisticated primitive variable reconstruction designed in order to enjoy low dissipation properties thanks to a model of sixth derivative. In the case of advection with uniform velocity, those schemes present superconvergence properties, in the sense that, when applied to a Cartesian subregion of mesh, these second-order schemes are of higher order (up to sixth-order).

With the NLV6 option, flux functions are reconstructed, instead of primitive variables. Upwinding is made through sign-based Riemann solvers for using only flux values. The proposed schemes still apply to unstructured triangulations where they stay essentially second-order accurate. In contrast to LV6 schemes, the NLV6 schemes enjoy superconvergence properties also for *nonlinear arbitrary fluxes.* Superconvergent order of accuracy is between 4 and 6.

7.4 Positivity of mixed element-volume formulations

7.4.1 Introduction

Some of the most useful properties of upwind approximation schemes for hyperbolic equations are their monotonicity and positivity properties. For example, for the Euler equations, and by combining flux splitting and limiters, Perthame and co-workers [Perthame and Khobalate, 1992], [Perthame and Shu, 1996] have proposed second-order accurate schemes that maintain a density and a temperature positive. We refer also to [Linde and Roe, 1998] and to the workshop [Venkatakrisnan, 1998]. Non-oscillating schemes [Harten et al., 1987], [Cockburn and Shu, 1989] propose high accuracy approximations applicable to many problems. However, they do not enjoy a strict satisfaction of positivity or monotony, which remains an important issue for stiff simulations, particularly in relation with highly heterogeneous fluid flows (see for example [Abgrall, 1996], [Murrone and Guillard, 2005]).

The purpose of the present section is to analyze the conditions of positivity for the MEV scheme. We first examine the maximum principle for a scalar conservation law, then we introduce a flux splitting that preserves density positivity for the Euler equations. This provides a basis to construct multidimensional schemes that ensure density positivity and maximum principle for convected species. This is of paramount importance for most flows of industrial interest for two reasons. A direct one is that most industrial flows are at medium Mach number and they generally do not induce negative pressures but more often negative densities that can arise at after bodies (negative pressures are more often obtained in high Mach number detached shocks). The second reason is that in many Reynolds-Averaged Navier-Stokes flows, limiters are not necessary for the mean flow itself, but robustness problems arise in the computation of turbulence closure variables such as k and ε.

This section is then organised as follows. First, a scalar nonlinear model is considered and allows us to introduce the main features of the new scheme. Second, well-known positivity 1D statements for the Euler equations are reformulated in such a way that we can derive the extension of the new scheme to density-positive treatment of the 2D and 3D Euler equations. Lastly, we present an example of numerical application.

7.4.2 Positive schemes and LED schemes for nonlinear scalar conservation laws

This section recalls some useful existing results. We keep the usual TVD/LED (TVD: Total Variation Diminishing, LED: Local Extremum diminishing) criterion preferably to weaker positivity criteria in order to deal also with the maximum principle.

7.4.2.1   Nonlinear scalar conservation laws in fixed and moving domains

Fixed domains   We consider a nonlinear scalar conservation law form for the unknown $U(\mathbf{x}, t)$:

$$\frac{\partial U}{\partial t}(\mathbf{x}, t) + \nabla_{\mathbf{x}} \cdot \mathcal{F}\Big(U(\mathbf{x}, t)\Big) = 0 \qquad (7.50)$$

where t denotes the time, $\mathbf{x} = (x_1, x_2, x_3)^t$ is the space coordinate and $\mathcal{F}(U) = (F(U), G(U), H(U))^t$ (3D case). Under some classical assumptions [Godlewski and Raviart, 1996], the solution U satisfies a *maximum principle* that, for simplicity, we write in the case of the whole space:

$$\min_{\mathbf{x}} U(\mathbf{x}, 0) \leq U(\mathbf{x}, t) \leq \max_{\mathbf{x}} U(\mathbf{x}, 0) \qquad (7.51)$$

Moving domains   We consider a nonlinear scalar conservation law in ALE form for the unknown $U(\mathbf{x}, t)$. Similar notations to [Farhat et al., 2001] are used. We denote an instantaneous configuration by $\Omega(\mathbf{x}, t)$ and a reference configuration by $\Omega(\xi, \tau = 0)$ where $\xi = (\xi_1, \xi_2, \xi_3)^t$ denotes the space coordinate and τ the time. We have a map function $\mathbf{x} = \mathbf{x}(\xi, \tau), t = \tau$, from $\Omega(\xi, \tau = 0)$ to $\Omega(\mathbf{x}, t)$, and $J = det(\partial \mathbf{x}/\partial \xi)$ denotes its determinant. Then, the nonlinear scalar conservation law in ALE form can be written as:

$$\frac{\partial JU}{\partial t}|_\xi(\mathbf{x}, t) + J \nabla_{\mathbf{x}} \cdot \Big(\mathcal{F}(U(\mathbf{x}, t)) - \mathbf{w}(\mathbf{x}, \mathbf{t})U(\mathbf{x}, t)\Big) = 0 \qquad (7.52)$$

where $\mathcal{F}(U) = (F(U), G(U), H(U))^t$ $\;and\;$ $\mathbf{w} = \frac{\partial \mathbf{x}}{\partial t}|_\xi$. As for fixed domain problems, the solution U satisfies a *maximum principle*.

7.4.2.2   Positivity/LED criteria

Assuming that the mesh nodes are numbered, we call U_i the value at mesh node i that can move in time for moving grids. We recall now the classical positivity statement for an explicit time-integration (the proof is immediate).

LEMMA 7.1

A positivity criterion: suppose that an explicit first-order time-integration of equation (7.50) or (7.52) can be expressed in the form:

$$\frac{U_i^{n+1} - U_i^n}{\Delta t} = b_{ii}U_i^n + \sum_{j \neq i} b_{ij}U_j^n , \qquad (7.53)$$

where all the b_{ij}, $j \neq i$, are non-negative and $b_{ii} \in \mathbb{R}$. Then it can be shown that the above explicit scheme preserves positivity under the following condition on the time-step Δt: $b_{ii} + \dfrac{1}{\Delta t} \geq 0$.

REMARK 7.3   Under a possibly different restriction on Δt, a high-order explicit time discretization can still preserve the positivity, see [Shu and Osher, 1988].   ▯

Another scheme formulation relies on the so-called incremental form that dates back to Harten [Harten, 1983]. It was used by Jameson in [Jameson, 1987] for defining LED schemes (see also [Godlewski and Raviart, 1996]). We recall now the theory introduced by Jameson [Jameson, 1987] on LED schemes in the case of an explicit time-integration:

LEMMA 7.2

A LED criterion [Jameson, 1987]: suppose that an explicit first-order time-integration of equation (7.50) or (7.52) can be written in the form:

$$\frac{U_i^{n+1} - U_i^n}{\Delta t} = \sum_{k \in V(i)} c_{ik}(U^n)\,(U_k^n - U_i^n),\qquad (7.54)$$

with all the $c_{ik}(U^n) \geq 0$, and where $V(i)$ denotes the set of the neighbours of node i. Then the previous scheme verifies that a local maximum cannot increase and a local minimum cannot decrease, and under an appropriate condition on the time-step the positivity and the maximum principle are preserved.

PROOF   Given U_i^n a local maximum, we deduce that $(U_k^n - U_i^n) \leq 0$ for all $k \in V(i)$. Therefore equation (7.54) implies that $\dfrac{U_i^{n+1} - U_i^n}{\Delta t} \leq 0$, and the local maximum cannot increase. Likewise, we can prove that a local minimum cannot decrease. On the other hand, the reader can easily check that the positivity and the maximum principle are preserved when the time-step satisfies $\dfrac{1}{\Delta t} - \displaystyle\sum_{k \in V(i)} c_{ik}(U^n) \geq 0.$ ▯

7.4.2.3   First-order space-accurate MEV schemes on fixed and moving grids

Fixed grids   We integrate (7.50) over a cell C_i, integrating by parts the resulting convective fluxes and using a conservative approximation leads to the following semi-discretization of (7.50):

$$a_i \frac{dU_i}{dt} + \sum_{j \in V(i)} \Phi(U_i, U_j, \nu_{ij}) = 0 \qquad (7.55)$$

where a_i is the measure of cell C_i. In the above semi-discretization, the values U_i and U_j correspond to a constant per cell *interpolation* of the variable

U, and Φ is a *numerical flux function* so that $\Phi(U_i, U_j, \nu_{ij})$ approximates $\int_{\partial C_{ij}} \mathcal{F}(U) \cdot \mathbf{n}_i(\sigma) d\sigma$. In general, the numerical flux function $\Phi : (u, v, \nu) \mapsto \Phi(u, v, \nu)$ is assumed to be Lipschitz continuous, monotone increasing with respect to u, monotone decreasing with respect to v, and consistent:

$$\Phi(u, u, \nu) = \mathcal{F}(u) \cdot \nu \ . \tag{7.56}$$

Moving grids In the case of moving grids, these cells can move and deform with time according to vertices motion. Integrating (7.52) over a cell $C_i(0)$ of the ξ space, switching to the $\mathbf{x}$ space, integrating by part the resulting convective fluxes and using a conservative approximation lead to the following semi-discretization of (7.52):

$$\frac{d\Big(a_i(t)U_i\Big)}{dt} + \sum_{j \in V(i)} \Phi(U_i, U_j, \nu_{ij}(t), \kappa_{ij}(t)) = 0 \tag{7.57}$$

where $a_i(t)$ is the measure of cell $C_i(t)$, $\nu_{ij}(t) = \int_{\partial C_{ij}(t)} \mathbf{n}_i(\sigma, t) \, d\sigma$ and

$\kappa_{ij}(t) = \int_{\partial C_{ij}(t)} \mathbf{w}(\sigma, t) \cdot \mathbf{n}_i(\sigma, t) \, d\sigma$. In the above semi-discretized scheme, Φ is a *numerical flux function* so that $\Phi(U_i, U_j, \nu_{ij}(t), \kappa_{ij}(t))$ approximates $\int_{\partial C_{ij}(t)} (\mathcal{F}(U) - \mathbf{w}U) \cdot \mathbf{n}_i d\sigma$. As previously, $\Phi : (u, v, \nu, \kappa) \mapsto \Phi(u, v, \nu, \kappa)$ is assumed to be Lipschitz continuous, monotone increasing with respect to u, monotone decreasing with respect to v, and consistent:

$$\Phi(u, u, \nu, \kappa) = \mathcal{F}(u) \cdot \nu - \kappa u \ . \tag{7.58}$$

7.4.2.4 Limited high-order space-accurate MEV scheme on fixed and moving grids

According to section 7.3, we replace the values U_i and U_j by "better" interpolations U_{ij} and U_{ji} at the interface ∂C_{ij}. More precisely, the first-order MEV scheme becomes:

$$a_i \frac{dU_i}{dt} + \sum_{j \in V(i)} \Phi(U_{ij}, U_{ji}, \nu_{ij}) = 0 \quad \text{in the case of *fixed grids*}, \tag{7.59}$$

and

$$\frac{d\Big(a_i(t)U_i\Big)}{dt} + \sum_{j \in V(i)} \Phi(U_{ij}, U_{ji}, \nu_{ij}(t), \kappa_{ij}(t)) = 0 \quad \text{for *moving grids*},$$

$$\tag{7.60}$$

where U_{ij} and U_{ji} are left and right values of U at the interface ∂C_{ij}. Our purpose is to build a scheme that is non oscillatory and positive. It has been early proved that high-order positive schemes must be necessarily built with a nonlinear process, often called a *limiter*. A limiter locally chooses between the first-order monotone version of the schemes and a higher order one. In the case of unstructured meshes and scalar models, second-order positive schemes were derived using a two-entry symmetric limiter by Jameson in [Jameson, 1987]. Here, instead of the Jameson symmetric limiter, we choose to work in a MUSCL formulation, involving two limiters per edge. The adaptation to triangulations is close to the one proposed in [Fezoui and Dervieux, 1989] and [Stoufflet et al., 1996]. Following [Debiez, 1996], we extend it to a three-entry limiter which allows to design a positive scheme of third- (or even fifth-) order far from extrema when U varies smoothly. Then (7.59) becomes for *fixed grids*:

$$a_i \frac{dU_i}{dt} + \sum_{j \in V(i)} \Phi(U_i + \frac{1}{2}L_{ij}(U), U_j - \frac{1}{2}L_{ji}(U), \nu_{ij}) = 0, \qquad (7.61)$$

and (7.60) becomes for *moving grids*:

$$\frac{d(a_i(t)U_i)}{dt} + \sum_{j \in V(i)} \Phi(U_i + \frac{1}{2}L_{ij}(U), U_j - \frac{1}{2}L_{ji}(U), \nu_{ij}(t), \kappa_{ij}(t)) = 0.$$
$$(7.62)$$

In order to define $L_{ij}(U)$ and $L_{ji}(U)$ we use the upstream and downstream triangles (or tetrahedra) T_{ij} and T_{ji} (see Figures 7.12 and 7.11), as introduced in [Fezoui and Dervieux, 1989]. Element T_{ij} is *upstream* to vertex i with respect to edge ij if for any small enough real number η the vector $-\eta \vec{ij}$ is inside element T_{ij}. Symmetrically, element T_{ji} is *downstream* to vertex i with respect to edge ij if for any small enough real number η the vector $\eta \vec{ji}$ is inside element T_{ij}. Let $\epsilon_{ri}, \epsilon_{si}, \epsilon_{ti}, \epsilon_{jn}, \epsilon_{jp}$ and ϵ_{jq} be the components of vector $\vec{ji}$ (resp. $\vec{ij}$) in the oblique system of axes $(\vec{ir}, \vec{is}, \vec{it})$ (resp. $\vec{jn}, \vec{jp}, \vec{jq})$):

$$\vec{ji} = \epsilon_{ri}\,\vec{ir} + \epsilon_{si}\,\vec{is} + \epsilon_{ti}\,\vec{it},$$

$$\vec{ij} = \epsilon_{jn}\,\vec{jn} + \epsilon_{jp}\,\vec{jp} + \epsilon_{jq}\,\vec{jq}.$$

Then T_{ij} and T_{ji} are upstream and downstream elements means that they have been chosen in such a way that the components ϵ_{ri}, etc. are all nonnegative: T_{ij} upstream and T_{ji} downstream $\Leftrightarrow$ $\epsilon_{ri}, \epsilon_{si}, \epsilon_{ti}, \epsilon_{jn}, \epsilon_{jp}, \epsilon_{jq}$ are all non-negative. Let us introduce the following notations:

$$\Delta^- U_{ij} = \nabla U|_{T_{ij}} \cdot \vec{ij} \quad , \quad \Delta^0 U_{ij} = U_j - U_i \quad and \quad \Delta^- U_{ji} = \nabla U|_{T_{ji}} \cdot \vec{ij},$$

where the gradients are those of the P1 (continuous and linear) interpolation of U. Jameson in [Jameson, 1987] has noted that (for the 3D case):

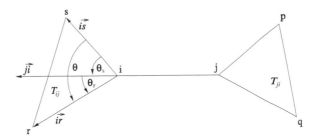

FIGURE 7.12: Two-dimensional case: splitting of vector $\vec{ji}$ in the directions of downstream and upstream triangles edges.

$$\Delta^- U_{ij} = \epsilon_{ri}\,(U_i - U_r) + \epsilon_{si}\,(U_i - U_s) + \epsilon_{ti}\,(U_i - U_t),$$

and

$$\Delta^- U_{ji} = \epsilon_{jp}\,(U_p - U_j) + \epsilon_{jq}\,(U_q - U_j) + \epsilon_{jn}\,(U_n - U_j),$$

with the same non-negative ϵ_{ri}, ϵ_{si}, ϵ_{ti}, ϵ_{jn}, ϵ_{jp} and ϵ_{jq}.

Now, we introduce a family of continuous limiters with three entries, satisfying:

$(P1)$ $L(u,v,w) = L(v,u,w)$

$(P2)$ $L(\alpha\,u, \alpha\,v, \alpha\,w) = \alpha\,L(v,u,w)$

$(P3)$ $L(u,u,u) = u$

$(P4)$ $L(u,v,w) = 0$ if $uv \leq 0$

$(P5)$ $0 \leq \frac{L(u,v,w)}{v} \leq 2$ if $v \neq 0$.

Note that there exists K^- and K^0 depending on (u,v,w) such that:

$$L(u,v,w) = K^- u = K^0 v\,, \ \text{with } 0 \leq K^- \leq 2\,,\ 0 \leq K^0 \leq 2\,. \quad (7.63)$$

A function verifying (P1) to (P5) exists, and in the numerical examples, we shall use the following version of the Superbee method of Roe:

$$\begin{cases} L_{SB}(u,v,w) = 0 & \text{if} \ \ uv \leq 0 \\[2mm] \qquad = Sign(u)\,\min(\,2\,|u|\,,\,2\,|v|\,,\,|w|\,) \ \text{otherwise.} \end{cases} \quad (7.64)$$

We define:

$$L_{ij}(U) = L(\Delta^- U_{ij}\,,\,\Delta^0 U_{ij}\,,\,\Delta^{HO} U_{ij})\ ;\ \ L_{ji}(U) = L(\Delta^- U_{ji}\,,\,\Delta^0 U_{ij}\,,\,\Delta^{HO} U_{ji}).$$
$$(7.65)$$

where $\Delta^{HO} U_{ji}$ is a third way of evaluating the variation of U which we can introduce for increasing the accuracy of the resulting scheme (see the following remark).

REMARK 7.4 Let assume that the option $L_{ij}(U) = \Delta^{HO} U_{ij}$ gives a high order approximation with order ω. For U smooth enough and assuming that the mesh size is smaller than α (small), there exists $\epsilon(\alpha)$ such that if $\dfrac{|\nabla U.\vec{ij}|}{||\vec{ij}||} > \epsilon(\alpha)$ for an edge ij, the limiter L_{ij} is not active, *i.e.* $L_{ij}(U) = \Delta^{HO} U_{ij}$. Then, the scheme is locally of order $\omega.\Box$.   ⬚

REMARK 7.5 Second-order accuracy of MEV in case of arbitrary unstructured meshes is difficult to state in general. We refer to [Mer, 1998b] for proofs in simplified cases. $\Box$.   ⬚

The proposed analysis does not need any assumption concerning the terms $\Delta^{HO}U_{ij}$. For example, we can use the following flux:

$$\Delta^{HO3}U_{ij} = \frac{1}{3}\Delta^{-}U_{ij} + \frac{2}{3}\Delta^{0}U_{ij}, \qquad (7.66)$$

which gives us a third-order space-accurate scheme for linear advection on Cartesian triangular meshes. More generally, the high-order flux can use extra data in order to increase the accuracy. In [Debiez et al., 1998] and [Camarri et al., 2004], the following version is studied:

$$
\begin{aligned}
\Delta^{HO5}U_{ij} = {} & \Delta^{HO3}U_{ij} \\
& - \frac{1}{30}\left[\Delta^{-}U_{ij} - 2\Delta^{0}U_{ij} + \Delta^{-}U_{ji}\right] \\
& - \frac{2}{15}\left[(\nabla U)_M.\vec{ij} - 2(\nabla U)_i.\vec{ij} + (\nabla U)_j.\vec{ij}\right] . \qquad (7.67)
\end{aligned}
$$

For a vertex k, the notation $(\nabla U)_k$ holds for the following average of the gradients on tetrahedra T having the node k as a vertex:

$$(\nabla U)_k = - \frac{1}{Vol(C_k)} \sum_{T,k \in T} \frac{Vol(T)}{4} \nabla U|_T. \qquad (7.68)$$

The term $(\nabla U)_M$ is the gradient at point M, intersection of line ij with the face of T_{ij} that does not have i as vertex, see Figure 7.11 (3D case). It is computed by linear interpolation of the nodal gradient values at the vertices contained in the face opposite to i in the upwind tetrahedron T_{ij}. This option gives fifth-order accuracy for the linear advection equation on Cartesian meshes and has a dissipative leading error expressed in terms of sixth-order derivatives.

7.4.2.5 Maximum principle for first-order space-accurate MEV schemes on fixed and moving grids

Fixed grids the first-order explicit time-integration of (7.55) leads to the following scheme:

$$\frac{U_i^{n+1} - U_i^n}{\Delta t} = \sum_{j \in V(i)} c_{ij} \, (U_j^n - U_i^n) + d_i \, U_i^n \qquad (7.69)$$

where the coefficients:

$$c_{ij} = -\frac{1}{a_i} \frac{\Phi(U_i^n, U_j^n, \nu_{ij}) - \Phi(U_i^n, U_i^n, \nu_{ij})}{U_j^n - U_i^n} \qquad (7.70)$$

are always positive since the flux Φ is monotone decreasing with respect to the second variable. The last term in (7.69) involves:

$$d_i = -\frac{1}{a_i} \frac{\displaystyle\sum_{j \in V(i)} \Phi(U_i^n, U_i^n, \nu_{ij})}{U_i^n} \qquad (7.71)$$

which, due to the consistent condition (7.56), can be transformed as follows:

$$d_i = -\frac{1}{a_i} \frac{1}{U_i^n} \, \mathcal{F}(U_i^n) \cdot \sum_{j \in V(i)} \nu_{ij} \, . \qquad (7.72)$$

Therefore this term vanishes since the finite-volume cells are closed:

$$\sum_{j \in V(i)} \nu_{ij} = 0, \ \forall \ i \, . \qquad (7.73)$$

According to Lemma 4, we conclude classically that the scheme resulting from the first-order explicit time-integration of (7.55) is L^∞-stable under the CFL condition:

$$\frac{1}{\Delta t} - \sum_{j \in V(i)} c_{ij} \geq 0. \qquad (7.74)$$

Moving grids The maximum principle has been extended to *ALE formulations* by Farhat et al. in [Farhat et al., 2001] for first-order space-accurate schemes. We recall this in short. The time-integration of (7.57) is obtained by combining a given time-integration scheme for fixed grid with a procedure for evaluating the geometric quantities that arise from the ALE formulation. Since the mesh configuration changes in time, an important problem is the correct computation of the numerical flux function Φ through the evaluation of the geometric quantities $\nu_{ij}(t)$ and $\kappa_{ij}(t)$. In order to address this problem, the discrete scheme is required to preserve a constant solution. In the case of a first-order explicit time-integration of (7.57), imposing this condition to the discrete scheme leads to the following relation called the *Discrete Geometric Conservation Law (DGCL)*:

$$a_i^{n+1} - a_i^n = \Delta t \sum_{j \in V(i)} \bar{\kappa}_{ij} \tag{7.75}$$

with $a_i^n = |C_i^n|$ and $a_i^{n+1} = |C_i^{n+1}|$, and where $\bar{\kappa}_{ij}$ is a time-averaged value of $\kappa_{ij}(t)$. This gives a procedure for the evaluation of the geometric quantities in the numerical flux function based on a combination of suited mesh configurations.

Then the first-order space-accurate ALE scheme (7.57) combined with a first-order explicit time-integration can be written as follows:

$$\frac{a_i^{n+1} U_i^{n+1} - a_i^n U_i^n}{\Delta t} + \sum_{j \in V(i)} \Phi(U_i^n, U_j^n, \bar{\nu}_{ij}, \bar{\kappa}_{ij}) = 0, \tag{7.76}$$

where $\bar{\nu}_{ij}$ and $\bar{\kappa}_{ij}$ are averaged value of $\nu_{ij}(t)$ and $\kappa_{ij}(t)$ on suited mesh configurations so that the numerical scheme satisfies the DGCL. Given that $\sum_{j \in V(i)} \bar{\nu}_{ij} = 0$ since the cells C_i remain closed during the mesh motion, the consistency condition (7.58) implies that:

$$\Phi(U_i^n, U_i^n, \bar{\nu}_{ij}, \bar{\kappa}_{ij}) = \Phi(U_i^n, U_i^n, \bar{\nu}_{ij}, \bar{\kappa}_{ij}) - \sum_{j \in V(i)} \mathcal{F}(U_i^n) \cdot \bar{\nu}_{ij} = - \sum_{j \in V(i)} \bar{\kappa}_{ij} U_i^n.$$

Then, the above scheme can also be written as:

$$\frac{U_i^{n+1} - U_i^n}{\Delta t} = \sum_{j \in V(i)} c_{ij} (U_j^n - U_i^n) + e_i U_i^n, \tag{7.77}$$

$$c_{ij} = -\frac{1}{a_i^{n+1}} \frac{\Phi(U_i^n, U_j^n, \bar{\nu}_{ij}, \bar{\kappa}_{ij}) - \Phi(U_i^n, U_i^n, \bar{\nu}_{ij}, \bar{\kappa}_{ij})}{U_j^n - U_i^n}, \tag{7.78}$$

$$e_i \quad - \left(\frac{a_i^n - a_i^{n+1}}{a_i^{n+1} \Delta t} + \frac{1}{a_i^{n+1}} \sum_{j \in V(i)} \bar{\kappa}_{ij} \right). \tag{7.79}$$

We note again that the coefficients c_{ij} are always positive since the numerical flux function Φ is monotone decreasing with respect to the second variable. We observe also that the DGCL (7.75) is exactly the condition for which e_i defined by (7.79) vanishes. As for the fixed grids case, from Lemma 4 we conclude that the ALE scheme (7.76) is L^∞-stable under the CFL condition:

$$\frac{1}{\Delta t} - \sum_{j \in V(i)} c_{ij} \geq 0. \tag{7.80}$$

7.4.2.6 Maximum principle for limited high-order space-accurate MEV scheme on fixed and moving grids

Fixed grids A first-order explicit time-integration of the high-order upwind scheme given by equation (7.59) leads to the following equation:

$$\frac{a_i U_i^{n+1} - a_i U_i^n}{\Delta t} + \sum_{j \in V(i)} \Phi(U_{ij}^n, U_{ji}^n, \nu_{ij}) = 0. \tag{7.81}$$

LEMMA 7.3
The scheme defined by (7.81) combined with (7.64), (7.65) and (7.66) or (7.67) satisfies the maximum principle under an appropriate CFL condition.

PROOF Let us first introduce the following coefficients:

$$g_{ij} = -\frac{1}{a_i} \frac{1}{U_{ji}^n - U_i^n} \left(\Phi(U_{ij}^n, U_{ji}^n, \nu_{ij}) - \Phi(U_{ij}^n, U_i^n, \nu_{ij}) \right) \tag{7.82}$$

$$h_{ij} = -\frac{1}{a_i} \frac{1}{U_{ij}^n - U_i^n} \left(\Phi(U_{ij}^n, U_i^n, \nu_{ij}) - \Phi(U_i^n, U_i^n, \nu_{ij}) \right) \tag{7.83}$$

$$d_i = -\frac{1}{a_i} \frac{1}{U_i^n} \left(\sum_{j \in V(i)} \Phi(U_i^n, U_i^n, \nu_{ij}) \right). \tag{7.84}$$

Then, the scheme (7.81) can be written as:

$$\frac{U_i^{n+1} - U_i^n}{\Delta t} = \sum_{j \in V(i)} g_{ij} (U_{ji}^n - U_i^n) + \sum_{j \in V(i)} h_{ij} (U_{ij}^n - U_i^n) + d_i U_i^n. \tag{7.85}$$

We can notice that the coefficients g_{ij} and h_{ij} are respectively positive and negative since the numerical flux function Φ is monotone increasing with the first variable and monotone decreasing with the second variable. As in the previous section, the term d_i vanishes since the numerical flux function Φ satisfies the consistency condition (7.56) and the finite-volume cells C_i are closed so that identity (7.73) holds. On the other hand, according to (7.63) and (7.65) we can write $L_{ij}(U^n)$ as:

$$L_{ij}(U^n) = K_{ij}^- \Delta^- U_{ij}^n$$

where K_{ij}^- is a positive function of U^n, so that we have:

$$U_{ij}^n - U_i^n = \frac{1}{2} L_{ij}(U^n) = \frac{1}{2} K_{ij}^- \left(\epsilon_{ri}(U_i^n - U_r^n) + \epsilon_{si}(U_i^n - U_s^n) + \epsilon_{ti}(U_i^n - U_t^n) \right). \tag{7.86}$$

Likewise, from (7.63) and (7.65) we can write $L_{ji}(U^n)$ in the following form:

$$L_{ji}(U^n) = K_{ji}^0 \Delta^0 U_{ij}^n$$

where K_{ji}^0 is a positive function of U^n smaller than 2, so that we get:

$$U_{ji}^n - U_i^n = U_j^n - \frac{1}{2} L_{ji}(U^n) - U_i^n = \left(1 - \frac{K_{ji}^0}{2}\right)(U_j^n - U_i^n) \qquad (7.87)$$

in which the coefficient $1 - \dfrac{K_{ji}^0}{2}$ is positive. In the identity (7.85), we substitute $U_{ij}^n - U_i^n$ and $U_{ji}^n - U_i^n$ respectively by their expressions given by (7.86) and (7.87), so that we can write the discrete upwind scheme in the following form:

$$\frac{U_i^{n+1} - U_i^n}{\Delta t} = \sum_{j \in V(i)} \alpha_{ij}(1 - \frac{K_{ji}^0}{2})(U_j^n - U_i^n) + \sum_{j \in V(i)} \beta_{ij}(U_j^n - U_i^n) \quad (7.88)$$

where the coefficients α_{ij} and β_{ij} are positive, so that this scheme satisfies the maximum principle under a CFL condition according to Lemma 4. ☐

Moving grids: A first-order explicit time-integration of the high-order upwind scheme given by (7.60) leads to the following equation:

$$\frac{a_i^{n+1} U_i^{n+1} - a_i^n U_i^n}{\Delta t} + \sum_{j \in V(i)} \Phi(U_{ij}^n, U_{ji}^n, \bar{\nu}_{ij}, \bar{\kappa}_{ij}) = 0 \qquad (7.89)$$

where $\bar{\nu}_{ij}$ and $\bar{\kappa}_{ij}$ are averaged value of $\nu_{ij}(t)$ and $\kappa_{ij}(t)$ on suited mesh configurations so that the numerical scheme satisfies the DGCL (7.75).

LEMMA 7.4
The scheme defined by (7.89) combined with (7.64), (7.65) and (7.66) or (7.67) satisfies the maximum principle under an appropriate CFL condition.

PROOF   Let first introduce the following coefficients:

$$g_{ij} = -\frac{1}{a_i^{n+1}} \frac{\Phi(U_{ij}^n, U_{ji}^n, \bar{\nu}_{ij}, \bar{\kappa}_{ij}) - \Phi(U_{ij}^n, U_i^n, \bar{\nu}_{ij}, \bar{\kappa}_{ij})}{U_{ji}^n - U_i^n} \qquad (7.90)$$

$$h_{ij} = -\frac{1}{a_i^{n+1}} \frac{\Phi(U_{ij}^n, U_i^n, \bar{\nu}_{ij}, \bar{\kappa}_{ij}) - \Phi(U_i^n, U_i^n, \bar{\nu}_{ij}, \bar{\kappa}_{ij})}{U_{ij}^n - U_i^n}, \qquad (7.91)$$

$$e_i = \left(\frac{a_i^n - a_i^{n+1}}{a_i^{n+1} \Delta t} + \frac{1}{a_i^{n+1}} \sum_{j \in V(i)} \bar{\kappa}_{ij}\right). \qquad (7.92)$$

Then, the scheme defined by (7.89) becomes:

$$\frac{U_i^{n+1} - U_i^n}{\Delta t} = \sum_{j \in V(i)} g_{ij} (U_{ji}^n - U_i^n) + \sum_{j \in V(i)} h_{ij} (U_{ij}^n - U_i^n) + e_i U_i^n.$$
(7.93)

We observe again that the coefficients g_{ij} and h_{ij} are respectively positive and negative combinations of the unknown, and that e_i vanishes due to the DGCL. Using the expressions of $U_{ij}^n - U_i^n$ and $U_{ji}^n - U_i^n$ given by (7.86) and (7.87), we can rewrite (7.93) in the following form:

$$\frac{U_i^{n+1} - U_i^n}{\Delta t} = \sum_{j \in V(i)} \alpha_{ij}(1 - \frac{K_{ji}^0}{2})(U_j^n - U_i^n) + \sum_{j \in V(i)} \beta_{ij}(U_j^n - U_i^n) \quad (7.94)$$

where α_{ij} and β_{ik} are positive coefficients. As above, according to Lemma 4, we conclude that the ALE scheme satisfies the maximum principle under a CFL condition, provided that $K_{ij}^0 \leq 2$. This last condition is satisfied thanks to the property (P5) of the limiter L, which ends the proof. $\quad\Box$

7.4.3 Density-positive MEV schemes for the Euler equations

The building block for density positivity is flux splitting. We first consider the unidirectional Euler equations and the Godunov exact Riemann solver which is an example of flux difference splitting method that preserves the positivity of ρ under an appropriate CFL condition. Then we derive the multidimensional-CFL condition ensuring that the 2D and 3D schemes involving the unidirectional positive splitting still preserve the positivity of ρ.

7.4.3.1 Unidirectional Euler positive MEV scheme: Godunov's method

Let us consider the unidirectional formulation of the Euler equations for the usual five variables, applicable to fields which do not depend of y and z:

$$\frac{\partial U}{\partial t}(\mathbf{x}, t) + \frac{\partial F(U(\mathbf{x}, t))}{\partial x} = 0 \quad (7.95)$$

$$U = \begin{pmatrix} \rho \\ \rho u \\ \rho v \\ \rho w \\ e \end{pmatrix} \qquad F(U) = \begin{pmatrix} \rho u \\ \rho u^2 + P \\ \rho u v \\ \rho u w \\ (e + P)u \end{pmatrix} \quad (7.96)$$

where (u, v, w) is the velocity and e denotes the total energy per unit volume given by $e = \rho \epsilon + \frac{1}{2}\rho(u^2 + v^2 + w^2)$ in which ϵ is the internal energy per unit mass. We restrict the study to the case of an ideal gas, the pressure P being defined through an equation of state $P = (\gamma - 1)\epsilon\rho$ where γ is constant. The general form of an explicit conservative scheme for the Euler equations is:

$$\frac{U_i^{n+1} - U_i^n}{\Delta t} = -\frac{1}{\Delta x}\left(\Phi_{i+1/2}^n - \Phi_{i-1/2}^n\right) , \qquad (7.97)$$

where U_i^n is the average over the cell $[x_{i-1/2} , x_{i+1/2}]$ and $\Phi_{i+1/2}^n = \Phi(U_i^n, U_{i+1}^n)$ denotes the numerical flux approximation of $F(U^n)|_{i+1/2}$. Assuming that ρ and P are positive at time level n, we look for schemes which keep this still true for time level $n + 1$. There exists in the literature a lot of flux splittings which enjoy density and pressure positivity, some popular examples are the Boltzman splitting [Perthame and Khobalate, 1992] and the HLLE splitting [Einfeldt et al., 1991]. In the Godunov method, we solve two independent Riemann problems at each cell interfaces, that do not interact in the cell provided that:

$$\frac{|V_{max}|\,\Delta t}{\Delta x} \le \frac{1}{2} , \qquad (7.98)$$

where $|V_{max}|$ denotes the maximum absolute value of the Riemann problem wave speeds. We obtain U_i^{n+1} by averaging:

$$U_i^{n+1} =$$
$$\frac{1}{\Delta x}\int_{x_{i-\frac{1}{2}}}^{x_i} W_{RP}\left(\frac{x - x_{i-\frac{1}{2}}}{\Delta t}, U_{i-1}^n, U_i^n\right)\,dx + \qquad (7.99)$$
$$\frac{1}{\Delta x}\int_{x_i}^{x_{i+\frac{1}{2}}} W_{RP}\left(\frac{x - x_{i+\frac{1}{2}}}{\Delta t}, U_i^n, U_{i+1}^n\right)\,dx ,$$

where W_{RP} is the exact ρ-positive solution of the Riemann problem. This illustrates the well-known fact that under the CFL condition (7.98) the Godunov scheme preserves the positivity of density. Let us consider now any unidirectional scheme for the Euler equations, that is ρ-positive under a CFL condition:

$$\frac{\Delta t |V_{max}|}{\Delta x} \le \alpha_1 \qquad (7.100)$$

where α_1 is a coefficient which depends only on the scheme under study. For example, with Godunov scheme, positivity holds for Courant numbers smaller than $\alpha_1 = 0.5$.

For any arbitrary couple of initial states $U_L = U_i$ and $U_R = U_{i+1}$, the above positivity property can be expressed as a property of the flux between these states. For this we need a third state U_{i-1} to assemble the fluxes around node i. Let us choose it as the mirror state of U_i in x-direction, that is to say $\rho_{i-1}^n = \rho_i^n$, $u_{i-1}^n = -u_i^n$, $v_{i-1}^n = v_i^n$, $w_{i-1}^n = w_i^n$, and $e_{i-1}^n = e_i^n$. The first component Φ_1 of the numerical flux function $\Phi_{i-1/2,1}^n$ should ideally be zero:

$$\Phi_1(U_{i-1}, U_i) = 0 \quad \text{if} \quad U_{i-1} \text{ is the mirror state of } U_i . \qquad (7.101)$$

This property is verified by the three previous refered positive schemes. In particular for the Godunov scheme, due to symmetry reasons, we have:

$\Phi_{i-1/2,1}^n = W_{RP,2}(0, U_{i-1}^n, U_i^n) = 0$ where $W_{RP,2}$ denotes the second component of the exact solution W_{RP} of the Riemann problem.

We restrict our study to schemes which satisfy (7.101). Then the discretized equation (7.97) for the density at node i can be written as:

$$\frac{\rho_i^{n+1} - \rho_i^n}{\Delta t} = -\frac{1}{\Delta x}\Phi_{i+1/2,1}^n .$$

Let us denote $\phi_{i+1/2}^+ = Max(0, \Phi_{i+1/2,1}^n)$ and $\phi_{i+1/2}^- = Min(0, \Phi_{i+1/2,1}^n)$, we can write the previous equation as:

$$\rho_i^{n+1} = \left(1 - \frac{\Delta t}{\Delta x}\frac{\phi_{i+1/2}^+}{\rho_i^n}\right)\rho_i^n - \frac{\Delta t}{\Delta x}\phi_{i+1/2}^- .$$

For $\rho^n \geq 0$ and under condition (7.100), ρ^{n+1} is positive. This implies that either $\phi_{i+1/2}$ is negative, or the coefficient of ρ_i^n in right hand side is positive. Both cases are summed up as follows:

$$\frac{\Delta t\ \phi_{i+1/2}^+}{\rho_i^n\Delta x} \leq 1 .$$

Which should hold for the maximum allowed $\Delta t_{max} = \alpha_1\Delta x/|V_{max}|$. We finally get:

LEMMA 7.5
An unidirectional scheme that can be written in the form (7.97), that verifies (7.101) and that is $\rho-$positive under the CFL condition:

$$\frac{\Delta t|V_{max}|}{\Delta x} \leq \alpha_1$$

satisfies the following property:

$$\frac{\alpha_1\ \phi_{i+1/2}^+}{|V_{max}|\rho_i^n} \leq 1 . \quad\square \tag{7.102}$$

More generally, considering a Riemann problem between two states U_R and U_L, $|V_{max}|$ being the maximum wave speed between these two states, we will say that *a flux splitting Φ is ρ-positive* if there exists α_1 so that:

$$\frac{|V_{max}|\Delta t}{\Delta x} \leq \alpha_1 \quad \text{implies} \quad \frac{\Delta t\ \Phi_1^+(U_L, U_R)}{\rho_L\Delta x} \leq 1 . \tag{7.103}$$

where Φ_1 is the first component of Φ. For such a flux splitting, we have:

$$\frac{\alpha_1\ \Phi_1^+(U_L, U_R)}{|V_{max}|\rho_L} \leq 1 . \tag{7.104}$$

The Godunov, HLLE [Einfeldt et al., 1991] and Perthame [Perthame and Khobalate, 1992] schemes involve flux splittings that are ρ-positive according to the above definition. Further, although derived for the x-direction, the previous ρ-positivity statement extends to an arbitrary direction ν (by a rotation of moments and their equations for example). The positivity relation then writes:

$$\frac{\alpha_1 \; \Phi_1^+(U_L, U_R, \nu)}{|V_{max}|\rho_L} \leq 1 \; . \tag{7.105}$$

7.4.3.2 First-order positive MEV scheme for the 3D Euler equations

Let us come back to the unsteady Euler model. Combining a first-order space-accurate finite-volume discretization of the mass conservation equation with a first-order explicit time-integration gives:

$$a_i\rho_i^{n+1} = a_i\rho_i^n - \Delta t \sum_{j \in V(i)} \Phi_1(U_i^n, U_j^n, \nu_{ij}) = a_i\rho_i^n - \Delta t \sum_{j \in V(i)} \phi_{ij}\, l_{ij} \tag{7.106}$$

in which $l_{ij} = \left\|\nu_{ij}\right\| = \left\|\int_{\partial C_{ij}} \mathbf{n}_i(\sigma) \, d\sigma\right\|$ and Φ is defined as a ρ-positive flux splitting in direction ν_{ij}, and as previously Φ_1 represents the first component of Φ. Therefore, using (7.105), we successively get:

$$\phi_{ij} \leq \frac{|V_{max}|\rho_i^n}{\alpha_1} \quad , \quad \Delta t \sum_{j \in V(i)} \phi_{ij}l_{ij} \leq \frac{|V_{max}|\rho_i^n \Delta t}{\alpha_1} L_i$$

where $L_i = \sum_{j \in V(i)} l_{ij}$ is the measure of the cell boundary ∂C_i. From equation (7.106) we finally get that the positivity of ρ_i^{n+1} is ensured when the following condition on the time-step is satisfied:

$$\frac{|V_{max}|\Delta t}{\alpha_1\left(\frac{a_i}{L_i}\right)} \leq 1 \; .$$

LEMMA 7.6

The three-dimensional first-order space-accurate scheme built from a ρ-positive flux splitting is ρ-positive under the CFL condition:

$$\frac{|V_{max}|\Delta t}{\left(\frac{a_i}{L_i}\right)} \leq \alpha_1 \; . \tag{7.107}$$

REMARK 7.6   The above formula measures the loss with respect to 1D case in positively-stable time step: the ratio $\frac{a_i}{L_i}$ is in general only a fraction of mesh size Δx.   ☐

7.4.3.3 High-order positive MEV scheme for the 3D Euler equations

We now derive *high-order ρ-positive schemes* in several dimensions. For sake of clarity we restrict our proof to the 2D case, but it works in a similar way in 3D. We assume that all the ρ_j^n are positive. The general form of the explicit high-order space-accurate scheme governing ρ_i^{n+1} writes:

$$a_i \rho_i^{n+1} = a_i \rho_i^n - \Delta t \sum_{j \in V(i)} \Phi_1(U_{ij}^n, U_{ji}^n, \nu_{ij}) \tag{7.108}$$

$$= a_i \rho_i^n - \Delta t \sum_{j \in V(i)} \phi_{ij}^{HO} l_{ij}$$

$$= a_i \rho_i^n - \Delta t \sum_{j \in V(i)} \left(\frac{\phi_{ij}^{HO+} l_{ij}}{\rho_{ij}^n} \right) \rho_{ij}^n + \left(\frac{\phi_{ij}^{HO-} l_{ij}}{\rho_{ji}^n} \right) \rho_{ji}^n , \tag{7.109}$$

where Φ is a ρ-positive flux splitting in direction ν_{ij} and $\phi_{ij}^{HO} = \frac{1}{l_{ij}} \Phi_1(U_{ij}^n, U_{ji}^n, \nu_{ij})$ is a flux integration of "high-order" space-accuracy. Following the reconstruction of the solution at the interface ∂C_{ij} given by equations (7.86) and (7.87), ρ_{ij}^n and ρ_{ji}^n write as:

$$\rho_{ij}^n = \rho_i^n + \frac{K_{ij}^-}{2} \left(\epsilon_{ri}(\rho_i^n - \rho_r^n) + \epsilon_{si}(\rho_i^n - \rho_s^n) \right) \tag{7.110}$$
$$\rho_{ji}^n = \rho_j^n - \frac{K_{ji}^0}{2}(\rho_j^n - \rho_i^n)$$

where K_{ij}^-, K_{ji}^0, ϵ_{ri} and ϵ_{si} are positive. First, we can notice that ρ_{ij}^n and ρ_{ji}^n are positive. Indeed, according to (7.64) and (7.65) we can also write ρ_{ij}^n as:

$$\rho_{ij}^n = \rho_i^n + \frac{1}{2} L_{ij}(\rho^n) = \rho_i^n + \frac{K_{ij}^0}{2} \Delta^0 \rho_{ij}^n = \rho_i^n + \frac{K_{ij}^0}{2}(\rho_j^n - \rho_i^n) \tag{7.111}$$

where K_{ij}^0 is positive. As the property (P5) of the limiter implies $K_{ij}^0 \leq 2$ and $K_{ji}^0 \leq 2$, we deduce from equations (7.110) and (7.111) that

$$min(\rho_i^n, \rho_j^n) \leq \rho_{ij}^n, \rho_{ji}^n \leq max(\rho_i^n, \rho_j^n)$$

so that ρ_{ij}^n and ρ_{ji}^n are positive. Using (7.110), we can rewrite the discrete equation (7.109) as follows:

$$\rho_i^{n+1} = \sum_{j \in V(i)} \alpha_{ij} \rho_j^n +$$

$$\rho_i^n \left(1 - \frac{\Delta t}{a_i} \sum_{j \in V(i)} \left(\frac{\phi_{ij}^{HO+} l_{ij}}{\rho_{ij}^n} \right) (1 + K_{ij}^- \frac{\epsilon_{ri} + \epsilon_{si}}{2}) + \frac{\Delta t}{a_i} \sum_{j \in V(i)} \left(-\frac{\phi_{ij}^{HO-} l_{ij}}{\rho_{ji}^n} \right) \frac{K_{ji}^0}{2} \right) \tag{7.112}$$

where α_{ij} are positive since $K_{ji}^0 \leq 2$ as already seen. We deduce that the positivity of ρ is preserved under the following condition on Δt:

$$\frac{\Delta t}{a_i} \sum_{j \in V(i)} \left(\frac{\phi_{ij}^{HO+} l_{ij}}{\rho_{ij}^n} \right) \left(1 + K_{ij}^- \frac{\epsilon_{ri} + \epsilon_{si}}{2} \right) \leq 1 . \tag{7.113}$$

In the above equation, the mesh dependant quantity $M_{ij} = \epsilon_{ri} + \epsilon_{si}$ can be written as:

$$M_{ij} = \frac{l_j^2}{l_s l_r} \left(\frac{l_r \sin \theta_r + l_s \sin \theta_s}{l_j \sin \theta} \right)$$

where θ_r, θ_s and θ_t are defined as in Figure 7.12 and l_p denotes the length of the vector **ip** for $p = r, s$ or j. Given $M_i = \max_j M_{ij}$, the CFL condition (7.113) ensuring the positivity of ρ is satisfied when:

$$\frac{\Delta t}{a_i} \sum_{j \in V(i)} \left(\frac{\phi_{ij}^{HO+} l_{ij}}{\rho_{ij}^n} \right) \leq \frac{1}{1 + M_i} , \tag{7.114}$$

since $K_{ij}^- \leq 2$ thanks to the properties (P1) and (P5) of the limiter L. On the other hand, we have $\phi_{ij}^{HO} = \frac{1}{l_{ij}} \Phi_1(U_{ij}^n, U_{ji}^n, \nu_{ij})$ with Φ a ρ-positive flux splitting, so that we get according to (7.105):

$$\frac{\alpha_1 \phi_{ij}^{HO+}}{|V_{max}| \rho_{ij}} \leq 1 . \tag{7.115}$$

Given $L_i = \sum_{j \in V(i)} l_{ij}$ the measure of the cell boundary ∂C_i, from equations (7.114) and (7.115) we derive a positivity statement for a class of high-order schemes, that we write for simplicity for the previous third-order scheme:

LEMMA 7.7
The quasi third-order scheme introduced in section 2, see equations (7.64)-(7.65)-(7.66), based on a ρ-positive flux splitting (7.103) is ρ-positive under the CFL condition:

$$\frac{\Delta t |V_{max}|}{\left(\frac{ai}{L_i} \right)} \leq \frac{\alpha_1}{1 + M_i} . \tag{7.116}$$

In the ALE case, the model is modified in a manner that is similar to the scalar conservation law case. For the spatial discretization, the Riemann solver is modified by the term involving the mesh velocity, but its positivity property is unchanged. Also, the time derivative $d\Big(a_i(t) U_i(t) \Big)/dt$ is now approximated by $(a_i^{n+1} \rho_i^{n+1} - a_i^n \rho_i^n)/\Delta t$, and equation (7.112) becomes:

$$\frac{a_i^{n+1}}{a_i^n} \rho_i^{n+1} = \sum_{j \in V(i)} \alpha_{ij} \rho_j^n +$$

$$\rho_i^n \left(1 - \frac{\Delta t}{a_i^n} \sum_{j \in V(i)} \left(\frac{\phi_{ij}^{HO+} l_{ij}}{\rho_{ij^n}} \right) (1 + K_{ij}^- \frac{\epsilon_{ri} + \epsilon_{si}}{2}) + \frac{\Delta t}{a_i^n} \sum_{j \in V(i)} \left(-\frac{\phi_{ij}^{HO-} l_{ij}}{\rho_{ji}^n} \right) \frac{K_{ji}^0}{2} \right)$$

where α_{ij} are positive, so that we can derive the same ρ-positivity statement than for the fixed grids case.

REMARK 7.7 We observe that the DGCL is not necessary for the positivity of ρ with the ALE formulation. ⬚

REMARK 7.8 In the case of a viscous flow, a particular attention has to be paid to the discrete diffusion operators that should also preserve positivity of variables; for finite-element discretization, this is related to standard acute angle condition, see [Baba and Tabata, 1981]. ⬚

REMARK 7.9 All the previous results are extended to passive specie convection in [Cournède et al., 2006] and can be easily extended to boundary for various boundary conditions by applying mirror principles. ⬚

7.4.4 A numerical example

The theoretical results presented in this chapter have three types of consequence on software. Firstly, the extra robustness of the upwind-element method with respect to the nodal gradient method gets a theoretical confirmation. Secondly, for the explicit upwind-element scheme, a nonlinear stability condition is available for software in order to get more robustness than standard time-step length evaluations relying on unproved extensions of linear and scalar heuristics. Thirdly, with the proposed methodology, we can introduce and limit new higher-order schemes in such a way that density positivity holds. The resulting robustness has been abundantly illustrated by high Mach calculations in several papers, such as [Stoufflet et al., 1996] and papers referenced in it. The high Mach calculations presented in [Stoufflet et al., 1996] were not possible with previous versions of the schemes (*i.e.* without the upwind elements).

Instead of presenting more high Mach flow calculations, we show in this section that the new limited scheme is useful for transonic flow simulations since it represents a good compromise between robustness and accuracy.

Two 3D applications are then considered in order to illustrate the gain obtained in accuracy when the standard second-order scheme equipped with van Albada limiter as in [Stoufflet et al., 1996] is replaced by limiter (7.65) combined with the HO5 flux (see (7.67)). In both case an implicit BDF2 time advancing is used. The first 3D example concerns the calculation of the inviscid steady flow around a supersonic jet geometry at a farfield Mach

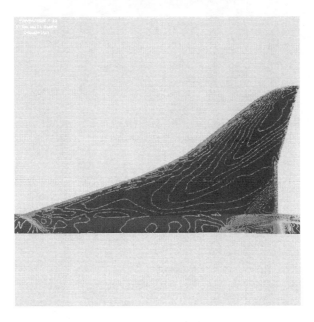

FIGURE 7.13: Flow around Dassault SSBJ geometry, Mach 1.6, incidence 0: Mach number contours on upper side obtained by applying the previous version of the scheme.

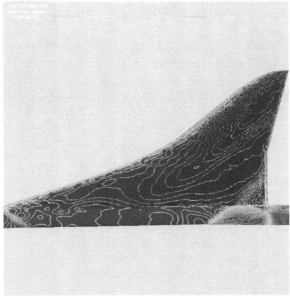

FIGURE 7.14: Flow around Dassault SSBJ geometry, Mach 1.6, incidence 0: Mach number contours on upper side obtained by applying the new version of the scheme.

number set to 1.6 and an angle of attack of 0 degree. The unstructured tetrahedral mesh involves 170000 vertices, which represents a medium-fine mesh for this geometry. In Figures 7.13 and 7.14, we depict the Mach number isolines on the upper surface of the wing-body set for the standard second-order scheme and the new higher-order one, respectively. The comparison of both results shows rather important differences, with extremas predicted at different locations on the geometry and more flow details captured by the higher-order scheme. Using scheme (7.65)-(7.67) produces an improvement of 4% for the drag (from 15940 to 15313) and of 2% for the lift (from 120690 to 123345).

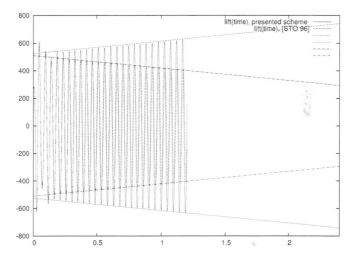

FIGURE 7.15: Flutter of the Agard Wing 445.6 underlined by the amplitude of the lift oscillation as a function of time. Dashes: the previous version [Stoufflet et al., 1996] of the scheme is applied, the initial oscillation is damped. Line: scheme (7.65)-(7.67) is applied, oscillations are consistently amplified with the experimental results.

The second 3D example concerns the simulation of an unsteady inviscid flow involving mesh deformations. We consider a rather standard flutter test case: the flutter of the AGARD Wing 445.6 which has been measured with various flow conditions by Yates [Yates, 1987]. We focus on a rather tricky transonic case, for which the farfield Mach number is 1.072. More precisely, the reference density is set to $9.838 \ 10^{-8}$ $slugs/inch^3$ and the reference pressure to 16.8 $slugs/(inch.sec^2)$. According to the experimental results detailed in [Yates, 1987], the flow conditions are inside the unstability domain and the wing flutter is already pronounced.

The three-dimensional unstructured tetrahedral CFD mesh contains 22014 vertices. For the aeroelastic analysis, the structure of the wing is discretized by a thin plate finite-element model which contains 800 triangular composite shell elements and is based on the informations given in [Yates, 1987].

This transient aeroelastic case is computed with the fluid-structure interaction methodology developed by Farhat and co-workers, see [Farhat, 1995]. We compare again the scheme defined in [Stoufflet et al., 1996] to scheme (7.65)-(7.67). The time step is defined according to a maximum fluid Courant number of 900. In Figure 7.15, we depict the wing lift coefficient as a function of time. With the new scheme, the flutter amplification is well predicted consistently with the experimental results while an important damping is predicted with the previous scheme leading to erroneous results.

7.4.5   Conclusion for positiveness

This section focuses on the positivity of Mixed-Element-Volume upwind schemes. We propose an analysis for defining a positive sub-family of MUSCL-based second-order accurate schemes. Three types of positivity are examined: maximum principle for a scalar nonlinear conservation law, density positivity for the Euler equations, and maximum principle for convected species in a multi-component flow. In each of the three cases, the proposed analysis is extended to moving mesh ALE approximations. In particular we extend to second-order accuracy the contribution of [Farhat et al., 2001]. We also *a posteriori* state, as a particular case, the robustness of the upwind-element scheme used in [Stoufflet et al., 1996] which was derived empirically for the special treatment of very stiff flows, involving strong bow shocks at large Mach numbers.

In the case of explicit time advancing, this analysis brings a rigorous time-step evaluation for positivity. With this analysis, we have also defined new schemes that are as robust but less dissipative than the previous basic upwind-element scheme, because the conditions for positiveness are accurately identified.

The numerical study presented in this chapter highlights the benefits that can be acquired when these new schemes are applied. With the new spatial discretization schemes presented in this work, steady and unsteady flow calculations show thin and monotone shock structures, and a lower amount of numerical dissipation compared to the previous MEV schemes. It should be noted that dissipation can be even more decreased by adding sensors dedicated to the inhibition of limiters in regions where the flow is regular. The improvement in control of both monotony and dissipation is finally demonstrated by computing a rather critical flutter case for which the scheme accuracy is a determining factor on qualitative outputs.

Let us specify some important limits of the above techniques. Limiters should be avoided or well locally controlled in calculations which need a very low level of dissipation, such as those addressed with the superconvergent V6 schemes (in practice, large eddy simulation, aeroacoutics). A second limitation of the theory of this chapter is that it does not apply to the case where a low Mach number preconditioner is used.

7.5   3D multi-scales anisotropic mesh adaptation

When dealing with real-life CFD problems, mesh adaptation is recognized as a complementary approach to high-order schemes classically used to solve the problem at hand.   It provides a way to control the accuracy of the numerical solution by modifying the domain discretization according to size and directional constraints.   Among mesh adaptation methods, anisotropic unstructured adaptation is of paramount influence on the accuracy of many CFD predictions [Castro-Diaz et al., 1997], [Frey and Alauzet, 2005], [Li et al., 2005], [Pain et al., 2001], [Tam et al., 2000].   In the particular context of flow with shocks, anisotropic mesh adaptation provides very accurate solutions by reducing considerably the numerical dissipation of shock capturing schemes.   This technique allows (i) to substantially reduce the number of degrees of freedom, thus impacting favorably the cpu time and (ii) to reduce (optimize) the numerical scheme dissipation by automatically taking into account the anisotropy of the physical phenomena inside the mesh.

If the previous features of unstructured mesh adaptation are now quite classical, it has been recently pointed out that mesh adaptation has further consequences impacting directly numerical schemes used to approximate the flow.   Indeed, a loss of convergence order generally occurs due to the presence of steep gradients (Naviers Stokes equations) or genuine discontinuities (Euler equations) in the flow, even if a probably spatially high order method is employed. The computed mesh convergence order on uniformly refined meshes is not the theoretical expected one. In [Dervieux et al., 2003], [Loseille et al., 2007], it has been demonstrated that the convergence order of numerical schemes can be recovered thanks to this mesh adaptation procedure.   This ability to approach the asymptotic convergence and, therefore, to obtain more easily an accurate prediction for complex flows is another key feature of mesh adaptation

In this section, we focus on a recent family of methods, often refered to as **metric analysis methods**, or **Hessian-based methods** that have shown a very fertile development, from the pioneering works in [Castro-Diaz et al., 1997], [Fortin et al., 1996]. Thanks to recent formalisms, see for instance [Loseille and Alauzet, 2009], these ideas turned into a clean set of functional analysis problematics relying on an ideal representation of the interpolation error and of a mesh. Getting rid of error iso-distribution and prefering L^p error minimization allow to take into account discontinuities with higher-order convergence [Dervieux et al., 2007], [Loseille et al., 2007]. This theory combines perfectly with unstructured mesh generation [Frey, 2000], [George and Borouchaki, 1998] and addressed applications are either steady or un-steady [Alauzet et al., 2007], [Frey and Alauzet, 2005], [Guégan, 2007]. Metric-based mesh adaptation efficiency and genericity have been proved by many successful applications for 3D complex problems [Alauzet et al., 2007], [Bottasso, 2004], [Dompierre et al., 1997], [Formaggia et al., 2004], [Frey and Alauzet, 2005], [Pain et al., 2001], [Schall et al., 2004], [Tam et al., 2000].

7.5.1   Anisotropic mesh generation

The generation of 3D anisotropic adapted meshes uses the notion of length in a metric space [Frey and George, 2008]. The idea is to introduce a metric tensor in the dot product definition to modify size evaluation in all directions. In 3D, a metric is a 3×3 symmetric definite-positive matrix. The mesh is automatically adapted by generating *a unit mesh* with respect to this metric, *i.e.,* the mesh is such that all edges have a length close to one in the metric and such that all elements are almost regular. The mesh is then uniform in prescribed metric space and, non-uniform and anisotropic in the Euclidean space.

Consequently, every mesh generator which is able to deal with a metric field can be utilized whatever the meshing technique it uses: Delaunay, local Delaunay, local refinements, ... Note that a lot of adaptive mesh generators are now able to interpret this metric concept. Let us mention [Frey, 2001] for discrete surface mesh adaptation and [Bottasso, 2004], [Coupez, 2000], [Dobrzynski and Frey, 2008], [George, 1999], [Jones et al., 2006], [Li et al., 2005], [Pain et al., 2001], [Tam et al., 2000] in 3D.

In the context of numerical simulation, the accuracy level of the solution depends on the current mesh used for its computation and the mesh adaptation prescription, *i.e.,* the metric field, is provided by the current solution. This points out the non-linearity of the anisotropic mesh adaptation problem. Therefore, an iterative process needs to be set up in order to converge both the mesh and the solution, or equivalently the metric field and the solution. For stationary simulations, an adaptive computation is carried out *via* a mesh adaptation loop inside which an algorithmic convergence of the couple mesh-solution is sought [Frey and Alauzet, 2005]. At each iteration, all components of the mesh adaptation loop are involved successively: the flow solver, the error estimate, the adaptive mesh generator [Dobrzynski and Frey, 2008], [George, 1999] and the solution interpolation step. This procedure is repeated until convergence of the couple mesh-solution is reached.

7.5.2   Continuous mesh model and optimality

Let u be a smooth given function. The problem of deriving a mesh that minimizes the $\mathbf{L}^p$ norm of the interpolation error is an active field of research [Castro-Diaz et al., 1997], [Agouzal et al., 1999], [Frey and Alauzet, 2005], [Huang, 2005], [Vassilevski and Agouzal, 2005], [Courty et al., 2006]. Our approach is based on a promising continuous mesh model that allows to predict effectively the interpolation error on a fictitious so called continuous mesh [Loseille and Alauzet, 2009]. In addition to this interpolation estimate, the main benefit is that we are practically able to generate a computational optimal discrete mesh both in two and three dimensions.

A *continuous mesh* $\mathbf{M} = (\mathcal{M}(\mathbf{x}))_{\mathbf{x}\in\Omega}$ of a domain Ω is a Riemannian metric field [Berger, 2003]. For all $\mathbf{x}$ of Ω, $\mathcal{M}(\mathbf{x})$ is a symmetric tensor having $(\lambda_i(\mathbf{x}))_{i=1,3}$ as eigenvalues along the principal directions $\mathcal{R}(\mathbf{x}) = (\mathbf{v}_i(\mathbf{x}))_{i=1,3}$. Sizes along these directions are denoted $(h_i(\mathbf{x}))_{i=1,3} = (\lambda_i^{-2}(\mathbf{x}))_{i=1,3}$. With this definition, $\mathbf{M}$ admits the more practical decomposition:

$$
\mathcal{M}(\mathbf{x}) = d^{\frac{2}{3}}(\mathbf{x})\, \mathcal{R}(\mathbf{x})
\begin{pmatrix}
r_1^{-\frac{2}{3}}(\mathbf{x}) & & \\
& r_2^{-\frac{2}{3}}(\mathbf{x}) & \\
& & r_3^{-\frac{2}{3}}(\mathbf{x})
\end{pmatrix}
{}^t\mathcal{R}(\mathbf{x}),
$$

where

- the *mesh density* d is equal to: $(h_1 h_2 h_3)^{-1} = (\lambda_1 \lambda_2 \lambda_3)^{\frac{1}{2}} = \sqrt{\det(\mathcal{M})}$,

- the three *anisotropic quotients* r_i are equal to: $h_i^3 (h_1 h_2 h_3)^{-1}$.

The anisotropic quotients represent the overall anisotropic ratio of a tetrahedron taking into account all the possible directions. It is a complementary measure to anisotropic ratio given by $\max_i(h_i)/\min_i(h_i)$. By integrating the node density, we define the *mesh complexity* $\mathcal{C}$ of a continuous mesh which is the continuous counterpart of the total number of vertices:

$$
\mathcal{C}(\mathbf{M}) = \int_\Omega d(\mathbf{x})\, d\mathbf{x} = \int_\Omega \sqrt{\det(\mathcal{M}(\mathbf{x}))}\, d\mathbf{x}.
$$

This real-value parameter is useful to quantify the global level of accuracy of the continuous mesh $\mathbf{M} = (\mathcal{M}(\mathbf{x}))_{\mathbf{x}\in\Omega}$.

It has been shown in [Loseille and Alauzet, 2009] that $\mathbf{M}$ defines a class of equivalence of discrete meshes. The equivalence relation is based on the notion of **unit mesh** with respect to $\mathbf{M}$. A mesh $\mathcal{H}$ is unit wih respect to $\mathbf{M}$ when each tetrahedron $K \in \mathcal{H}$ defined by its list of edges $(\mathbf{e}_i)_{i=1\ldots6}$ verifies:

$$
\forall i \in [1,6], \quad \ell_{\mathcal{M}}(\mathbf{e}_i) \in \left[\frac{1}{\sqrt{2}}, \sqrt{2}\right] \quad \text{and} \quad Q_{\mathcal{M}}(K) \in [\alpha, 1] \text{ with } \alpha > 0.
$$

A classical and admissible value of α is 0.8. The length of an edge $\ell_{\mathcal{M}}(\mathbf{e}_i)$ and the quality of an element $\mathcal{Q}_{\mathcal{M}}(K)$ are integrated to take into account the variations of $\mathbf{M}$ in Ω:

$$\mathcal{Q}_{\mathcal{M}}(K) = \frac{36}{3^{\frac{1}{3}}} \frac{|K|_{\mathcal{M}}^{\frac{2}{3}}}{\sum_{i=1}^{6} \ell_{\mathcal{M}}^{2}(\mathbf{e}_i)} \in [0,1], \quad \text{with } |K|_{\mathcal{M}} = \int_K \sqrt{\det(\mathcal{M}(\mathbf{x}))}\, d\mathbf{x},$$

$$\text{and } \ell_{\mathcal{M}}(\mathbf{e}_i) = \int_0^1 \sqrt{{}^t\mathbf{ab}\, \mathcal{M}(\mathbf{a} + t\, \mathbf{ab})\, \mathbf{ab}}\, dt, \quad \text{with } \mathbf{e}_i = \mathbf{ab}.$$

This model is also particularly well suited to the study of the interpolation error. Indeed, there exists a unique continuous interpolation error that models the (infinite) set of interpolation errors computed on the class of unit meshes. See [Loseille and Alauzet, 2009] for the proof along with equivalence between discrete and continuous formulations. In 3D, for a smooth function u, the continuous linear interpolate $\pi_{\mathcal{M}} u$ is a function of the Hessian H_u of u and verifies:

$$(u - \pi_{\mathcal{M}} u)(\mathbf{x}) = \frac{1}{10} \text{trace}(\mathcal{M}^{-\frac{1}{2}}(\mathbf{x})\, |H_u(\mathbf{x})|\, \mathcal{M}^{-\frac{1}{2}}(\mathbf{x})) \qquad (7.117)$$

$$= \frac{1}{10} d^{-\frac{2}{3}} \sum_{i=1}^{3} r_i^{\frac{2}{3}}\, {}^t\mathbf{v_i}\, |H_u|\, \mathbf{v_i},$$

where $|H_u|$ is deduced from H_u by taking the absolute values of its eigenvalues. $\pi_{\mathcal{M}}$ replaces the discrete operator Π_h in this continuous framework. Note that (7.117) does not require any assumption linking u and $\mathbf{M}$ as, for instance, any alignment condition.

In practice, returning to a unit mesh $\mathcal{H}$ with respect to $\mathbf{M}$, then the following estimate of the linear interpolation error $u - \Pi_{\mathcal{H}} u$ holds:

$$\|u - \Pi_{\mathcal{H}} u\|_{\mathbf{L}^p(\Omega_h)} \leq \left(\int_\Omega \left(\text{trace}(\mathcal{M}^{-\frac{1}{2}}(\mathbf{x}) |H_u(\mathbf{x})| \mathcal{M}^{-\frac{1}{2}}(\mathbf{x})) \right)^p d\mathbf{x} \right)^{\frac{1}{p}}. \quad (7.118)$$

The previous inequality turns out to be an equality for some p and some smoothness assumptions on u. Choosing the continuous formulation of [Alauzet et al., 2006], [Loseille, 2008], the right hand side of (7.118) can be then minimized under the constraint of a mesh complexity equal to a parameter N, giving the unique optimal continuous mesh $\mathbf{M}_{\mathbf{L}^p} = (\mathcal{M}_{\mathbf{L}^p}(\mathbf{x}))_{\mathbf{x}\in\Omega}$ minimizing the right hand side of (7.118):

$$\mathcal{M}_{\mathbf{L}^p} = D_{\mathbf{L}^p}\, (\det |H_u|)^{\frac{-1}{2p+3}}\, |H_u| \quad \text{with} \quad D_{\mathbf{L}^p} = N^{\frac{2}{3}} \left(\int_\Omega (\det |H_u|)^{\frac{p}{2p+3}} \right)^{-\frac{2}{3}},$$
$$(7.119)$$

where $N = \mathcal{C}(\mathbf{M})$ is the continuous mesh complexity fixing the accuracy (size) of the mesh. $D_{\mathbf{L}^p}$ is a global normalization term set to obtain a continuous mesh with complexity N and $(\det |H_u|)^{\frac{-1}{2p+3}}$ is a local normalization term accounting for the sensitivity of the $\mathbf{L}^p$ norm. Indeed, the choice of a $\mathbf{L}^p$ norm is essential in a mesh adaptation process regarding the type of problems solved. For instance in CFD, physical phenomena may involve large scale variations. Capturing weak phenomena is crucial for obtaining an accurate solution by taking into account all phenomena interactions in the main flow area. Intrinsically, metrics constructed with lower p norms are more sensitive to weaker variations of the solution whereas the $\mathbf{L}^\infty$ norm mainly concentrates on strong singularities (e.g. shocks).

Finally, the interpolation error (7.118) can be rewritten for $\mathbf{M_{L^p}}$, the following bound follows up for a unit mesh $\mathcal{H}_{\mathbf{L}^p}$ with respect to $\mathbf{M_{L^p}}$:

$$\|u - \Pi_{\mathcal{H}_{\mathbf{L}^p}} u\|_{\mathbf{L}^p(\Omega_h)} \leq 3N^{-\frac{2}{3}} \left(\int_\Omega (\det |H_u|)^{\frac{p}{2p+3}} \right)^{\frac{2p+3}{3p}} \leq \frac{Cst}{N^{2/3}} . \qquad (7.120)$$

A main result arises from the previous bound: a global second-order asymptotic mesh convergence is expected for the considered variable u. Indeed, a simple analogy with regular grids leads to consider that $N = O\left(h^{-3}\right)$ so that the previous estimate becomes: $\|u - \Pi_{\mathcal{H}_{\mathbf{L}^p}} u\|_{\mathbf{L}^p(\Omega_h)} \leq Cst'h^2$. The second order convergence property still holds even when singularities are present in the flow field for all $p \in [1, \infty[$, see [Loseille et al., 2007]. This theoretical result is verified numerically in our simulations and is used to assess the obtained numerical solutions.

7.5.3 Application to numerical computation

In our case, the numerical solution provides a continuous piecewise linear by elements representation of the solution. Consequently, our analysis cannot be applied directly to the numerical solution. The idea is to build a higher order solution approximation u^* of u from u_h which is twice continuously differentiable and to consider u^* in our error estimate. More precisely, the interpolation error is approximated as $\|u - \Pi_h u\|_\Omega \approx \|u^* - \Pi_h u^*\|_\Omega$. If u^* and u_h coincide at mesh vertices then we have $\|u^* - \Pi_h u^*\|_\Omega = \|u^* - u_h\|_\Omega$ illustrating that our estimate approximates the approximation error. Practically, only the Hessian of u^* is recovered. In the context of discontinuous flows, the numerical solution is also piecewise linear by elements even if it approximates a discontinuous solution. In this case, we still approximate the solution u with a continuous higher order representation and we still apply our error estimate.

We briefly present a Hessian recovery method based on a Green formulation. Let $\mathcal{H}$ be a mesh of a domain $\Omega_h \subset \mathbb{R}^3$. We denote by $\varphi_i \in V_h$ the basis function associated with vertex i, where V_h is the approximation space associated with the P^1 Lagrange finite-element. We denote by S_i the stencil

φ_i, *i.e.*, $S_i = supp\,\varphi_i$, which is in fact the ball of i. K represents a mesh element. The Hessian of the solution is recovered using a weak formulation, based on the Green formula, considering that the gradient of u_h is constant by element. For each vertex k of $\mathcal{H}$, we have for $1 \leq i, j \leq 3$:

$$\int_{\mathcal{H}} \frac{\partial^2 u_h}{\partial x_i \partial x_j}\, \varphi_k = \int_{S_k} \frac{\partial^2 u_h}{\partial x_i \partial x_j}\, \varphi_k = -\int_{S_k} \frac{\partial u_h}{\partial x_j} \frac{\partial \varphi_k}{\partial x_i} + \int_{\partial S_k} \frac{\partial u_h}{\partial n}\, \varphi_k\, d\sigma$$

$$= -\sum_{K \in S_k} \int_K \frac{\partial u_h}{\partial x_j} \frac{\partial \varphi_k}{\partial x_i},$$

as the shape function is zero on the boundary of the stencil ∂S_k. A specific treatment is done close to the boundary. Each component of the Hessian is then recovered with the relation:

$$\frac{\partial^2 u^*}{\partial x_i \partial x_j}(k) := \frac{-\displaystyle\int_{S_k} \frac{\partial u_h}{\partial x_j} \frac{\partial \varphi_k}{\partial x_i}}{\displaystyle\int_{S_k} \varphi_k} = -\frac{\displaystyle\sum_{K \in S_k} (\frac{\partial u_h}{\partial x_j})|_K \int_K \frac{\partial \varphi_k}{\partial x_i}}{\dfrac{|S_k|}{4}}.$$

7.5.4   Application to a supersonic business jet

In this section, we study a low-drag-shaped jet provided by Dassault Aviation. The aircraft geometry is shown in Figure 7.17 (top). The length of the jet is $L = 37$ meters and it has a wing span of 17 meters. The surface mesh accuracy varies between 1 millimeter and 30 centimeters. The computational domain is a cylinder of 2.25 kilometers length and 1.5 kilometers diameter. This represents a scale factor of 10^6 if the size of the domain is compared to the maximal accuracy of the jet surface mesh.

The flight conditions are a supersonic cruise speed of Mach 1.6 at an altitude of 13 680 meters (45 000 feet). In this study, the cruise lift has been set to $C_l = 0.115$ by Dassault Aviation. The angle of attack for the aircraft is set such that this lift is attained. The obtained angle of attack is near to 3 degrees.

The multi-scales mesh adaptation considers a control of the interpolation error in $\mathbf{L}^2$ norm of the local Mach number. The local Mach number has been selected as it is really representative of supersonic flows. We have deliberately avoided to adapt the aircraft surface mesh. Indeed, the provided surface mesh is very accurate and is of high quality for the computation. A mesh gradation of 2.5 has been set [Alauzet, 2009]. A total of 32 adaptations have been performed. The mesh adaptation loop is split into 4 steps of 8 adaptations. At each step, the couple mesh-solution is algorithmically converged at a fixed metric complexity. This complexity is multiplied by two between two steps. This strategy has two main advantages. First, it enables us to perform a convergence study of the whole solution as a series of couple mesh-solution with an increasing accuracy is obtained. Second, considering an increasing dynamic complexity level accelerates the convergence of the whole process. We have fixed the following complexities:

$$[100\,000, 200\,000, 400\,000, 800\,000],$$

which give meshes the size of which is almost 0.8, 1.7, 4 and 9 millions of vertices.

To validate the CFD computations, the solution global convergence order is analyzed with respect to a reference solution. The reference solution is the final solution obtained on the finest adapted mesh for a complexity equal to 800 000. The convergence is computed in $\mathbf{L}^2$ norm on the local Mach number which is the variable used for the mesh adaptation. A convergence order equal to 2 is obtained as predicted by the theory, see Figure 7.16, left.

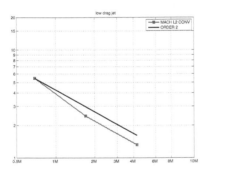

FIGURE 7.16: Left: global mesh convergence order for the SSBJ: a second order spatial convergence is obtained. Right: the Mach cone emitted by the SSBJ. The maximal Mach cone diameter is 1.25 kilometers.

As regards the obtained solution, a very accurate result in the whole computational domain is achieved for the final adapted meshes containing almost 9 millions vertices. In particular, the accuracy of the tetrahedral meshes has reached the surface meshes accuracy and the refinements have been propagated in the whole computational domain. This is illustrated in Figure 7.16, right, where the Mach cones of the low drag SSBJ is depicted. This result points out that the numerical dissipation of the flow solver has been drastically reduced thanks to the anisotropic mesh adaptation. The shock waves have been accurately propagated in the whole computational domain, more than one kilometer below the jet.

Figure 7.17 shows the final adapted mesh (middle, in the symmetry plane Oxz and, bottom, in the cut plane Oxy) and the associated local Mach number iso-values. We notice that the shock waves and the anisotropic refinements have been propagated in the whole domain without any dissipation. We also remark the very accurate capture inside the mesh of all the shock waves. This points out the multi-scales behavior of the mesh adaptation approach. Each element of the geometry emits its own shock waves. The focalization of shocks during their propagation is also illustrated inside the mesh. These pictures also point out the great complexity of the physical phenomenon and of the mesh refinement. This demonstrates the necessity to use a fully automatic mesh adaptive method.

7.6   3D goal-oriented anisotropic mesh adaptation

7.6.1   Introduction

So far, anisotropic features are mainly deduced from an interpolation error estimate as presented in the previous section. However, these methods are limited to the minimization of some interpolation errors for some solution fields. If for many applications, this standpoint is an advantage, there are also many applications where Hessian-based adaptation is no more optimal regarding the way the degrees of freedom are distributed in the computational domain. Indeed, metric-based methods aim at controlling the interpolation error but this goal is not often so close to the objective that consists in obtaining the best solution of a PDE. This is particularly true in many engineering applications where a specific functional needs to be accurately evaluated: lift, drag, heat flux, pressure field... Unfortunately, a Hessian-like anisotropic approach does not directly apply to the goal-oriented mesh adaptation methods that take into account both the solution and the PDE in the error estimation.

In contrast, the formulation of **goal-oriented** mesh adaptation [Giles, 1997], [Giles and Pierce, 1999], [Pierce and Giles, 2000], [Venditti and Darmofal, 2002], [Venditti and Darmofal, 2003] has brought many improvements in the formulation and the resolution of mesh adaptation for PDE approximations. Let us write the continuous PDE and the discrete one as:

$$\Psi(w) = 0 \quad \text{and} \quad \Psi_h(w_h) = 0. \tag{7.121}$$

In this context, we focus on deriving the best mesh to observe a given functional j depending of the solution w. To this end, we examine how to control

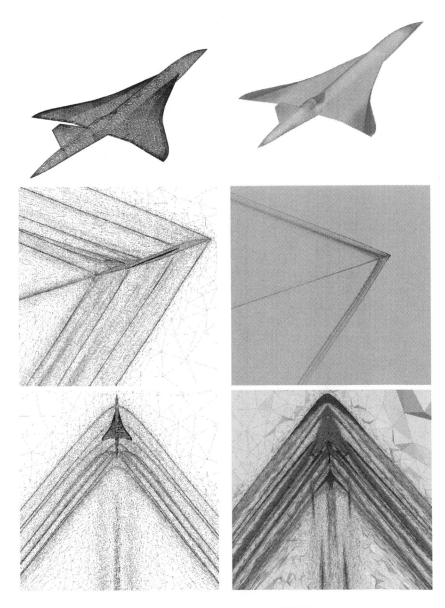

FIGURE 7.17: Left: final adapted mesh for the SSBJ computation containing 9 465 835 vertices and 56 568 966 tetrahedra, right: the associated local Mach number iso-value. From top to bottom: the surface mesh, view of the adapted volume mesh through the cut plane along $0xz$ and view of the adapted volume mesh through the cut plane along $0xy$.

the approximation error of the functional: $j(w) - j(w_h)$. Nevertheless, the objective of goal-oriented mesh adaptation is different from the one of deriving the optimal mesh to control the global approximation error $\|w - w_h\|$, see [Becker and Rannacher, 1996], [Verfürth, 1996] for *a posteriori* error estimate devoted to this task. The formulation of this problem is slightly different. Assuming that the functional j is enough regular to be observed through its jacobian g we simplifiy it as follows:

$$j(w) = (g, w).$$

We also assume that there is no discrete error evaluation on j, this means that $j_h(w_h) = j(w_h)$. On this basis, we seek for the mesh $\mathcal{H}$ which gives the smallest error for the evaluation of j from the solution w_h:

$$\min_{\mathcal{H}} |(g, w_h) - (g, w)|, \qquad (7.122)$$

where w and w_h verify (constraint) state equations (7.121). The mesh adaptation problem is recast with equation (7.122) as an optimization problem. In order to go a step forward in the analysis, we need to implicitly take into account constraints (7.121) in equation (7.122). The initial approximation error on the cost functional $|(g, w_h) - (g, w)|$ can be simplified as a local error thanks to the introduction of the adjoint state:

$$(g, \; w_h - w) \approx \left(g, \; \left(\frac{\partial \Psi}{\partial w}\right)^{-1} \Psi(w_h)\right) = (w^*, \; \Psi(w_h)) \,,$$

where the adjoint state w^* is solution of:

$$\left(\left(\frac{\partial \Psi}{\partial w}\right)^* w^*, \; \psi\right) = (g, \; \psi).$$

In practice, the exact adjoint w^* is not available. By introducing an approximate adjoint w_h^*, we get:

$$(g, \; w_h - w) \approx (w_h^*, \; \Psi(w_h)) \,.$$

The right hand side is a spatial integral the integrand of which can be used to decide where to refine the mesh. The iso-distribution of the error can be approximated by refining according to a tolerance, as in [Becker and Rannacher, 1996]. In [Giles and Suli, 2002], it is proposed to use this right hand side as a correction that importantly improves the quality (in particular the convergence order) of the approximation of j by setting:

$$j^{corrected} = (g, \; w_h) + (w_h^*, \; \Psi(w_h)) \,.$$

However, by substituting w^* by w_h^*, we introduce an error in $O(w_h^* - w^*)$, which results in being the main error term when we use $j^{corrected}$. In [Venditti and Darmofal, 2002], [Venditti and Darmofal, 2003], it is proposed to keep the corrector and to adapt the mesh to this higher order error term, *i.e.*:

$$j^{corrected} - j \approx (w_h^* - w^*, \Psi(w_h)),$$

or equivalently:

$$j^{corrected} - j \approx (w - w_h,\ g - \left(\frac{\partial \Psi}{\partial w}\right)^* w_h^*).$$

In order to evaluate numerically these terms, the authors chose to approach these approximation errors, *i.e.*, $w_h^* - w^*$ and $w_h - w$, by interpolation errors, by computing differences between the linear representation $L_{h/2}^h$ and a quadratic representation $Q_{h/2}^h$ reconstructed on a finer mesh.

The adjoint-L^1 approach In this approach, metric analysis and goal-oriented analysis are complementary. Indeed, a metric-based method specifies the object of our search through an accurate description of the ideal mesh while a goal-oriented method specifies precisely the purpose of the search in terms of which error will be reduced. It is then very motivating to seek for a combination of both methods, with the hope of obtaining a metric-based specification of the best mesh for reducing the error committed on a target functional. A few works address this purpose. In [Venditti and Darmofal, 2003], an anisotropic step relying on the Hessian of the Mach number is introduced into the *a posteriori* estimate. In [Rogé and Martin, 2008], an adhoc formula gives a better impact to the anisotropic component. This section presents a different contribution to the combination of both methods.

The first key point of this work is to use a metric-based parameterization of meshes. This means to work in a continuous (non-discrete) formulation by following the continuous interpolation analysis proposed in the previous section. Usually, metric-based methods use an interpolation error, the deviation between the exact solution and its linear interpolation on the mesh. This assumes the knowledge of the solution, this is an *a priori* standpoint.

In contrast, goal-oriented methods are generally envisaged from an *a posteriori* standpoint, we refer to [Apel, 1999], [Becker and Rannacher, 1996], [Formaggia et al., 2004], [Giles and Suli, 2002], [Picasso, 2003], [Verfürth, 1996]. With this option, the error committed is known on an existing mesh element-wise. Therefore, mesh refinement scheme based on a such *a posteriori* estimations depends on an equi-distribution principle and is thus intrinsically isotropic. Fortunately, goal-oriented methods do not need to be systematically associated with an *a posteriori* analysis. Now, according to [Babuška and Strouboulis, 2001] *a priori* analysis can bring many useful informations. Anisotropy is often one of these informations [Formaggia and Perotto, 2001]. Further, the goal-oriented error can also be easily expressed by an *a priori* analysis, as we will demonstrate. This is the second key point of this work.

The third key point results from working with a numerical scheme that expresses the difference $\Psi_h(w) - \Psi(w)$ in term of interpolation errors. This can be done in a straightforward way by considering finite-element variational formulations.

The theoretical abstract framework is first introduced. Within this framework, a first *a priori* goal-oriented error estimate, equation (7.126), is derived. Its application to the compressible Euler equations is then studied in section 7.6.3 for a class of specific Galerkin-equivalent numerical schemes. From this study, a generic anisotropic error estimate, equation (7.129), is expressed. The estimate is then minimized globally on the abstract space of continuous meshes, section 7.6.4. Finally, the numerical part gives some details on the main modifications of the adaptive loop as compared to classical Hessian-based mesh adaptation. The pratical optimal metric field minimizing the goal-oriented error estimate is then exhibited, equation (7.132). 3D detailed examples conclude this last section by providing a numerical validation of the theory.

7.6.2   A more accurate nonlinear error analysis

An accurate error analysis cannot be done without specifying the operator which permits to pass from continuous to discrete and *vice versa*. Since the P^1 interpolate is the pivot of today metric analysis, this operator is naturally used in our analysis. It is denoted by Π_h in the sequel.

7.6.2.1   Assumptions and definitions

Let V be a space of functions (at least a Banach space). We write the state equation under a variational statement:

$$w \in V \ , \ \forall \varphi \in V \ , \ (\Psi(w), \varphi) = 0, \qquad (7.123)$$

where the operator $(,)$ holds for a $V' \times V$ product, V' is the topological dual of V and w is the solution of this equation. Symbol Ψ holds for a functional that is linear with respect to test function φ but *a priori* nonlinear with respect to w. The continuous adjoint w^* is solution of:

$$w^* \in V \ , \ \forall \psi \in V \ , \ \left(\frac{\partial \Psi}{\partial w}(w)\psi, w^* \right) = (g, \psi), \qquad (7.124)$$

where g is the jacobian of a given functional j. Let V_h be a subspace of $\mathcal{V} = V \cap \mathcal{C}^0$ of finite dimention N, we write the discrete state equation as follows:

$$w_h \in V_h \ , \ \forall \varphi_h \in V_h \ , \ (\Psi_h(w_h), \varphi_h) = 0.$$

Then, we can write:

$$(\Psi_h(w), \varphi_h) - (\Psi_h(w_h), \varphi_h) = (\Psi_h(w), \varphi_h) - (\Psi(w), \varphi_h) = ((\Psi_h - \Psi)(w), \varphi_h).$$
(7.125)

For the *a priori* analysis, we assume that the solutions w and w^* are sufficiently regular:

$$w \in V \cap C^0 \ , \ w^* \in V \cap C^0,$$

and that we have an interpolation operator:

$$\Pi_h : V \cap C^0 \ \rightarrow \ V_h.$$

7.6.2.2 *A priori* estimation

We start from a functional defined as: $j(w) = (g, w)$, where g is a function of V. Our objective is to estimate the following approximation error on the functional:

$$\delta j = j(w) - j(w_h),$$

as a function of continuous solutions, of continuous residuals and of discrete residuals. The error δj is split as follows:

$$\delta j = j(w) - j(w_h) = (g, w - \Pi_h w) + (g, \Pi_h w - w_h).$$

δj is now composed of an **interpolation error** and of an **implicit error** which involves only discrete terms. Let us introduce the discrete adjoint system:

$$w_h^* \in V_h \ , \ \forall \psi_h \in V_h \ , \ \left(\frac{\partial \Psi_h}{\partial w}(\Pi_h w)\psi_h, w_h^* \right) = (g, \psi_h).$$

We can derive the following extension of δj with the choice $\psi_h = \Pi_h w - w_h$:

$$\delta j = (g, w - \Pi_h w) + \left(\frac{\partial \Psi_h}{\partial w}(\Pi_h w)(\Pi_h w - w_h), w_h^* \right).$$

This new right hand side second term is simplified by using an extension of Ψ_h. According to (7.125), we have:

$$(\Psi_h(\Pi_h w), w_h^*) - (\Psi_h(w_h), w_h^*) = (\Psi_h(\Pi_h w), w_h^*) - (\Psi_h(w), w_h^*) + ((\Psi_h - \Psi)(w), w_h^*),$$

which gives by using a Taylor extension:

$$\left(\frac{\partial \Psi_h}{\partial w}(\Pi_h w)(\Pi_h w - w_h), w_h^* \right) = (\Psi_h(\Pi_h w), w_h^*) - (\Psi_h(w), w_h^*) + ((\Psi_h - \Psi)(w), w_h^*) + R_1,$$

where the remainder R_1 is:

$$R_1 = \left(\frac{\partial \Psi_h}{\partial w}(\Pi_h w)(\Pi_h w - w_h), w_h^* \right) - (\Psi_h(\Pi_h w), w_h^*) + (\Psi_h(w_h), w_h^*).$$

Thus, we get the following expression of δj:

$$\delta j = (g, w - \Pi_h w) + (\Psi_h(\Pi_h w), w_h^*) - (\Psi_h(w), w_h^*) + ((\Psi_h - \Psi)(w), w_h^*) + R_1,$$

We now apply a second Taylor extension to get:

$$(\Psi_h(\Pi_h w), w_h^*) - (\Psi_h(w), w_h^*) = \left(\frac{\partial \Psi_h}{\partial w}(w)(\Pi_h w - w), w_h^*\right) + R_2,$$

with remainder term

$$R_2 = \left(\Psi_h(\Pi_h w), w_h^*\right) - \left(\Psi_h(w), w_h^*\right) - \left(\frac{\partial \Psi_h}{\partial w}(w)(\Pi_h w - w), w_h^*\right).$$

This implies:

$$\delta j = (g, w - \Pi_h w) + \left(\frac{\partial \Psi_h}{\partial w}(w)(\Pi_h w - w), w_h^*\right) + ((\Psi_h - \Psi)(w), w_h^*) + R_1 + R_2.$$

In contrast to an *a posteriori* analysis, this analysis starts with a discrete adjoint w_h^*. However, our purpose is to derive a continuous description of the main error term. Thus, we get rid of the discrete solutions in the dominating terms. To this end, we re-write δj as follows:

$$\delta j = (g, w - \Pi_h w) + \left(\frac{\partial \Psi}{\partial w}(w)(\Pi_h w - w), w^*\right)$$
$$+ ((\Psi_h - \Psi)(w), w^*) + R_1 + R_2 + D_1 + D_2 + D_3,$$

where,

$$D_1 = \left(\left(\frac{\partial \Psi_h}{\partial w} - \frac{\partial \Psi}{\partial w}\right)(w)(\Pi_h w - w), w_h^*\right),$$

$$D_2 = \left(\frac{\partial \Psi}{\partial w}(w)(\Pi_h w - w), w_h^* - w^*\right),$$

$$D_3 = ((\Psi_h - \Psi)(w), w_h^* - w^*).$$

The latter expression of δj can be even more simplified thanks to the continuous adjoint of equation (7.124), leading to:

$$\delta j = ((\Psi_h - \Psi)(w), w^*) + R_1 + R_2 + D_1 + D_2 + D_3. \qquad (7.126)$$

At least formally, the R_i and the D_k are higher order terms, and the first term in the right hand side of (7.126) is the dominating one. It remains to give the studied context and to exhibit from (7.126) a formulation specifying the optimal mesh.

7.6.3 The case of the steady Euler equations

In this section, we study how equation (7.126) can be applied in the context of the steady Euler equations. To this end, we restrict to the particular discretization of these equations given in this Chapter and we consider a variational analysis.

7.6.3.1 Variational analysis

Equation (7.4) will play the role of equation (7.123) of the abstract analysis of the previous section. For the discretization, we consider a discrete domain Ω_h and a discrete boundary Γ_h which are not necessarily identical to the continuous ones. The MUSCL scheme introduced in the first sections can be written as a variational Galerkin-type scheme complemented by a higher-order term which can be assimilated to a numerical viscosity term D_h. This gives: $\forall \phi_h \in V_h$,

$$\int_{\Omega_h} \phi_h \nabla.\mathcal{F}_h(W_h) \, \mathrm{d}\Omega_h - \int_{\Gamma_h} \phi_h \hat{\mathcal{F}}_h(W_h).\mathbf{n} \, \mathrm{d}\Gamma_h = -\int_{\Omega_h} \phi_h \, D_h(W_h) \mathrm{d}\Omega_h.$$

$$(7.127)$$

According to [Mer, 1998a], the diffusion term is of higher order as soon as it is applied to the interpolation of a smooth enough field W on a sufficiently regular mesh:

$$\left| \int_{\Omega_h} \phi_h \, D_h(W_h) \mathrm{d}\Omega_h \right| \leq h^3 K(W)|\phi_h|_{L^2} .$$

As a result, the dissipation term will be neglected in the same way we neglect the remainders R_i and D_k of Relation (7.126). In the case of a flow with shocks, we have chosen to follow the strategy of the Hessian-based study in [Loseille et al., 2007] which consists in avoiding to introduce the error term from artificial dissipation.

7.6.3.2 Approximation error estimation

Returning to the output functional $j(W) = (g, W)$ and according to Estimate (7.126), the main term of the *a priori* error estimation of δj becomes:

$$\delta j = (g, W - W_h) \approx ((\Psi_h - \Psi)(W), W^*) \quad \text{with} \quad \frac{\partial \Psi}{\partial W} W^* = g,$$

where W^* is the continuous adjoint state. Using the exact solution W in equations (7.4) and (7.127) while neglecting the dissipation D_h leads to:

$$(g, W - W_h) \approx \int_{\Omega_h} W^* \left(\nabla.\mathcal{F}_h(W) - \nabla.\mathcal{F}(W)\right) \mathrm{d}\Omega_h$$

$$- \int_{\Gamma_h} W^* \left(\hat{\mathcal{F}}_h(W) - \hat{\mathcal{F}}(W)\right).\mathbf{n} \, \mathrm{d}\Gamma_h.$$

By integrating by parts the previous estimate, it comes:

$$(g, W - W_h) \approx \int_{\Omega_h} \nabla W^* \left(\mathcal{F}(W) - \mathcal{F}_h(W) \right) d\Omega_h$$

$$- \int_{\Gamma_h} W^* \left(\bar{\mathcal{F}}(W) - \bar{\mathcal{F}}_h(W) \right).\mathbf{n} \, d\Gamma_h,$$

where fluxes $\bar{\mathcal{F}}$ are given by:

$$\bar{\mathcal{F}}(W).\mathbf{n} = \mathcal{F}(W).\mathbf{n} - \hat{\mathcal{F}}(W).\mathbf{n}.$$

By definition, $\mathcal{F}_h$ is the linear interpolate of $\mathcal{F}$, *i.e.*, $\Pi_h \mathcal{F} = \mathcal{F}_h$, thus we have:

$$\delta j \approx \int_{\Omega_h} \nabla W^* \left(\mathcal{F}(W) - \Pi_h \mathcal{F}(W) \right) d\Omega_h - \int_{\Gamma_h} W^* \left(\bar{\mathcal{F}}(W) - \Pi_h \bar{\mathcal{F}}(W)) \right).\mathbf{n} \, d\Gamma_h.$$

$$(7.128)$$

We observe that this estimate of δj is expressed in terms of interpolation errors for the fluxes and in terms of the continuous functions W and W^*.

7.6.3.3   Error bound with a safety principle

The integrands in (7.128) contain positive and negative parts which can compensate for some particular meshes. In our strategy, we prefer to avoid these parasitic effects. To this end, all integrands are bounded by their absolute values:

$$(g, W_h - W) \leq \int_{\Omega_h} |\nabla W^*| \, |\mathcal{F}(W) - \Pi_h \mathcal{F}(W)| \, d\Omega_h +$$

$$\int_{\Gamma_h} |W^*| \, |(\bar{\mathcal{F}}(W) - \Pi_h \bar{\mathcal{F}}(W)).\mathbf{n}| \, d\Gamma_h. \, (7.129)$$

In other words, we prefer to locally over-estimate the error.

7.6.4   Error model minimization

Starting from Bound (7.129), several options are possible to derive an optimal mesh for the observed functional. We propose to work in the continuous mesh framework by adopting a complete continuous view which is possible with *a priori* estimates. It allows us to define proper differentiable optimization [Absil et al., 2008], [Arsigny et al., 2006] or to use the calculus of variations that is undefined on the class of discrete meshes. This framework lies in the class of metric-based methods. Working in this framework enables us, as in the previous section, to write Estimate (7.129) in a continuous form:

$$(g, W_h - W) \approx E(\mathbf{M}) = \int_{\Omega} |\nabla W^*| \, |\mathcal{F}(W) - \pi_{\mathcal{M}} \mathcal{F}(W)| \, d\Omega +$$

$$\int_{\Gamma} |W^*| \, |(\bar{\mathcal{F}}(W) - \pi_{\mathcal{M}} \bar{\mathcal{F}}(W)).\mathbf{n}| \, d\Gamma, (7.130)$$

where $\mathbf{M} = (\mathcal{M}(\mathbf{x}))_{\mathbf{x} \in \Omega}$ is a continuous mesh defined by a Riemannian metric field and $\pi_{\mathcal{M}}$ is the continuous linear interpolate defined hereafter. We are now focusing on the following (continuous) mesh optimization problem:

$$\text{Find } \mathbf{M}_{opt} = \text{Argmin}_{\mathbf{M}} \, E(\mathbf{M}). \tag{7.131}$$

A constraint is added to the previous problem in order to bound mesh fineness. In this continuous framework, we impose the total number of nodes to be equal to a specified positive integer N.

The same reasoning as section 7.5 can be applied for the three following particular cases.

7.6.4.1 Weighted interpolation error

Let u be a twice continuously differentiable function and g be a strictly positive function. We consider the following optimization problem in the continuous framework:

$$\text{Find } \mathbf{M}_{wgt} = \text{Argmin}_{\mathbf{M}} E_{wgt}(\mathbf{M}) \quad \text{with} \quad E_{wgt}(\mathbf{M}) = \int_{\Omega} g \, |u - \pi_{\mathcal{M}} u| \, d\Omega,$$

under the equality constraint $\mathcal{C}(\mathbf{M}) = N$. The continuous interpolation error related to continuous mesh $\mathbf{M}$, can be expressed (up to a constant negligible thanks to the constraint $\mathcal{C}(\mathbf{M}) = N$) in terms of the Hessian H_u of function u, see equation (7.117). Following the reasoning of [Loseille et al., 2007] used for Hessian-based mesh adaptation, we get the expression of the optimal metric:

$$\mathcal{M}_{wgt}(g, u) = D(g, u) \det(g \, |H_u|)^{-\frac{1}{5}} g \, |H_u|$$

$$\text{where} \quad D(g, u) = N^{\frac{2}{3}} \left(\int_{\Omega} (\det(g \, |H_u|))^{\frac{2}{5}} \right)^{-\frac{2}{3}}.$$

7.6.4.2 Sum of interpolation errors

The previous variational calculus extends to a linear combination of interpolation errors. Let u, v, α, $\beta > 0$ be four twice continuously differentiable functions. We aim at finding the metric which optimizes the L^1 norm of the weighted sum of interpolation errors:

$$\text{Find } \mathbf{M}_{sum} = \text{Argmin}_{\mathbf{M}} \, E_{sum}(\mathbf{M})$$

with

$$E_{sum}(\mathbf{M}) = \int_{\Omega} \alpha \, |u - \pi_{\mathcal{M}} u| \, d\Omega + \int_{\Omega} \beta \, |v - \pi_{\mathcal{M}} v| \, d\Omega,$$

under the constraint $\mathcal{C}(\mathbf{M}) = N$. It can be shown that minimizing E_{sum} is equivalent to minimizing a single interpolation error of a function having as

Hessian the linear combination $\alpha\,|H_u| + \beta\,|H_v|$. The optimal metric is then

$$\mathcal{M}_{sum}(\alpha, u, \beta, v) = D(\alpha, u, \beta, v)\det(\alpha\,|H_u| + \beta\,|H_v|)^{-\frac{1}{5}}(\alpha\,|H_u| + \beta\,|H_v|),$$

$$\text{where} \quad D(\alpha, u, \beta, v) = N^{\frac{2}{3}}\left(\int_\Omega \det(\alpha\,|H_u| + \beta\,|H_v|)^{\frac{2}{5}}\right)^{-\frac{2}{3}}.$$

7.6.4.3   Mixing boundary and volume error contributions

We consider now an optimization problem involving at the same time volume and surface interpolation error terms. Let u and $\bar{u}$ be two functions that are defined on Ω and $\Gamma = \partial\Omega$, respectively. The function $\bar{u}$ is simply the trace of u on the boundary. We consider g and $\bar{g}$ two positive functions. The problem reads:

$$\text{Find } (\mathbf{M}_{vol}, \bar{\mathbf{M}}_{surf}) = \text{Argmin}_{\mathbf{M}, \bar{\mathbf{M}}}\, E_{surf}(\mathbf{M}, \bar{\mathbf{M}}),$$

with

$$E_{surf}(\mathbf{M}, \bar{\mathbf{M}}) = \int_\Omega g\,|u - \pi_\mathcal{M} u|\,\mathrm{d}\Omega + \int_\Gamma \bar{g}\,|\bar{u} - \pi_{\bar{\mathcal{M}}}\bar{u}|\,\mathrm{d}\Gamma,$$

under the constraint $\mathcal{C}(\mathbf{M}) + \mathcal{C}(\bar{\mathbf{M}}) = N$. The optimal solution is then sought as a couple of two 3D metric fields: $\mathbf{M}_{vol}$ defined in the whole domain Ω and $\mathbf{M}_{surf}$ defined only on the boundary Γ. A similar calculus as previously dealing with both terms separately is applied. The global optimal metric $\mathbf{M}_{opt}$ is then defined by:

$$\mathcal{M}_{opt}(\mathbf{x}) = \begin{cases} \mathcal{M}_{vol}(\mathbf{x}) & \text{for } \mathbf{x} \in \Omega \\ \mathcal{M}_{vol}(\mathbf{x}) \cap \mathcal{M}_{surf}(\mathbf{x}) & \text{for } \mathbf{x} \in \Gamma \end{cases}.$$

7.6.5   Adaptive strategy

The adaptive strategy for the proposed goal-oriented mesh adaptation criterion (7.130) is quite similar to any anisotropic metric-based mesh adaptation. As both the solution and the mesh are changing during the computation, a non-linear loop is set up in order to converge toward a fixed point for the couple mesh-solution. From an initial couple mesh-solution $(\mathcal{H}_0, \mathcal{S}_0)$, it is composed of the following sequences. At step i, the flow is first converged on the current mesh $\mathcal{H}_i$ to get the solution $\mathcal{S}_i$. Then, a metric tensor field $\mathcal{M}_i$ is deduced from $(\mathcal{H}_i, \mathcal{S}_i)$ thanks to an anisotropic error estimate. The latter is used by the adaptive mesh generator which generates a unit mesh with respect to $\mathcal{M}_i$. The previous solution is then linearly interpolated on the new mesh. This procedure is repeated until convergence of the couple mesh-solution. We refer to [Loseille et al., 2007] for more details.

We now investigate the differences when dealing with the adjoint-based anisotropic error estimate. The main modifications concern the flow solver and the remeshing stage. In this section, the following notations are used. $\mathcal{H}$ denotes the mesh of the domain Ω_h, $\partial\mathcal{H}$ the mesh of the boundary Γ_h of Ω_h, W_h is the state provided by the flow solver and $j(W_h)$ the observed functional defined on $\gamma \subset \Omega_h$.

7.6.5.1 Flow solver and adjoint state

As compared to Hessian-based mesh adaptation, the new step in the solver is the resolution of the linear system providing the adjoint state: $A_h^* W_h^* = g_h$, where g_h is the approximated jacobian of $j(W_h)$ with respect to the conservative variables and W_h^* is the adjoint state. A_h^* is the adjoint matrix of order one deduced by linearizing the numerical scheme. Thus, the main over-cost is memory. The linear system is solve with the iterative GMRES method coupled with an incomplete $BILU(0)$ preconditioner [Saad, 2003].

Once W_h^* is computed, its point-wise gradient is recovered by using a $\mathbf{L}^2$ projection from the neighboring element-wise constant gradients [Alauzet and Loseille, 2009]. We summarize the final couples of variables made available by the flow solver:

- the gradients of the adjoint state $\left(\dfrac{\partial W_h^*}{\partial x}, \dfrac{\partial W_h^*}{\partial y}, \dfrac{\partial W_h^*}{\partial z}\right)$ associated with fluxes $(\mathcal{F}_1(W_h), \mathcal{F}_2(W_h), \mathcal{F}_3(W_h))$.

- the adjoint state associated with the boundary fluxes $\bar{\mathcal{F}}$ on γ.

7.6.5.2 Optimal goal-oriented metric

The optimal metric found in section 7.6.4 is composed of a volume tensor field $\mathcal{M}_{go}$ defined in Ω_h and a surface one $\bar{\mathcal{M}}_{go}$ defined on Γ_h. We then compute:

- for each vertex $\mathbf{x}$ of $\mathcal{H}$, the Hessian matrix arising from the volume contribution of each component of the Euler fluxes:

$$H(\mathbf{x}) = \sum_{j=1}^{5} \left([\Delta x]_j + [\Delta y]_j + [\Delta z]_j\right),$$

where
$$[\Delta x]_j = \left|\frac{\partial W_h^*}{\partial x}\right| \left|H_R([\mathcal{F}_1(W_h)]_j)\right|, \quad [\Delta y]_j = \left|\frac{\partial W_h^*}{\partial y}\right| \left|H_R([\mathcal{F}_2(W_h)]_j)\right|,$$

$$[\Delta z]_j = \left|\frac{\partial W_h^*}{\partial z}\right| \left|H_R([\mathcal{F}_3(W_h)]_j)\right| \quad \text{with} \quad [\mathcal{F}_i(W_h)]_j \text{ denoting the } j^{th}$$
component of the vector $\mathcal{F}_i(W_h)$

- for each vertex $\mathbf{x}$ of $\partial\mathcal{H}$, the Hessian matrix arising from the surface contribution:

$$\bar{H}(\mathbf{x}) = \sum_{j=1}^{5} \left|W_h^*\right| \left|H_R\left(\sum_{i=1}^{3} \bar{\mathcal{F}}_i(W_h)n_i\right)\right|,$$

where $\mathbf{n} = (n_1, n_2, n_3)$ is the outward normal of Γ.

H_R stands for the operator that recovers numerically the second order derivatives of an initial piecewise linear by element solution field. The recovery method is based on the Green formula [Alauzet and Loseille, 2009]. The standard $\mathbf{L}^1$ norm normalization is then applied independently on each metric tensor field:

$$\mathcal{M}^{go}(\mathbf{x}) = C \ \det(|H(\mathbf{x})|)^{-\frac{1}{5}} |H(\mathbf{x})| \qquad (7.132)$$

$$\text{and} \quad \bar{\mathcal{M}}^{go}(\mathbf{x}) = \bar{C} \ \det(|\bar{H}(\mathbf{x})|)^{-\frac{1}{4}} |\bar{H}(\mathbf{x})|.$$

Constants C et $\bar{C}$ depends on the desired complexity N.

7.6.5.3   Mesh adaptation

Goal-oriented mesh adaptation requires to adapt the surface mesh of the surface γ on which the functional is observed. This standpoint is needed in order to ensure a valid coupling between the volume mesh and the surface mesh. This constraint implies numerous complications for the re-meshing phase. In our case, a global re-meshing is carried out after re-meshing the surface γ. We use Yams [Frey, 2001] for the adaptation of the surface and an anisotropic extension of Gamhic [George, 1999] for the volume mesh. When the surface is not adapted, we use Mmg3d [Dobrzynski and Frey, 2008].

7.6.6   Some examples

7.6.6.1   High-fidelity pressure prediction of an aircraft

We consider the flow around a supersonic business jet (SSBJ). The geometry provided by Dassault-Aviation is depicted in Figure 7.17 (top left). Flight conditions are Mach 1.6 with an angle of attack of 3 degrees. As for a body flying at a supersonic speed, each geometric singularity generates a shock wave having a cone shape; a multitude of conic shock waves are emitted by the aircraft geometry. They generally coalesce around the aircraft while propagating to the ground. The goal, here, is to compute accurately the pressure signature only on a plane located 100m below the aircraft. The observation plane has a length of 40m and a width of only 2m whereas the wing span is about 17m. The scope of this test case is to evaluate the ability of the adjoint to prescribe refinements only in areas that impact the observation region. The functional is given by:

$$j(W) = \frac{1}{2} \int_\gamma \left(\frac{p - p_\infty}{p_\infty} \right)^2 \mathrm{d}\gamma,$$

with $\gamma = \{(x, y, z) \in \mathbb{R}^3 \mid 100 \leq x \leq 140, -1 \leq y \leq 1, z = -100\}$. Observation area γ and its position with respect to the aircraft is shown in Figure 7.18 (top right).

In order to exemplify how adjoint-based mesh adaptation gives an optimal distribution of the degrees of freedom to evaluate the functional, this adaptation is compared to the Hessian-based mesh adaptation presented in section 7.5. The adaptation is done on the local Mach number and the interpolation error is controlled in $\mathbf{L}^2$ norm.

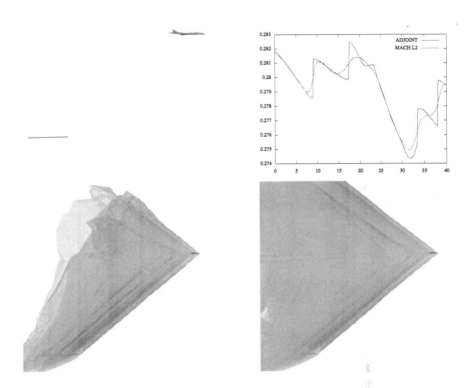

FIGURE 7.18: Top left: location of the observation plane below the aircraft. Top right: pressure signature along x axis in the observation plane. Bottom, local Mach number iso-surface for the adjoint-based (left) and Hessian-based (right) methods.

The adaptive loop is divided into 5 steps, each step is composed of 6 sub-iterations having a constant complexity for a total of 30 adaptations. The complexity sequence is $[10\,000, 20\,000, 40\,000, 80\,000, 160\,000]$. Hessian-based and adjoint-based final adapted meshes are composed of almost 800 000 vertices. They are represented in Figure 7.19 where several cuts in the final adapted meshes for the adjoint-based (top) and Hessian-based (bottom) adaptations are displayed.

For the Hessian-based adaptation, the mesh is adapted in the whole computational domain along the Mach cones and in the wake, see Figure 7.19 (right). Such an anisotropic adapted mesh provides an accurate solution everywhere in the domain, cf. Figure 7.18 (bottom right). But, if the aim is to only compute an accurate pressure signature on surface γ then we clearly notice that a large amount of degrees of freedom is wasted in the upper part of the domain and in the wake where accuracy is not needed.

As regards the adjoint-based adaptation, the mesh is mainly adapted below the aircraft in order to capture accurately all the shock waves that impacts the observation plane Figure 7.19 (left). On the contrary, areas that do not impact the functional are ignored with this new approach: the region over the aircraft, the wake of the SSBJ and in the region just behind the aircraft, only the lower half of the conic shock waves are refined and the angular amplitude of the refined part keeps on decreasing along with the distance to the aircraft.

Another point of main interest, which is more technical, is that the Hessian-based adaptation in $\mathbf{L}^2$ norm prescribes a mesh size that depends on the shock intensity. A stronger shock is then more refined than a weaker one. In this simulation, shocks directly below the aircraft have a lower intensity than lateral or upper shocks emitted by the wings, see Figure 7.19 (bottom). Consequently, the adapted meshes with the Hessian-based method are less accurate in regions that directly affect the observation plane. On the contrary, shocks below the aircraft are not uniformly refined with the adjoint-based strategy. The shock waves are all the more refined as they are influent on the functional independently of their amplitudes. This has a drastic consequence on the accuracy of the observed functional, see Figure 7.18 (top left).

7.6.6.2   Wing tip vortices capture

In this example, we study the accurate prediction of wing tip vortices at large distance in the wake for transonic flow conditions. We consider the Falcon business jet geometry provided by Dassault Aviation, see Figure 7.20. The jet is flying at transonic cruise speed with Mach number 0.8 and an angle of attack of 3 degrees. The computational domain is a cylinder of radius 250 m and of length 700 m. The Hessian-based adaptation on the local Mach number is compared to the adjoint-based adaptation on the vorticity functional:

$$j(W) = \frac{1}{2} \int_{\gamma} \|\nabla \times (\mathbf{u} - \mathbf{u}_{\infty})\|_2^2 \, d\gamma,$$

where γ is a plane located 400 m behind the aircraft orthogonal to the aircraft path, $\mathbf{u}$ the velocity field and $\mathbf{u}_{\infty}$ the velocity field at infinity.

As the aircraft is flying at a transonic speed, the flows is composed of both shocks and smooth vortices. These phenomena have different magnitudes and mathematical properties. Across a shock, all the variables become discontinuous whereas a vortex corresponds to a smooth variation of the variables while having a very small amplitude. These features are exemplified in Figure 7.20 (right). An extraction of the pressure across the

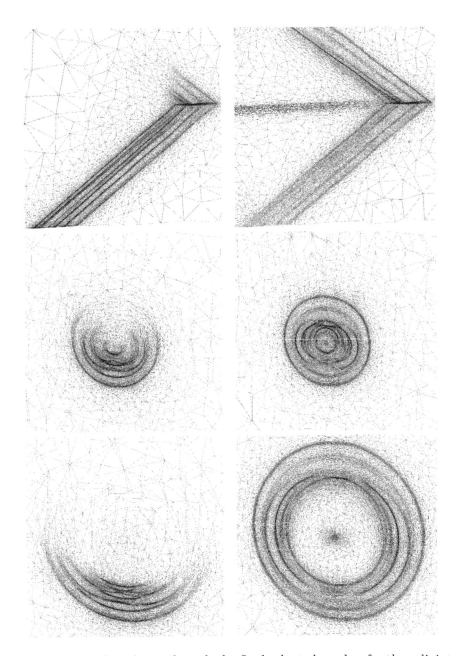

FIGURE 7.19: Cut planes through the final adapted meshes for the adjoint-based (left) and Hessian-based (right) methods. A cut in the symmetry plane (top) and, two cuts with an increasing distance behind the aircraft orthogonal to its path (middle and bottom).

wing extrados where a shock occurs (light gray curve) is superposed to the pressure variation in the wake across a vortex located 400 m behind the aircraft (dark gray curve). The amplitude of the vortex is less than 2% of the amplitude of the shock. Moreover, the smoothness property of the vortex is a supplementary difficulty as its derivatives involved in our estimate are also smooth. Consequently, vortices are difficult to detect and to not diffuse. Detecting and preserving these vortices are still a challenge in the field of CFD.

Final adapted meshes are composed of almost 1.5 million vertices for both methods. Vorticity iso-values are visualized in Figure 7.22 for planes $x = 100$ m, $x = 200$ m and $x = 400$ m, that are located behind the aircraft and orthogonal to the aircraft path. As regards Hessian-based adaptation, the vortex is accurately captures up to 100 m behind the aircraft and then it is diffused with the distance to the Falcon. On the contrary, with adjoint-based method, the vortex keeps a constant size and its core is not diffused when increasing the distance to the aircraft, see Figure 7.21. When looking at the cuts in $x = cte$ planes, the meshes are almost isotropic. Indeed, as the vortex iso-values are circular the ideal mesh is isotropic. In fact, the anisotropic gains are along the x-axis as illustrated in Figure 7.23. We also observe that the wing shock is merely refined in the adjoint-based mesh contrary to the Hessian-based mesh. In the observation plane 400 m behind the aircraft, the adjoint-mesh is strongly anisotropic whereas Hessian-based mesh has already lost the vortex and is poorly anisotropic, see Figure 7.23 (bottom).

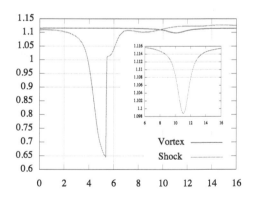

FIGURE 7.20: Left: surface mesh of the Falcon business jet geometry, right: pressure variation through a section across the wing (light gray curve) and across the wake 400 m behind the aircraft (dark gray curve). For the light gray curve, the jump represents the shock on the wing extrados. For the dark gray curve, the small Gaussian curve represents the vortex amplitude.

FIGURE 7.21: Pressure iso-surfaces along a 400 m path behind the Falcon obtained with the adjoint-based mesh adaptation.

7.6.6.3 Conclusions

As a conclusion, it is clear that in this context, the Hessian-based adaptation gives a non-optimal result with an inappropriate distribution of the degrees of freedom for the evaluation of the functional. It also demonstrates how the adjoint defines an optimal distribution of the degrees of freedom for the specific target. However, it is important to note that the mesh obtained with the Hessian-based strategy is optimal to evaluate globally the local Mach number.

It is worth mentioning that this method is completely automatic and gives an optimal result. It seems quite difficult to find a manual adaptation strategy to obtain an accurate evaluation of the functional while reducing the number of degrees of freedom. For instance, one may consider an approach that would consist in ignoring the upper part of the flow, however such a mesh would not be optimal as there exists at each distance a specific angle for refinement depending on the width of the observation plane γ. Moreover, considering the amplitude on the physical phenomenon to deduce a size prescription is not optimal as it gives weights that are not optimal to each physical phenomena involved in the flow.

A similar conclusion as previous example also applied here: the utilization of anisotropic meshes prevents the numerical diffusion of the physical phenomena while an appropriate weighting of the physical variables captures the small scale phenomena.

We have proposed a new method providing the anisotropic adapted mesh optimizing the first error term in the approximation of a functional depending of the solution of a flow problem. This method is based on a new formal *a priori* estimation of the functional approximation error and its resolution in an abstract continuous framework. It has been applied successfully to the compressible Euler equations. This new method exploits two advanced technologies and their good synergy:

- up-to-date anisotropic mesh generators that contribute to build optimal anisotropic adapted meshes

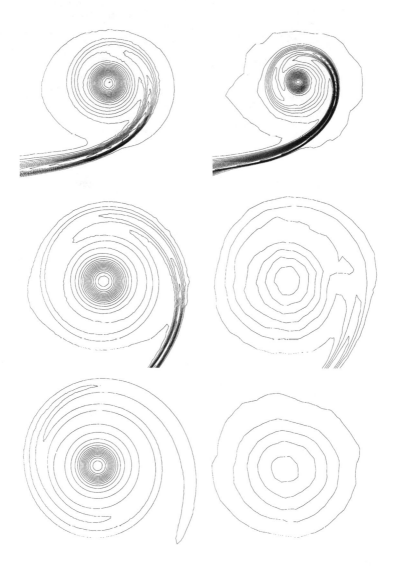

FIGURE 7.22: Comparison between the solutions (vorticity) obtained with the goal-oriented (left) and the Hessian-based (right) strategies in several planes behind the Falcon and orthogonal to the aircraft path. From top to bottom: planes 100 m, 200 m and 400 m behind the jet.

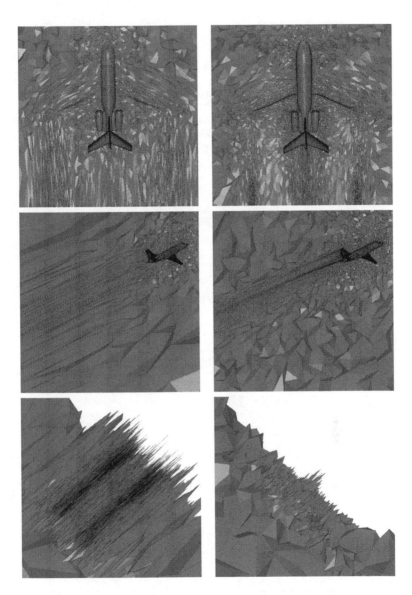

FIGURE 7.23: Comparison between the adapted meshes obtained with the goal-oriented (left) and the Hessian-based (right) strategies. Cuts through the volume along the planes $z = 0$ (top), $y = 0$ (middle) and $x = 400$ (bottom).

- the flow solver which is basically a P^1 Galerkin one relying on a continuous piecewise linear representation of the solution. It satisfies P^1-exactness property allowing a P^1-interpolation based error estimate to be used. Moreover, it is stabilized with a modern shock capturing method enabling the accurate computation of thin numerical shocks in strongly anisotropic adapted meshes.

The method has the following features:

- it produces an optimal anisotropic metric uniquely specified as the optimum of a functional and explicitly given by variational calculus from the continuous state and the adjoint state. The coupled system of the metric and of the two states is the object of the discretization. This should be put in contrast with the usual process of starting from a (discrete) mesh and then improving it

- to apply it, there is no need to choose in a more or less arbitrary way any local refinement "criterion" and no need to fix any parameter except the total number of vertices which represents the error threshold

- mesh convergence is performed in a natural way by increasing the total number of prescribed vertices at each stage of the mesh adaptation process

The new method has been applied to a sample of challenging 3D problems. Numerical experiments show that the new method enjoys at best level the advantages of Hessian-based anisotropic methods and of goal-oriented methods. As compared to the Hessian-based method, the anisotropic stretching of the meshes is not lost but even more strengthened and better distributed along shocks. As compared with goal-oriented methods, the new method behaves like a goal-oriented method, but also naturally takes the anisotropy related to functional into account.

Some issues have not been addressed in this work such as the asymptotic convergence order of the error on the functional. The authors are addressing some of them, together with the issues of extending the above method to viscous and unsteady flows.

7.7   Concluding remarks

In this chapter, we have presented several important questions that rise when a numerical model for inviscid compressible flows needs to be extended to a large set of applications. Each of the proposed methods relies on a mathematical analysis.

The question of **numerical dissipation** is the central point in the use of these *collocated* (in contrast to *staggered*) approximation. This dissipation takes different forms which depend on the problem to deal with: low Mach number flows, highly stretched meshes, applications requiring very low levels of dissipation. For the latter case, applications to Large Eddy Simulation (see the companion chapter in this book dealing with Large Eddy Simulation) and aeroacoustic simulations (see for example [Abalakin et al., 2004]) have been performed with good results.

Density positivity is the key strategy for robustness of this kind of numerical models. The study proposed in this chapter gives a complete theory for positivity relying on TVD-type limiters.

Then these approximation schemes are used for developing new anisotropic mesh optimization methods. The central role of P^1 approximation allows to apply two efficient strategies. The continuous mesh method combined with L^p interpolation error minimization offers high quality predictions, with higher order convergence properties for non-regular flows. The continuous mesh method adapts also to goal-oriented adaptation, offering an optimal anisotropy.

All the above methods extend or are being extended to Navier-Stokes models. In particular Navier-Stokes numerical models based on the numerics presented in this chapter are addressed in the companion chapter on Large Eddy Simulation.

Chapter 8

LES, variational multiscale LES, and hybrid models

Hilde Ouvrard
Université Montpellier 2, Département de Mathématiques
Place Eugène Bataillon
34095 Montpellier cedex, France

Maria-Vittoria Salvetti
Università di Pisa, Dipartimento di Ingegneria Aerospaziale
Via G. Caruso 8
56122 Pisa, Italy

Simone Camarri
Università di Pisa, Dipartimento di Ingegneria Aerospaziale
Via G. Caruso 8
56122 Pisa, Italy

Stephen Wornom
Société Lemma, Les Algorithmes
Bât. Thalès, 2000 Route des Lucioles
06410 Biot, France

Alain Dervieux
Institut National de Recherche en Informatique et Automatique
2004 Route des Lucioles
06902 Sophia-Antipolis, France

Bruno Koobus
Université Montpellier 2, Département de Mathématiques
Place Eugène Bataillon
34095 Montpellier cedex, France

8.1 Introduction

The numerical simulation of turbulent flows (for an introduction, see e.g. [Pope, 2000], [Tennekes and Lumley, 1972], [Mathieu and Scott, 2000], [Davidson, 2004]) is one of the great challenges in Computational Fluid Dynamics (CFD). It is commonly accepted that the physics of the flow of a continuous fluid is well represented by the Navier-Stokes equations. The direct numerical simulation (DNS) (for a review, see [Moin and Mahesh, 1998]) discretizes directly the three-dimensional Navier-Stokes equations. The basic requirement for such a simulation to succeed is the use of numerical schemes of high-order accuracy and meshes fine enough to capture the smallest scales of motion, to the order of the Kolmogorov scales. However, when the ratio of inertial forces to viscous ones, quantified by the Reynolds number, increases the smallest scales become smaller, and the amount of information (handled and processed) necessary for a Navier-Stokes based prediction becomes enormous. For homogeneous isotropic turbulence, for instance, the number of grid points needed is typically proportional to the 9/4 power of the Reynolds number. Then DNS applies well to rather academic problems characterized by simple geometry and low Reynolds numbers, and is a wonderful tool for the understanding of turbulence and validation of models. Nonetheless, the prediction of most engineering flows cannot be done by DNS with today's computers, and probably not before computing power have been increased during a couple of decades.

So far, the amount of information carried by DNS in a single computation is not only large but also not necessary in practice. Only statistics derived from it are used. In order to deal with the complex flows associated with higher Reynolds numbers and complex geometries, as those encountered in practical engineering applications, turbulence modeling was introduced. The principle is to try to drastically reduce the simulation unknowns by solving a mathematical model different from Navier-Stokes. The Reynolds-Averaged Navier-Stokes equations (**RANS**) approach, in which only the time or ensemble averaged flow is solved, is widely used for the simulation of complex turbulent flows in engineering applications. Many extensive reviews on these models are provided in the literature (see e.g. [Launder and Spalding, 1972], [Rodi, 1982], [Jones and Launder, 1972], [Hanjalic and Launder, 1976], [Newman and Launder, 1981], [Ahmadi, 1984], [Pope, 2000], [Wilcox, 2006], [Peyret and Krause, 2000]). Due to nonlinearities, the pure mathematical averaging of the Navier-Stokes system introduces new terms. The closure of the new system needs to be obtained from phenomenological information provided by the study of simplified flows. Ensemble averaging and the need of extra information for closure are indeed two limitations of RANS:

- When the mean flow is steady, ensemble averaging reduces to time averaging. Then an important part of the unsteady dynamics is not described by the RANS. The idea of phase averaging as in [Martinat et al., 2008] can improve this issue.

- Many closure models have been proposed in the literature. Each of them is known to give satisfactory results only for a particular class of problems and it is almost impossible to devise a model of general validity. However, the RANS models made it possible to predict high Reynolds number flows.

In the large eddy simulation (LES) approach, the reduction of the simulation unknowns is obtained through the application of a spatial filter to the Navier-Stokes equation. In most cases, the filter size is strictly related to the typical size of the computational grid (grid scale). Only the set of scales larger than the grid-scale, which we also call globally "grid-scale", are computed explicitly, while the small scales (subgrid-scale, **SGS**) are modeled. This concept lies on two presumptions:

- most global flow features are essentially governed by the largest scales,
- small-scale turbulence tends to local isotropy, and thus their modeling in a universal way can be found.

For solving complex unsteady flows as the flow around bluff-bodies, the LES approach gives generally more accurate predictions than the computationally cheaper RANS models, LES can also deliver an increased level of details. While RANS methods provide averaged results, LES is able to predict some instantaneous flow characteristics and to resolve important turbulent flow structures. Recent books and reviews on this technique can be found in [Sagaut, 2001], [Lesieur et al., 2005], [Mason, 1994], [Fureby, 2008], [Geurts, 2006].

In this chapter, we present a synthesis of our research activities carried out towards the accurate prediction of complex turbulent flows through the use of (i) LES, (ii) a variant of LES, *i.e.* the Variational Multiscale LES (**VMS-LES**) method, and (iii) a hybrid RANS/LES technique. We will consider the simulation of flows around bluff-body *i.e.* massively separated flows which contain very important physical features that are encountered in many engineering flows, including flow around offshore platforms, landing aircraft, cars, buildings, etc. In the perspective of the complex geometries of real life applications, we have used a mixed finite volume/finite element numerical method applicable to unstructured grids.

The success of a large eddy simulation depends on the combination and interaction of different factors, *viz.* the numerical discretization, which also provides filtering when no explicit one is applied, the grid refinement and quality and the closure model. On the other hand, all these aspects can be seen as possible sources of error in LES, especially in the simulation of complex flows. Usually, LES SGS modeling is based on the assumption of an universal behavior of the subgrid scales. Due to this assumption, most energy-containing eddies should not be filtered. Then large Reynolds numbers cannot be addressed with reasonable coarse meshes, except, in particular regions of large detached eddies. Even in the case of low Reynolds number or detached eddies, a particular attention must be paid to energetic eddies. Note however that for practical reasons LES simulations oriented to industrial or engineering applications are often characterized by coarse and very irregular grids. Classical eddy-viscosity SGS models, such as, for instance the Smagorinsky model [Smagorinsky, 1963], are purely dissipative, and thus they are unable to model backscatter of energy from small scales to large ones. They often apply, instead, a too large damping to the resolved energetic eddies. Indeed, although damping of the smallest resolved scale is needed in order to have a stable and accurate computation, this kind of models apply damping to the whole range of resolved scales and this may lead to a too large dissipation of energetic eddies. This situation is particularly critical in the case of irregular meshes, in which large elements may be close to very small ones. A cure for this can be the application of sophisticated adaptive SGS models as e.g. the dynamic eddy-viscosity model [Germano et al., 1991] or the dynamic mixed models [Meneveau and Katz, 2000]. Unfortunately, on unstructured grids, the increase in complexity and in computational costs introduced by the dynamic procedure is rather large.

Following the previous remarks, we have investigated the application of an alternative approach to large eddy simulation. In the particular case of spectral methods, the principle of Spectral Vanishing Viscosity consists in using the phase space for restricting the action of dissipation [Pasquetti, 2005], [Pasquetti, 2006b], [Pasquetti, 2006a]. In a more general standpoint, the Variational Multiscale (VMS) concept was proposed by Hughes et al. [Hughes et al., 2000]. Collis [Collis, 2001] proposed a new interpretation. The main idea of VMS-LES is to decompose the resolved scales in an LES into the largest and smallest ones and to add the SGS model only to the smallest ones. The VMS approach was originally introduced by Hughes [Hughes et al., 2000], [Hughes et al., 2001b] for the LES of incompressible flows and implemented in a Fourier spectral framework. In this original approach, the separation between largest and smallest scales was obtained through variational projection of the Navier-Stokes equations and no filtering is introduced, although a filter analog of the VMS approach may be also devised [Vreman, 2003]. Since the initial work of Hughes, several VMS-LES methods have been proposed, mainly in a spectral or finite element framework [Ramakrishnan and Collis, 2004], [Collis, 2002], [Gravemeier et al., 2004], [Gravemeier et al., 2005], [Gravemeier et al., 2006], [John and Kaya, 2005], [Munts et al., 2007]. These works have shown that the VMS-LES approach together with simple subgrid scale model, as the Smagorinsky one, can give results as accurate as traditional LES combined with dynamic formulation, but the former is less computationally expensive and does not require any *ad hoc* treatment (smoothing and clipping of the dynamic constant, as usually required with dynamic LES models) in order to avoid stability problems. In this chapter, we describe the VMS-LES implementation presented in [Koobus and Farhat, 2004] for the simulation of compressible turbulent flows on unstructured grids within a mixed finite volume/finite element framework.

Another major difficulty for the success of LES for the simulation of complex flows is the fact that the cost of LES increases as the flow Reynolds number increases. Indeed, the grid has to be fine enough to resolve a significant part of the turbulent scales, and this becomes particularly critical in the near-wall region.

Initiated by a few pioneering papers like [Speziale, 1998], a new class of models has recently been proposed in the literature which combines RANS and LES approaches. The purpose is to obtain simulations as accurate as in the LES case in some part of the flow but at reasonable computational costs (for recent reviews, see [Frohlich and von Terzi, 2008], [Labourasse and Sagaut, 2004], [Vengadesan and Nithiarasu, 2007]). These so-called hybrid methods can be divided in *zonal* approaches, in which RANS and LES are used in a-priori defined regions, and the so called *universal* models, which should be able to automatically switch from RANS to LES throughout the computational domain. In the perspective of the simulation of massively separated unsteady flows in complex geometry, as occur in many cases of engineering or industrial interest, we are primarily interested in universal hybrid models. Among the universal hybrid models described in the literature, the Detached Eddy Simulation (DES) has received the largest attention.

The DES approach [Spalart et al., 1997] is generally based on the Spalart-Allmaras RANS model modified in such a way that far from solid walls and with refined grids, the simulation switches to the LES mode with a one-equation SGS closure. Another hybrid approach has been recently proposed, the *Limited Numerical Scales* (LNS) [Batten et al., 2004], in which the blending parameter depends on the values of the eddy-viscosity given closure. In practice, the minimum of the two eddy-viscosities is used. This should ensure that, where the grid is fine enough to resolve a significant part of the turbulence scales, the model works in the LES mode, while elsewhere the RANS closure is recovered. An example of validation of LNS for the simulation of bluff-body flows is given in [Camarri et al., 2005].

A major difficulty in combining a RANS model with a LES one is due to the fact that RANS does not naturally allow for fluctuations, due to its tendency to damp them and to "perpetuate itself", as explained in [Spalart et al., 1997]. On the other hand, LES needs a significant level of fluctuations in order to accurately model the flow. The abrupt passage from a RANS region to a LES one may produce the so-called "modeled stress depletion" [Spalart et al., 1997]. Another limitation of DES-type approaches is that the two components need be of similar construction. We shall describe here a more general strategy for blending RANS and LES approaches in a hybrid model [Salvetti et al., 2007], [Pagano et al., 2006]. To this purpose, as in [Labourasse and Sagaut, 2002], the flow variables are decomposed in a RANS part (*i.e.* the averaged flow field), a correction part that takes into account the turbulent large-scale fluctuations, and a third part made of the unresolved or SGS fluctuations. The basic idea is to solve the RANS equations in the whole computational domain and to correct the obtained averaged flow field by adding, where the grid is adequately refined, the remaining resolved fluctuations. We search here for a hybridization strategy in which the RANS and LES models are blended in the computational domain following a given criterion. To this aim, a blending function is introduced, θ, which smoothly varies between 0 and 1. The correction term which is added to the averaged flow field is thus damped by a factor $(1 - \theta)$, obtaining a model which coincides with the RANS approach when $\theta = 1$ and recovers the LES approach in the limit of $\theta \to 0$. In particular, two different definitions of the blending function θ will be presented in this chapter. They are based on the ratios between (i) two eddy-viscosities and (ii) two characteristic length scales. The RANS model used in the proposed hybrid approach is a low-Reynolds version [Goldberg et al., 1998] of the standard $k - \varepsilon$ model [Launder and Spalding, 1979], while for the LES part the VMS approach is adopted [Hughes et al., 2000].

Finally, the last key ingredient of an LES (or VMS-LES or hybrid RANS-LES simulation) is the numerical discretization. As previously mentioned, we use a mixed finite-volume/finite-element discretization on unstructured grids, second-order accurate in space. The most critical point with this type of co-located schemes on unstructured grids is the need of numerical viscosity in order to obtain stable solutions. Due to the local topological irregularity, dissipative methods built on unstructured meshes generally produce more dissipation than those for structured ones. Further, for both cases, structured or unstructured meshes, the dissipation needed for stabilization increases with mesh irregularity, and in particular with mesh size variation. The introduction of a *numerical dissipation* and its possible interaction with the dissipation provided by the closure model is an important controversial point. Indeed, the most frequent practice in LES relies on the addition to the usual Navier-Stokes equations of a sub-grid scale (SGS) term, generally an eddy-viscosity term, and assumes that this same term is rather optimal for both turbulence modeling and numerical scheme stabilization. If dissipative numerical schemes are combined with a classical LES model, they can interact unfavorably with it, and significantly deteriorate the results as pointed out, for instance, by Garnier and co-workers [Garnier et al., 1999]. Conversely, in a different approach, a purely numerical stabilization term may fulfill the role of SGS. A typical example is the MILES method [Grinstein and Fureby, 2006] in which both subgrid modeling and numerical stabilization rely on monotonic dissipation by a second-order derivative (e.g. Flux-Corrected Transport (FCT) or Total-Variation Diminishing (TVD) schemes). However this family of model-free monotone methods seems to need a more refined grid than the classical approach for a given prediction quality. Our proposition was to dedicate the subgrid modeling to a physics-based model and to use for numerics a second-order accurate MUSCL upwind scheme equipped with a tunable dissipation made of *sixth-order* [Camarri et al., 2004] spatial derivatives of all flow variables. Fourier analysis clearly shows that such a dissipation has a damping effect which is much more localized on high frequencies than the one of stabilizations based on second-order derivatives. In this way we can reduce the interaction between numerical dissipation, which damps in priority the highest frequencies, and the SGS modeling. Moreover, a key coefficient (γ_s) permits to tune numerical dissipation to the smallest amount required to stabilize the simulation.

This chapter is organized as follows: in section 2, the numerical model, which is a key component for the success of a turbulent flow simulation, specially when LES type models are involved, is summarized. In section 3, the LES approach is presented and applied to the prediction of flows around a square cylinder at Reynolds number 22000 and around a forward-swept wing at high angle of attack. Effects of numerical viscosity, SGS modeling, and grid refinement are investigated. Section 4 reports on the description of the VMS-LES model and on its application for the prediction of a flow around a circular cylinder at Reynolds number 3900 and of vortex-induced motion of a complex spar geometry maintained by elastic moorings. Effects of SGS models, numerical viscosity, and grid resolution are discussed, and results obtained by the VMS approach are contrasted with those predicted by various LES models. In section 5, the previously described hybrid model is presented and its application to the flow around a circular cylinder at Reynolds number 140000 is shown. Effects of the blending function, SGS models and grid resolution are investigated. Finally, concluding remarks are presented in section 6.

8.2   Numerical model

8.2.1   Navier-Stokes equations

Let us consider the compressible Navier-Stokes equations given by

$$
\begin{cases}
\dfrac{\partial \rho}{\partial t} + \nabla \cdot (\rho \mathbf{u}) = 0 \\[2ex]
\dfrac{\partial \rho \mathbf{u}}{\partial t} + \nabla \cdot (\rho \mathbf{u} \otimes \mathbf{u} + P\mathbf{Id}) = \nabla \cdot \boldsymbol{\tau} \\[2ex]
\dfrac{\partial E}{\partial t} + \nabla \cdot [(E+P)\mathbf{u}] = \nabla \cdot (\tau u) + \nabla \cdot (\lambda \nabla T)
\end{cases}
\tag{8.1}
$$

where ρ is the density, $\mathbf{u}$ the velocity, P the pressure, τ the viscous stress tensor,

$$\tau_{ij} = 2\mu S_{ij}^* \; ; S_{ij}^* \equiv \frac{1}{2}\left(\frac{\partial u_i}{\partial x_j} + \frac{\partial u_j}{\partial x_i}\right) - \frac{1}{3}\frac{\partial u_k}{\partial x_k}\delta_{ij} \ ,$$

E the total energy, λ the thermal conductivity, and T the temperature of the fluid. This system is written in short:

$$W_t + \nabla \cdot \mathcal{F}(W) = \mathcal{R}(W), \tag{8.2}$$

with $W = (\rho, \rho u_1, \rho u_2, \rho u_3, E)$.

The spatial discretization is based on a mixed finite-volume/finite-element method applied to unstructured tetrahedrizations. The adopted scheme is upwind of MUSCL type as mentioned previously, and vertex centered, *i.e.* all degrees of freedom are located at the vertices. P1 Galerkin finite elements are used to discretize the diffusive terms $\mathcal{R}(W)$. In this section, we will briefly describe the numerical framework for the convective terms $\nabla.\mathcal{F}(W)$ and time scheme. More details on this numerical approach can be found in the companion chapter entitled *Numerical algorithms for unstructured meshes*.

8.2.2   Discretization of hyperbolic fluxes

A dual finite-volume grid is obtained by building a cell C_i around each vertex i. Different ways of constructing the finite-volume cells can be considered. The classical one consists in building cells by the rule of medians (*median cells*): the boundaries between cells are made of triangular interface facets. Each of these facets has a mid-edge, a facet centroid, and a tetrahedron centroid as vertices. The convective fluxes are discretized on this tessellation by a finite-volume approach, *i.e.* in terms of the fluxes through the common boundaries between each couple of neighboring cells:

$$\sum_{j \in V(i)} \int_{\partial C_{ij}} \mathcal{F}(W, \vec{n}) \, d\sigma \ , \tag{8.3}$$

where $V(i)$ is the set of neighboring nodes to vertex i, ∂C_{ij} is the boundary between cells C_i and C_j, $\vec{n}$ is the outer normal to the cell C_i and $\mathcal{F}(W, \vec{n})$ the Euler flux in the direction of $\vec{n}$. In all the schemes considered herein, the unknowns are discontinuous along the cell boundaries and this allows an approximate Riemann solver to be introduced. The Roe scheme [Roe, 1981] (with Turkel preconditioning) represents the basic upwind component for the numerical evaluation of the convective fluxes $\mathcal{F}$:

$$\int_{\partial C_{ij}} \mathcal{F}(W, \vec{n}) \, d\sigma \simeq \Phi^R(W_i, \, W_j, \, \vec{n}) = \frac{\mathcal{F}(W_i, \, \vec{n}) + \mathcal{F}(W_j, \, \vec{n})}{2} - \gamma_s d^R(W_i, \, W_j, \, \vec{n})$$

$$\tag{8.4}$$

$$d^R(W_i, \, W_j, \, \vec{n}) = P^{-1}|P\mathcal{R}(W_i, \, W_j, \, \vec{n})| \, \frac{W_j - W_i}{2} \tag{8.5}$$

in which W_i is the unknown vector at the i-th node and $\mathcal{R}$ is the Roe Matrix. The matrix $P(W_i, W_j)$ is the Turkel-type preconditioning term, introduced to avoid accuracy problems at low Mach numbers [Turkel, 1993]. Finally, the parameter γ_s multiplies the upwind part of the scheme and permits a direct control of the numerical viscosity, leading to a full upwind scheme (the usual Roe scheme) for $\gamma_s = 1$ and to a centered scheme when $\gamma_s = 0$.

The spatial accuracy of this scheme is only first order. The MUSCL linear reconstruction method ("Monotone Upwind Schemes for Conservation Laws"), introduced by Van Leer [Van Leer, 1977b], is employed to increase the order of accuracy of the Roe scheme. The basic idea is to express the Roe flux as a function of a reconstructed value of W at the boundary between the two cells centered respectively at nodes i and j: $\Phi^R(W_{ij}, W_{ji}, \vec{n}_{ij})$. W_{ij} and W_{ji} are extrapolated from the values of W at the nodes, as follows:

$$W_{ij} = W_i + \frac{1}{2}\left(\vec{\nabla}W\right)_{ij} \cdot \vec{ij} \tag{8.6}$$

$$W_{ji} = W_j - \frac{1}{2}\left(\vec{\nabla}W\right)_{ji} \cdot \vec{ij} \tag{8.7}$$

Schemes with different properties can be obtained by different numerical evaluation of the *slopes* $\left(\vec{\nabla}W\right)_{ij} \cdot \vec{ij}$ and $\left(\vec{\nabla}W\right)_{ji} \cdot \vec{ij}$. All the considered reconstructions can be written in the following general form:

$$\begin{aligned}(\vec{\nabla}W)_{ij} \cdot \vec{ij} = {} &(1-\beta)(\vec{\nabla}W)^C_{ij} \cdot \vec{ij} + \beta(\vec{\nabla}W)^U_{ij} \cdot \vec{ij} \\ &+\xi_c\left[(\vec{\nabla}W)^U_{ij} \cdot \vec{ij} - 2(\vec{\nabla}W)^C_{ij} \cdot \vec{ij} + (\vec{\nabla}W)^D_{ij} \cdot \vec{ij}\right] \\ &+\xi_d\left[(\vec{\nabla}W)_M \cdot \vec{ij} - 2(\vec{\nabla}W)_i \cdot \vec{ij} + (\vec{\nabla}W)_j \cdot \vec{ij}\right]\end{aligned} \tag{8.8}$$

With reference to Figure 8.1, $(\vec{\nabla}W)^U_{ij}$ is the gradient on the upwind tetrahedron T_{ij}, $(\vec{\nabla}W)^D_{ij}$ is the gradient on the downwind tetrahedron T_{ji}, $(\vec{\nabla}W)_i$ is the nodal gradient computed over the finite-volume cell around node i, $(\vec{\nabla}W)_j$ is the nodal gradient computed over the finite-volume cell around node j, $(\vec{\nabla}W)^C_{ij}$ is the centered gradient $((\vec{\nabla}W)^C_{ij} \cdot \vec{ij} = W_j - W_i)$ and $(\vec{\nabla}W)_M$ is the gradient at the point M. This last gradient is computed by interpolation of the nodal gradient values at the nodes contained in the face opposite to i in the upwind tetrahedron T_{ij}. The reconstruction of W_{ji} is analogous. In choosing a particular set of free coefficients (β, ξ_c, ξ_d) in equation (8.8) attention has been dedicated to the dissipative properties of the resulting scheme which is a key point for its successful use in LES simulations. Two schemes can be applied: the first one is characterized by $\beta = 1/3$, $\xi_c = \xi_d = 0$. It has a dissipative leading error proportional to the fourth-order derivatives of unknowns. We therefore denote it the **V4** scheme. It has been studied in details in [Camarri et al., 2002a] The latter one is obtained by putting $\beta = \frac{1}{3}$, $\xi_c = -\frac{1}{30}$ and $\xi_d - -\frac{2}{15}$. It has a dissipative leading error proportional to the

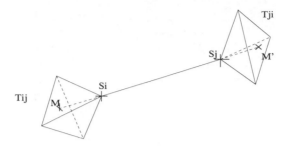

FIGURE 8.1: Sketch of points and elements involved in the computation of gradients.

fourth-order derivatives of unknowns. We therefore denote it the **V6** scheme. It has been introduced in [Camarri et al., 2004].

The numerical dissipation in the schemes V4 and V6 is made of fourth- and sixth-order space derivatives, respectively, and, thus, it is concentrated on a narrow-band of the highest resolved frequencies. This is important in LES type simulations in order to limit as far as possible the interactions between numerical and SGS dissipation, which could deteriorate the accuracy of the results.

8.2.3   Time advancing

Simulations can be advanced in time both with explicit or implicit schemes. In the first case, low-storage Runge-Kutta schemes are used. As for the implicit time advancing, a second-order time-accurate backward difference scheme is adopted. For a constant time step it writes:

$$\frac{3W^{n+1} - 4W^n + W^{n-1}}{2\Delta t} + \psi^{n+1} = 0 \qquad (8.9)$$

in which ψ^{n+1} denotes the discretized convective fluxes, diffusive and SGS terms evaluated at time step $n + 1$.

To avoid the solution of a non-linear system at each time step, the scheme (8.9) can be linearized by using the Jacobian of ψ^{n+1} with respect to the unknown variables. However, the evaluation of the Jacobian of ψ^{n+1} for the second-order accurate spatial discretization previously described and the solution of the resulting linear system implies significant computational costs and memory requirements. Thus, a defect-correction technique [Martin and Guillard, 1996] is used here, which consists in iteratively solving simpler problems obtained through an approximate linearization of (8.9). Thus, the following approximation is introduced:

$$\psi^{n+1} \simeq \psi^n + J_1^n(W^{n+1} - W^n)$$

in which J_1^n is the exact Jacobian of the spatial discretization terms when the convective fluxes are evaluated at first order. Then, the defect-correction iterations write as:

$$
\begin{cases}
\mathcal{W}^0 = W^n \\[2mm]
\left(\frac{3}{2\Delta t}Vol + J_1^s\right)\left(\mathcal{W}^{s+1} - \mathcal{W}^s\right) = -\dfrac{3\mathcal{W}^s - 4W^n + W^{n-1}}{2\Delta t} - \psi^s \\[2mm]
\qquad \text{for } s = 0, \cdots, M-1 \\[2mm]
W^{n+1} = \mathcal{W}^M
\end{cases}
$$

where Vol is the diagonal matrix containing the cell volumes and M is typically equal to 2. Indeed, it can be shown [Martin and Guillard, 1996] that only 2 defect-correction iterations are needed to reach a second-order accuracy. Information concerning the parallel implementation of the global algorithm can be found in [Koobus et al., 2007].

8.3 Large eddy simulation (LES)

8.3.1 Smagorinsky and dynamic models

The LES approach consists in filtering in space the Navier-Stokes equations in order to get rid of the high frequency fluctuations, and in simulating directly only the filtered flow. Due to the non-linearity of the problem, the filtered equations contain some unknown terms which represent the effect of the eliminated fluctuations on the filtered flow.

The filtered Navier-Stokes equations for compressible flows and in conservative form are considered in this subsection. In our simulations, filtering is implicit, *i.e.* the numerical discretization of the equations is considered as a filter operator (grid filter). In the first series of LES simulations shown in this section, two SGS models are used: the Smagorinsky model for compressible flows [Lesieur and Comte, 1997] and its dynamic version [Germano et al., 1991].

8.3.1.1 Smagorinsky model

The extension of the Smagorinsky model [Smagorinsky, 1963] to compressible flows [Lesieur and Comte, 1997] adopted here is intended to be used to study flows at high Reynolds numbers and such that low compressibility effects are present in the SGS fluctuations. In addition, we assume that heat transfer and temperature gradients are moderate. Thus, the retained SGS term in the momentum equation is the classical SGS stress tensor:

$$M_{ij} = \overline{\rho u_i u_j} - \overline{\rho} \tilde{u}_i \tilde{u}_j \ , \tag{8.10}$$

where the over-line denotes the grid filter and the tilde the density-weighted Favre filter ($\tilde{f} = \overline{(\rho f)} / (\overline{\rho})$). The isotropic part of M_{ij} can be neglected under the assumption of low compressibility effects in the SGS fluctuations [Erlebacher et al., 1992]. The deviatoric part, T_{ij}, is expressed by an eddy-viscosity term:

$$T_{ij} = -2\mu_{sgs} \left(\widetilde{S}_{ij} - \frac{1}{3} \widetilde{S}_{kk} \right) \ , \tag{8.11}$$

$$\mu_{sgs} = \overline{\rho} \, (C_s \Delta)^2 \left| \widetilde{S} \right| \ , \tag{8.12}$$

$\widetilde{S}_{ij}$ being the resolved strain tensor, μ_{sgs} the SGS viscosity, Δ the filter width, C_s a constant that must be *a priori* assigned and $\left| \widetilde{S} \right| = \sqrt{2 \widetilde{S}_{ij} \widetilde{S}_{ij}}$. To complete the definition of the SGS viscosity, the grid filter width must be specified. Note that no reliable criterion exist to define the width of the filter corresponding to the numerical discretization on unstructured grids. Nevertheless, the following expression has been employed here for each grid element l:

$$\Delta^{(l)} = \max_{i=1,..,6} \left(\Delta_i^{(l)} \right) \tag{8.13}$$

in which $\Delta_i^{(l)}$ is the length of the i-th side of the l-th element.

In the total energy equation, the effect of the SGS fluctuations has been modeled by the introduction of a constant SGS Prandtl number to be *a priori* assigned:

$$Pr_{sgs} = C_p \frac{\mu_{sgs}}{K_{sgs}} \tag{8.14}$$

where K_{sgs} is the SGS conductivity coefficient; it takes into account the diffusion of total energy caused by the SGS fluctuations and is added to the molecular conductivity coefficient.

8.3.1.2 Dynamic model

The dynamic version of the Smagorinsky model is now described. We follow the dynamic procedure proposed in [Germano et al., 1991] and extended to compressible flows in [Moin et al., 1991]. Another adaptation of the Germano dynamic model can be found in [Le Ribault et al., 2006]. Our adaptation of the dynamic procedure is now briefly described. A test filter (denoted by a hat) of larger width than the grid one is applied to the governing equations. Thus, a sub-test stress tensor appears in the momentum equation, which is modeled as the SGS stress tensor:

$$\mathcal{M}_{ij} = \widehat{\overline{\rho u_i u_j}} - \left(\frac{\widehat{\overline{\rho u_i}} \, \widehat{\overline{\rho u_i}}}{\widehat{\overline{\rho}}} \right) = - C \widehat{\Delta} 2 \widehat{\overline{\rho}} |\widehat{\widetilde{S}}| \widehat{\widetilde{P}}_{ij} \tag{8.15}$$

where $\widehat{\Delta}$ is the test filter width.

The test filter used here consists in evaluating the value of a flow variable on a given node by averaging on all the elements having this node as a vertex with a linear weighting function, that is the relative base function used in the P1 finite-element method.

It can be shown [Germano et al., 1991] that the SGS and the sub-test stress tensors are related by the following identity:

$$\mathcal{L}_{ij} = \widehat{\overline{\rho}\tilde{u}_i\tilde{u}_j} - \frac{1}{\widehat{\overline{\rho}}}(\widehat{\overline{\rho}\tilde{u}_i}\widehat{\overline{\rho}\tilde{u}_j}) = \mathcal{M}_{ij} - \widehat{M_{ij}^{(1)}} \qquad (8.16)$$

Then, by injecting equations (8.11) and (8.15) in the identity (8.16), the following tensorial equation is obtained:

$$L_{ij} = (C\Delta^2)\, B_{ij} \qquad (8.17)$$

in which $L_{ij} = \mathcal{L}_{ij} - \frac{1}{3}\mathcal{L}_{hh}\delta_{ij}$ and $B_{ij} = \widehat{\overline{\rho}|\tilde{S}|\tilde{P}_{ij}} - \left(\dfrac{\widehat{\Delta}}{\Delta}\right) 2\widehat{\overline{\rho}}|\widehat{\tilde{S}}|\widehat{\tilde{P}}_{ij}$. The only

unknown in equation (8.17) is $C\Delta 2$ and it can be determined by a least-square method. This gives:

$$(C\Delta^2) = \frac{B_{ij}L_{ij}}{B_{ij}B_{ij}} \qquad (8.18)$$

Note that we chose to compute $C\Delta^2$ instead of C to avoid the indetermination in the definition of the filter width.

The parameter Pr_t is computed by an analogous procedure, which is omitted here for sake of brevity. This results in:

$$Pr_t = \frac{Q_j Z_j}{Q_j Q_j} \qquad (8.19)$$

where

$$Q_i = \left[\left(\widehat{\overline{\rho}\tilde{e}} + \widehat{\overline{p}}\right)\frac{\widehat{\overline{\rho}\tilde{u}_i}}{\widehat{\overline{\rho}}}\right] - [(\widehat{\overline{\rho}\tilde{e} + \overline{p}})\,\tilde{u}_i] \text{ and } Z_i = C_p\left[\left(\frac{\widehat{\Delta}}{\Delta}\right) 2\widehat{\overline{\rho}}|\widehat{\tilde{S}}|\frac{\partial\widehat{\tilde{T}}}{\partial x_j} - \left(\widehat{\overline{\rho}|\tilde{S}|\frac{\partial\tilde{T}}{\partial x_j}}\right)\right].$$

The dynamic procedure proposed in the present section is usually unstable due to the oscillating behavior of $C\Delta^2$ with negative peaks and a large auto-correlation time. In order to avoid this problem, a local smoothing is applied by averaging over neighboring grid cells. A clipping procedure is also applied, setting $C\Delta^2$ to zero when the sum of the SGS and the molecular viscosity is negative.

8.3.2 Comparison of Smagorinsky and dynamic LES models

This application part has two main objectives: first to show the relative performance of the classical Smagorinsky model and of its dynamic version for the simulation of bluff-body flows on unstructured grids, and second to show the improvement brought by the upwind scheme providing a numerical dissipation based on sixth-order space derivatives, described in section 8.2.

8.3.2.1   Flow around a square cylinder (LES)

The flow around a square cylinder at $Re = 2.2 \times 10^4$ has been simulated. This flow was investigated experimentally [Bearman and Obasaju, 1982], [Luo et al., 1994], [Lyn et al., 1995], [Lyn and Rodi, 1994], [Norberg, 1993]. LES results are also available in the literature [Fureby et al., 2000], [Rodi et al., 1997], [Sohankar et al., 2000]. The simulations presented here are performed at a Mach number $M = 0.1$, in order to have negligible compressibility effects (experiments were performed for incompressible flow). The computational domain is sketched in Figure 8.2. Boundary conditions

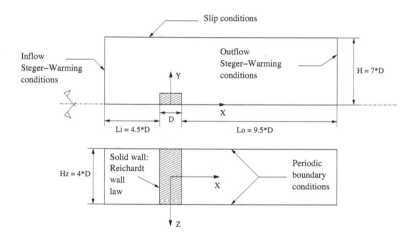

FIGURE 8.2: Flow past a square cylinder at Rey=22×10^3. Computational domain.

based on Steger-Warming decomposition [Steger and Warming, 1981a] are used at the inflow and at the outflow surfaces. On the side surfaces, free-slip is imposed and the flow is assumed to be periodic in the spanwise direction. No-slip boundary conditions at solid walls may lead in LES to inaccurate results, if very fine meshes are not employed. The use of such fine meshes is not in accordance with the standpoint adopted here, *i.e.* to reduce as much as possible the complexity of the simulation. We use instead the Reichardt wall-law described in the sequel. For details see [Camarri and Salvetti, 1999] and [Camarri et al., 2002a].

A first set of results has been obtained on a relatively coarse grid (GR1) with about 10^5 nodes and 6×10^5 elements (see also [Camarri et al., 2002a]); a more refined grid (GR2), having approximately 2×10^5 nodes and 1.1×10^6 elements has also been used. The average distance of the first layer of nodes from the cylinder surface is around $6 \times 10^{-2}D$ for GR1 and $4.5 \times 10^{-2}D$

for GR2. For both grids, approximately 32 nodes are used in the spanwise direction within the wake region, which corresponds to a spanwise resolution $\Delta z \simeq 0.125D$.

Simulations were performed on GR1 and GR2 using both the previously described SGS models, different values of the upwinding parameter, γ_s, and the schemes V4 and V6 for the discretization of convective fluxes. The simulation parameters are summarized in Table 8.1. In all cases, time advancing is carried out by a 4-stage explicit Runge-Kutta algorithm at CFL=1.

Simulation	grid	SGS model	γ_s	Convective fluxes
CSV4G05GR1	GR1	Smag. ($C_s = 0.1$)	0.05	V4
CSV6G05GR1	GR1	Smag. ($C_s = 0.1$)	0.05	V6
CSV6G05GR2	GR2	Smag. ($C_s = 0.1$)	0.05	V6
CDV4G05GR1	GR1	Dynamic	0.05	V4
CDV6G05GR1	GR1	Dynamic	0.05	V6
CDV4G10GR1	GR1	Dynamic	0.1	V4
CDV6G10GR1	GR1	Dynamic	0.1	V6
CDV6G05GR2	GR2	Dynamic	0.05	V6

Table 8.1: Flow past a square cylinder at Rey=22 × 10³. Summary of the simulations for the square cylinder test-case.

Results from simulations using the schemes V4 and V6 are evaluated in terms of accuracy of predictions in comparison with the experiments, sensitivity to numerical viscosity (γ_s) and effects of SGS modeling.

The main bulk coefficients obtained in the simulations are presented in Table 8.2, together with results from other LES simulations and experimental data.

Comparison with the experiments For a comparison in terms of accuracy between V4 and V6, it is possible to consider simulations that differ only for the scheme used for the convective fluxes. This is the case for (CSV4G05GR1, CSV6G05GR1), when the Smagorinsky model is used, and for (CDV4G05GR1, CDV6G05GR1) and (CDV4G10GR1, CDV6G10GR1), when the dynamic SGS model is used. Table 8.2 shows that the V6 scheme systematically gives more accurate results than V4 for all the bulk coefficients with the exception of C_d', whose value does not show significant variations in the different simulations considered. Moreover, the advantage of V6 over V4 increases as the upwinding parameter (γ_s) of the scheme is increased. This tendency is also confirmed by the comparison of time averaged flow fields (not shown here for sake of brevity).

It is also interesting to compare the accuracy obtained by V4 and V6 in the description of local flow fluctuations in time. This has been done by con-

Computational Fluid Dynamics

LES	C_l'	$\overline{C_d}$	C_d'	l_r
CSV4G05GR1	0.79	1.84	0.10	1.45
CSV6G05GR1	0.84	1.89	0.09	1.41
CSV6G05GR2	1.10	2.2	0.18	1.15
CDV4G05GR1	0.91	2.03	0.12	1.24
CDV6G05GR1	0.94	2.06	0.10	1.33
CDV4G10GR1	0.84	1.94	0.09	1.53
CDV6G10GR1	0.86	2.02	0.09	1.47
CDV6G05GR2	1.09	2.10	0.15	1.15
[Rodi et al., 1997]	[0.38,1.79]	[1.66,2.77]	[0.10,0.27]	[0.89,2.96]
[Sohankar et al., 2000]	[1.23,1.54]	[2.0,2.32]	[0.16,0.20]	[1.29-1.34]
Experimental data	C_l'	$\overline{C_d}$	C_d'	l_r
[Lyn and Rodi, 1994]	-	2.1	-	1.4
[Bearman and Obasaju, 1982]	1.2	2.28	-	-
[Norberg, 1993]	-	2.16	-	-
[Luo et al., 1994]	1.21	2.21	0.18	-

Table 8.2: Flow past a square cylinder at Rey=22×10^3. Bulk coefficients; comparison with experimental data and with other simulations in the literature. $\overline{C_d}$ is the mean drag coefficient, C_d' and C_l' are the r.m.s. of the drag and lift coefficients and l_r is the length of the mean recirculation bubble.

sidering the Reynolds stresses obtained in the simulations, reported together with the experimental data in Figure 8.3 for a vertical section in the wake.

The Reynolds stresses are averaged in the spanwise direction ($x_j = x_3$) and, consequently, only three components of the full Reynolds tensor are meaningful: $\langle u_1' u_1' \rangle$, $\langle u_2' u_2' \rangle$ and $\langle u_1' u_2' \rangle$. The symbol $\langle \ \rangle$ stands for average in time and in the homogeneous spanwise direction and the prime for the fluctuating part of the velocity components, according to the Reynolds decomposition: $u_j' = u_j - \langle u_j \rangle$. Note that experimental results should be filtered in space before being compared with numerical results, and the filter should be as similar as possible to the one used in the LES simulations. However, in our case it was not possible to filter the experimental data, available only in a few points. This explains the underestimate of the diagonal Reynolds stresses observed in all the simulations. Nevertheless, it is evident that V6 gives systematically more accurate results than V4, confirming that the more intense flow fluctuations obtained with V6 are not the effect of an incipient numerical instability but they are physically meaningful. In particular, V4 always gives definitely lower values of $\langle u'u' \rangle$ and $\langle v'v' \rangle$, indicating that velocity fluctuations are more damped than in V6 and the combination of V4 with the Smagorinsky model, which is more dissipative than the dynamic one, results in a strong damping of the velocity fluctuations.

Sensitivity to the numerical viscosity parameter γ_s Recall that γ_s multiplies the upwind part of the scheme leading to a fully upwind scheme (Roe scheme) for $\gamma_s = 1$ and to a centered scheme when $\gamma_s = 0$. As far as the sensitivity to γ_s is concerned, let us compare simulations CDV4G05GR1 with

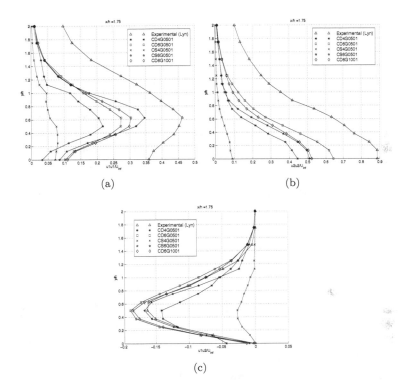

(a)

(b)

(c)

FIGURE 8.3: Flow past a square cylinder at Rey=22×10^3. Reynolds stresses, averaged in spanwise direction, on a vertical section at x/D = 1.75 (with reference to Figure 8.2).

CDV4G10GR1 for the V4 scheme and CDV6G05GR1 with CDV6G10GR1 for V6. The qualitative variation of the bulk coefficients (Table 8.2) with γ_s is the same for both V4 and V6, *i.e.* the agreement with the experiments improves as γ_s is decreased, as already pointed out in previous studies [Camarri and Salvetti, 1999], [Camarri et al., 2002a]. However, all the bulk coefficients obtained by V6 (except for C_l') are remarkably less sensitive to γ_s. The sensitivity of the Reynolds stresses to γ_s is rather low with V6, while they are more sensitive to the SGS model used, as shown in Figure 8.3.

Effect of SGS modeling and interaction with numerical viscosity model Let us compare now V4 and V6 in terms of sensitivity of the results to the SGS model. For this purpose, the simulation couples (CSV4G05GR1, CDV4G05GR1) and (CSV6G05GR1, CDV6G05GR1) are considered. Table 8.2 shows that the qualitative variations of the bulk coefficients with the SGS model are the same for V4 and V6, *i.e.* the dynamic model generally improves predictions with respect to the Smagorinsky model. However, quantitatively, the sensitivity to SGS modeling of the results obtained with V6 is lower than with V4. This is particularly evident from the comparison of the Reynolds stresses obtained in the simulations, as reported, for instance in Figure 8.3.

To better understand the reasons of this behavior, let us analyze more in details how numerical and SGS viscosities depend on the employed scheme. We expect that V6 introduces a numerical dissipation which is more localized on the highest resolved frequencies than V4. An *a posteriori* support to this speculation is indeed obtained in the present simulations, by analyzing, for instance, the time velocity signals recorded in the cylinder wake. Independently of the SGS model, in the simulations with V6 a larger energy content than with V4 is found in all the resolved frequencies, and especially in the highest ones, as shown for instance by the Fourier spectra in Figure 8.4.

The larger value of $\langle u_1' u_1' \rangle$ and $\langle u_2' u_2' \rangle$ obtained with V6 in all the simulations (as shown, for instance, in Figure 8.3) also confirms that in V6 the velocity fluctuations have a larger energy content than in V4. The mean convection velocity in the considered points within the wake is large enough to justify the Taylor hypothesis of frozen turbulence which allows us to assume that high time frequencies correspond to small scale in space; thus, one might conclude that small scales are less damped by the V6 scheme.

As far as SGS viscosity is concerned, for the Smagorinsky model it is practically insensitive to the scheme used for the convective fluxes. Conversely, for the dynamic model it is significantly higher when V6 is used, as shown, for instance, in Figure 8.5 for simulations CDV4G10GR1 and CDV6G10GR1.

This is again related to the lower damping of small scales introduced by V6. Indeed, in the dynamic procedure the parameter entering in the definition of the SGS viscosity is obtained as *the difference* between the LES field

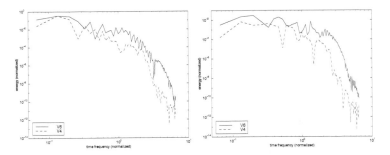

FIGURE 8.4: Flow past a square cylinder at Rey=22×10^3. Simulations CDV4G050GR1 and CDV6G050GR1. Fourier energy spectra of the velocity components recorded at $x = 3$, $y = 0.5$ and $z = 0$. (a) Transversal velocity v; (b) spanwise velocity w.

and a field twice filtered, with the second filter having a larger width than the grid one; thus, since with V6 the smallest resolved scales contain more energy than with V4, the dynamic procedure will compensate this and give a higher SGS viscosity level. Since, with the value of C and the definition of Δ used in the present study, the Smagorinsky model always gives in average a higher SGS viscosity than the dynamic one (see also [Camarri et al., 2002a]), in simulations with V6 the differences in μ_{sgs} given by the two models are reduced and this may explain why the results are more similar.

Effect of grid refinement   In spite of the global gain in accuracy obtained with the V6 scheme, in all simulations carried out on GR1, we observe a systematic underestimate of the drag coefficient even if the base pressure is predicted rather well, as in CDV4G05GR1, CDV6G05GR1 and CDV6G10GR1 (see Table 8.2). This is due to the underestimate of the pressure coefficient on the upwind face of the cylinder (see [Camarri et al., 2002a]), which is in turn plausibly caused by the inadequacy of GR1. Indeed, as shown in Figure 8.6(a), large elements are located too close to face AB, where gradients are high, and the grid refinement is too sharp to effectively increase resolution.

A better designed grid (GR2), shown in Figure 8.6(b), has been used in order to verify the previous conjecture; GR2 has approximately 2×10^5 nodes and 1.1×10^6 elements. Two simulations have been carried out on GR2, using the Smagorinsky (CSV6G05GR2) and the dynamic models (CDV6G05GR2) (see Table 8.1). The base pressure is well predicted and the mean drag coefficient is very close to the experimental value in both cases, as shown in Table 8.2; indeed, in those simulations the pressure coefficient on the upwind face of the cylinder matches very well the experiments (not shown here for sake of brevity). The same applies for the rms of the lift and drag coefficients, that are in both cases remarkably larger and closer to the experiments than the ones predicted by simulations on GR1.

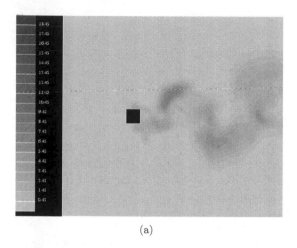

(a)

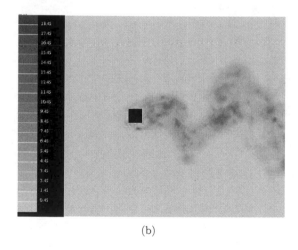

(b)

FIGURE 8.5: Flow past a square cylinder at Rey=22×10^3. Iso-contours of the SGS viscosity at a time corresponding to a peak in the C_L time history. (a) Simulation CDV4G10GR1 (V4 scheme); (b) simulation CDV6G10GR1 (V6 scheme).

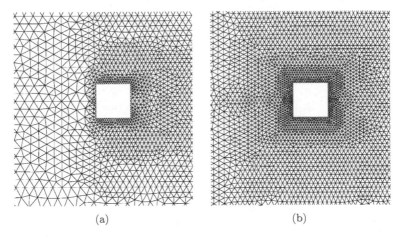

(a)   (b)

FIGURE 8.6: Flow past a square cylinder at Re=22 × 10³. Node distribution in grids GR1 (a) and GR2 (b) in the symmetry section in the spanwise direction.

The grid refinement improves results particularly when the Smagorinsky model is used, while the simulations with the dynamic model give acceptable results also on the grid GR1. This suggests that, with coarse grids, a dynamic model is to be prefered, even if significantly more expensive for unstructured grids (see [Camarri et al., 2002a]). Conversely, for rather resolved simulations the dynamic model loses its advantages over the Smagorinsky model, at least as far as the prediction of global flow quantities, such as bulk coefficients or the time-averaged flow field, is concerned.

8.3.2.2   Forward-swept wing at high angle of attack (LES)

This example is less academic, although still with a rather low Reynolds. The geometry of the considered wing is summarized in Table 8.3; a detailed description can be found in [Lombardi, 1993].

aspect ratio	taper ratio	twist
5.7	0.4	0 deg
sweep angle	wing section	mean aerodynamic chord
-25 deg	NACA0012	0.133 m

Table 8.3: Forward-swept wing at high angle of attack and Rey=19³. Geometry.

Simulations have been carried at an angle of attack of 20 deg. and at a very low Mach number (0.014), in order to obtain negligible effects of compressibility. The Reynolds number, based on the mean aerodynamic chord, is equal to 1.9×10^4. The computational domain is a circular cylinder: the diameter is equal to $20c_r$ and the height to $5c_r$, c_r being the length of the chord at the wing root. The wing root is located on a cylinder base, at which symmetry boundary conditions are imposed. On the remaining cylinder surfaces, inflow/outflow boundary conditions based on the Steger-Warming flow decomposition are used [Steger and Warming, 1981a]. The employed grid is unstructured and has 2.3×10^6 elements. Approximately 4.5×10^3 elements have a face laying on the wing surface and the normal distance from the wing of the first layer of nodes roughly varies from 0.01 to 0.02 of the local chord. The main parameters of the different simulations are summarized in Table 8.4. Time advancing of the equations has been carried out implicitly with a max-

Simulation	SGS model	γ_s	Convective fluxes
WSV4G05	Smag. $(C_s = 0.1)$	0.05	V4
WSV6G05	Smag. $(C_s = 0.1)$	0.05	V6
WSV4G10	Smag. $(C_s = 0.1)$	0.1	V4
WSV6G10	Smag. $(C_s = 0.1)$	0.1	V6
WDV4G05	Dynamic	0.05	V4
WDV6G05	Dynamic	0.05	V6
WDV4G10	Dynamic	0.1	V4
WDV6G10	Dynamic	0.1	V6
WNV4G05	None	0.05	V4

Table 8.4: Forward-swept wing at high angle of attack and Rey=19^3. Summary of the simulations.

imum CFL number of 75. It has been verified that the corresponding time step was not too large and that no significant information was lost in the time history of aerodynamic forces.

The values of the aerodynamic coefficients obtained in the different simulations are reported in Table 8.5, together with experimental values available for $Re = 10^5$ [Lombardi, 1993]. Aerodynamic forces are averaged in time, since unsteadiness are present in the flow due to the large separated flow on the wing upper surface and the corresponding coefficients are defined as follows:

$$C_L = \frac{L}{1/2\rho U^2 S} \quad ; \quad C_D = \frac{D}{1/2\rho U^2 S} \tag{8.20}$$

in which L and D are respectively the global lift and drag, ρ and U are the free-stream flow density and velocity and S is the wing surface.

LES and experiments	$\overline{C_L}$	$\overline{C_D}$
WSV4G05	0.641	0.219
WSV6G05	0.648	0.222
WSV4G10	0.644	0.219
WSV6G10	0.643	0.222
WDV4G05	0.748	0.251
WDV6G05	0.740	0.248
WDV4G10	0.757	0.260
WDV6G10	0.735	0.254
WNV4G05	0.771	0.257
[Lombardi, 1993]	0.714 ± 0.017	0.305 ± 0.005

Table 8.5: Forward-swept wing at high angle of attack and Rey=19^3. Aerodynamic coefficients; $\overline{C_L}$ and $\overline{C_D}$ are the mean lift and drag coefficients.

Note that, in this case, a meaningful comparison with the experiments is difficult because of the different Reynolds number and of the lack of detailed information in experiments (only global coefficients are available). The wing test-case, however, is useful to investigate the effects of numerical dissipation and SGS modeling for a more complex flow than the one around the square cylinder; indeed, here the boundary layer separation is not fixed by the geometry and no homogeneous directions is present.

Effect of SGS modeling   The global lift values obtained in the simulations with the Smagorinsky model are significantly lower than those given by the dynamic model; the largest value is found in the simulation with no SGS model. This behavior is due to the differences observed in the separated flow region, shown, for instance, by the isocontours of the pressure coefficient on the upper surface of the wing in Figure 8.7.

Indeed, independently of the numerical scheme, in the simulations with the Smagorinsky model the flow is found to be completely separated (*i.e.*, separation occurs practically at the leading edge) in a region ranging from the wing root to approximately the 80% of the wing span. In the rest of the upper surface the flow is attached. Conversely, in the simulations with the dynamic model, moving from the wing root to the tip, flow separation progressively moves downstream the leading edge. The simulation with no SGS model has the same qualitative behavior as the ones with the dynamic model, but boundary layer separation tends to occur more downstream than in those with the dynamic model. As far as the spanwise extent of the separated zone is concerned, it can be seen from the spanwise C_l distribution reported in Figure 8.8 (C_l is the lift coefficient of the single wing sections) that for the dynamic model and in the simulation without any model the reattachment point is located at the 80% of the wing span, while the Smagorinsky model predicts a slightly larger extent.

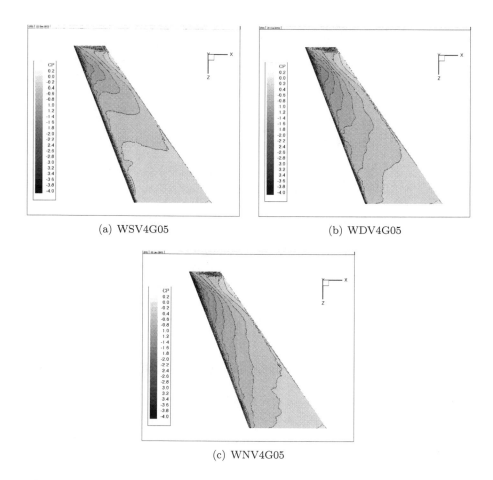

(a) WSV4G05   (b) WDV4G05

(c) WNV4G05

FIGURE 8.7: Forward-swept wing at high angle of attack and Rey=19^3. Iso-contours of the pressure coefficient on the wing upper surface. (a) Simulation WSV4G05; (b) simulation WDV4G05; (c) simulation WNV4G05.

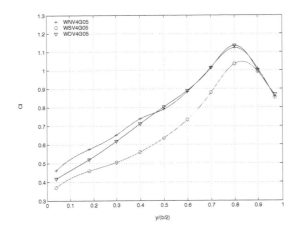

FIGURE 8.8: Forward-swept wing at high angle of attack and Rey=19^3. Effect of SGS modeling on the spanwise distribution of C_l.

It is also evident how in the separated zone the Smagorinsky model gives a C_l much lower than in the other simulations (due to the larger streamwise extent of separation) and this explains the underestimate of global lift. The picture obtained in the simulations with the dynamic model and without SGS model is certainly more accurate, at least from a qualitative point of view. Indeed, the same separation mechanism was observed in experiments at higher Reynolds numbers and different angles of attack [Lombardi et al., 1998] and is confirmed also from the flow visualizations carried out with wool tufts for the flow conditions considered here [Camarri et al., 2001b]. Moreover, this stall behavior is typical of forward-swept wings and is due to three-dimensional effects (strong cross-flow in the inboard direction) [Lombardi, 1993], [Lombardi et al., 1998]. If these effects are not well captured, one expects to obtain a picture similar to that of the simulations with the Smagorinsky model, *i.e.* stall occurring immediately close to the leading-edge, which is peculiar of NACA0012 airfoils. Thus, as already found in the cylinder test-case, it appears that the Smagorinsky model introduces excessive dissipation which damps 3D effects.

The previous considerations also explain the significant quantitative differences in the prediction of the aerodynamic coefficients given by the Smagorinsky and the dynamic model. Indeed, in the simulations with the Smagorinsky model the differences (with respect of the experiments) in the prediction of C_L and C_D range between -9% and -10% and between -27% and -28% respectively, while with the dynamic model they are reduced to [3%,6%] and [-15%,-19%].

Sensitivity to the numerical viscosity parameter γ_s Let us analyze first the effect of the scheme, by comparing simulations differing only for the discretization of convective fluxes, viz. (WSV6G05, WSV4G05), (WSV6G10, WSV4G10), (WDV6G05, WDV4G05) and (WDV6G10, WDV4G10). The effect of the scheme used for the discretization of the convective fluxes is much lower than that of SGS modeling; differences between simulations using V4 and V6 in the prediction of the aerodynamic coefficients are only of 1% for the Smagorinsky model and of 1%-3% for the dynamic one (see Table 8.5). For the wing flow the V6 scheme does not always improve the agreement with the experimental data; however, as discussed previously, in this case comparison with the experiments is not completely meaningful.

For both V4 and V6 schemes and for both SGS models, the predictions of the aerodynamic coefficients are also scarcely sensitive to the value of γ_s; differences between the aerodynamic coefficients obtained in the simulations with (WSV4G05, WSV4G10) and (WSV6G05, WSV6G10), using the Smagorinsky model, are lower than 1%. When the dynamic SGS model is employed, they are not higher than 3% for V4 and of 2% for V6.

Some remarks on computational costs As for the cost of the SGS models, when the time advancing is explicit, the CPU per time-step is increased of 180% for the dynamic model with respect to that corresponding to the Smagorinsky model. Indeed, the dynamic procedure on unstructured grids requires, at each node and at each time step, averages of several quantities over the elements containing the node. This implies a significant increase not only in CPU time, but also in memory requirements and communications costs in parallel simulations. We did not carry out a systematic optimization of the dynamic procedure implementation; however, it seems not trivial to find an algorithm which could significantly reduce the computational costs without further increasing the memory requirements.

As for the numerics, the V6 scheme requires approximately 100% more CPU time than V4 for the discretization of the convective fluxes. When explicit time advancing is used (square cylinder test-case), in the simulations with the Smagorinsky model this implies about 60% increase in the global CPU time, while with the dynamic model it reduces to 23%. For implicit time advancing, as in the wing test-case , the increase of global CPU time for the V6 simulations is practically negligible.

8.3.3   WALE and Vreman's models

The Smagorinsky model is a powerful model based on the simplified assumption of eddy-viscosity. Two more recent eddy-viscosity models have also been considered. Both use first-order derivatives of the velocity in the definition of eddy-viscosity so that they are rather simple and well-suited for engineering applications, and they are designed in order to have a better behavior than the Smagorinsky model for near-wall, shear and transitional flows.

WALE model In the wall-adapting local eddy-viscosity (WALE) SGS model, proposed by Nicoud and Ducros [Nicoud and Ducros, 1999], the eddy-viscosity term μ_{sgs} of the model is defined by:

$$\mu_{\text{sgs}} = \overline{\rho}(C_W \Delta)^2 \frac{(\widetilde{S}_{ij}^d \widetilde{S}_{ij}^d)^{\frac{3}{2}}}{(\widetilde{S}_{ij} \widetilde{S}_{ij})^{\frac{5}{2}} + (\widetilde{S}_{ij}^d \widetilde{S}_{ij}^d)^{\frac{5}{4}}} \tag{8.21}$$

with $\widetilde{S}_{ij}^d$ being the symmetric part of the tensor $g_{ij}^2 = g_{ik}g_{kj}$, where $g_{ij} = \partial \tilde{u}_i / \partial x_j$:

$$\widetilde{S}_{ij}^d = \frac{1}{2}(g_{ij}^2 + g_{ji}^2) - \frac{1}{3}\delta_{ij}g_{kk}^2$$

The filter width Δ has been defined as done for the Smagorinsky model. As recommended in [Nicoud and Ducros, 1999], the constant C_W is set to 0.5.

Vreman's model The model proposed by Vreman [Vreman, 2004a] defines the eddy-viscosity μ_{sgs} as follows:

$$\mu_{\text{sgs}} = \overline{\rho}C_V \left(\frac{B_\beta}{\alpha_{ij}\alpha_{ij}}\right)^{\frac{1}{2}} \tag{8.22}$$

with $\alpha_{ij} = \partial \tilde{u}_j / \partial x_i$, $\beta_{ij} = \Delta^2 \alpha_{mi}\alpha_{mj}$ and $B_\beta = \beta_{11}\beta_{22} - \beta_{12}^2 + \beta_{11}\beta_{33} - \beta_{13}^2 + \beta_{22}\beta_{33} - \beta_{23}^2$. As in [Vreman, 2004a], the constant is set to $C_V \approx 2.5C_s^2$ where C_s denotes the Smagorinsky constant. Again the filter width Δ has been defined as done for the Smagorinsky model.

The Smagorinsky model, the WALE model, and Vreman's model will also be used inside the novel method referred as Variational multiscale LES.

8.4 Variational multiscale large eddy simulation (VMS-LES)

8.4.1 Model features and description

A new approach to LES based on a variational multiscale (VMS) framework was recently introduced by Hughes and his co-workers [Hughes et al., 2000], [Hughes et al., 2001a], [Hughes et al., 2001b]. The VMS-LES differs from the traditional LES in a number of ways. In this approach, the scales are *a priori* separated — that is, before considering numerical discretization. And most importantly, a model for the effect of the unresolved scales is added only in the equations representing the smallest resolved scales, and not in the equations for the largest ones. Consequently, in the VMS-LES, energy is extracted from the fine resolved scales by a traditional model such as the Smagorinsky eddy-viscosity model, but no energy is directly extracted from the large structures in the flow. For this reason, one can reasonably hope to obtain a better behavior near walls, and less dissipation in the presence of large coherent structures. Furthermore, in the original formulation, the Navier-Stokes equations are not filtered but a variational projection is used instead. This is an important difference because as performed in the traditional LES, filtering works well with periodic boundary conditions but may raise mathematical issues in wall-bounded flows. The variational projection avoids these issues.

In this subsection, we present the VMS-LES method used in our works. This one has been developed for compressible flow equations, unstructured grids and a mixed finite volume/finite element framework, so that this VMS-LES method is well suited for engineering applications. More details concerning the derivation of the following VMS-LES governing equations can be found in [Farhat et al., 2006].

If we discretize the Navier-Stokes equations (8.1) by a mixed finite volume/finite element approach, we obtain the following set of equations:

$$
\begin{cases}
A(\mathcal{X}_i, \mathbf{W}) = \displaystyle\int_\Omega \frac{\partial \rho}{\partial t} \mathcal{X}_i \, d\Omega + \int_{\partial Sup\mathcal{X}_i} \rho \mathbf{u} \cdot \mathbf{n} \mathcal{X}_i \, d\Gamma = 0 \\[2ex]
\mathbf{B}(\mathcal{X}_i, \Phi_i, \mathbf{W}) = \displaystyle\int_\Omega \frac{\partial \rho \mathbf{u}}{\partial t} \mathcal{X}_i \, d\Omega + \int_{\partial Sup\mathcal{X}_i} \rho \mathbf{u} \otimes \mathbf{u} \, \mathbf{n} \mathcal{X}_i \, d\Gamma \\[1ex]
\qquad + \displaystyle\int_{\partial Sup\mathcal{X}_i} P \mathbf{n} \mathcal{X}_i \, d\Gamma + \int_\Omega \sigma \nabla \Phi_i \, d\Omega = \mathbf{0} \\[2ex]
C(\mathcal{X}_i, \Phi_i, \mathbf{W}) = \displaystyle\int_\Omega \frac{\partial E}{\partial t} \mathcal{X}_i \, d\Omega + \int_{\partial Sup\mathcal{X}_i} (E + P) \mathbf{u} . \mathbf{n} \mathcal{X}_i \, d\Gamma \\[1ex]
\qquad + \displaystyle\int_\Omega \sigma \mathbf{u} \cdot \nabla \Phi_i \, d\Omega + \int_\Omega \lambda \nabla T \cdot \nabla \Phi_i \, d\Omega = 0
\end{cases}
$$

where $\mathcal{X}_i$ is the characteristic function corresponding to the control volume C_i associated with node i and Φ_i the P1 shape function associated with node i.

Let $\mathcal{V}_{FV}$ and $\mathcal{V}_{FE}$ denote the spaces spanned by $\mathcal{X}_i$ and Φ_i, respectively. The decomposition *a priori* of these spaces into large resolved scales, small resolved scales and unresolved scales [Collis, 2001] can be written as

$$\mathcal{V}_{FV} = \overline{\mathcal{V}}_{FV} \oplus \mathcal{V}'_{FV} \oplus \widehat{\mathcal{V}}_{FV}$$

$$\mathcal{V}_{FE} = \overline{\mathcal{V}}_{FE} \oplus \mathcal{V}'_{FE} \oplus \widehat{\mathcal{V}}_{FE}$$

where the notation "¯", "′" and "^" denotes the large resolved scales, small resolved scales and unresolved scales, respectively.

Let $\mathbf{W} = \overline{\mathbf{W}} + \mathbf{W}' + \widehat{\mathbf{W}}$ be the decomposition of $\mathbf{W}$ into large resolved scales, small resolved scales and unresolved scales. Substituting this decomposition into the previous semi-discretized equations and by variational projection onto the large resolved scales space, small resolved scales space and unresolved scales space, we obtain the following set of equations governing the large resolved scales:

$$
\begin{cases}
A(\overline{\mathcal{X}}_i, \overline{\mathbf{W}} + \mathbf{W}') + A^*(\overline{\mathcal{X}}_i, \overline{\mathbf{W}}, \mathbf{W}', \widehat{\mathbf{W}}) & = 0 \\[2mm]
\mathbf{B}(\overline{\mathcal{X}}_i, \overline{\Phi}_i, \overline{\mathbf{W}} + \mathbf{W}') + \mathbf{B}^*(\overline{\mathcal{X}}_i, \overline{\Phi}_i, \overline{\mathbf{W}}, \mathbf{W}', \widehat{\mathbf{W}}) = 0 \\[2mm]
C(\overline{\mathcal{X}}_i, \overline{\Phi}_i, \overline{\mathbf{W}} + \mathbf{W}') + C^*(\overline{\mathcal{X}}_i, \overline{\Phi}_i, \overline{\mathbf{W}}, \mathbf{W}', \widehat{\mathbf{W}}) = 0,
\end{cases}
$$

the following set of equations governing the small resolved scales:

$$
\begin{cases}
A(\mathcal{X}'_i, \overline{\mathbf{W}} + \mathbf{W}') + A^*(\mathcal{X}'_i, \overline{\mathbf{W}}, \mathbf{W}', \widehat{\mathbf{W}}) & = 0 \\[2mm]
\mathbf{B}(\mathcal{X}'_i, \Phi'_i, \overline{\mathbf{W}} + \mathbf{W}') + \mathbf{B}^*(\mathcal{X}'_i, \Phi'_i, \overline{\mathbf{W}}, \mathbf{W}', \widehat{\mathbf{W}}) = 0 \\[2mm]
C(\mathcal{X}'_i, \Phi'_i, \overline{\mathbf{W}} + \mathbf{W}') + \mathbf{C}^*(\mathcal{X}'_i, \Phi'_i, \overline{\mathbf{W}}, \mathbf{W}', \widehat{\mathbf{W}}) = 0,
\end{cases}
$$

and the following set of equations governing the unresolved scales:

$$
\begin{cases}
A(\widehat{\mathcal{X}}_i, \overline{\mathbf{W}} + \mathbf{W}' + \widehat{\mathbf{W}}) & = 0 \\[2mm]
\mathbf{B}(\widehat{\mathcal{X}}_i, \widehat{\Phi}_i, \overline{\mathbf{W}} + \mathbf{W}' + \widehat{\mathbf{W}}) = 0 \\[2mm]
C(\widehat{\mathcal{X}}_i, \widehat{\Phi}_i, \overline{\mathbf{W}} + \mathbf{W}' + \widehat{\mathbf{W}}) = 0,
\end{cases}
$$

where terms A^*, B^* and C^* represent the effect of the unresolved scales on the large and small resolved ones.

Since the unresolved scales can not be captured by the numerical computation, the equations governing the unresolved scales are dropped from the system to be solved. This leads to reinterpreting the three-level formalism as a two-level framework:

$$
\begin{aligned}
\mathcal{V}_{FE_h} &= \overline{\mathcal{V}}_{FE_h} \oplus \mathcal{V}'_{FE_h} \\
\mathcal{V}_{FV_h} &= \overline{\mathcal{V}}_{FV_h} \oplus \mathcal{V}'_{FV_h}
\end{aligned}
$$

$$
\Rightarrow \mathbf{W}_h = \overline{\mathbf{W}}_h + \mathbf{W}'_h
$$

where the subscript h emphasizes the resolved aspects of the scales. This leads also to model the effect of the unresolved scales on the large and small resolved ones represented by terms A^*, B^* and C^*. Turbulence features a cascade process in which the kinetic energy transfers from larger eddies to smaller ones. Therefore, it can be assumed that energy transfer occurs mostly between neighboring scales. For this reason, we neglect the effect of the unresolved scales on the large resolved ones, and model only the effect of the unresolved scales on the small resolved ones. This effect, which is of energy dissipation type, is modeled here by any of the three non dynamic eddy-viscosity models previously presented (section 8.3.1.1 and 8.3.3). Furthermore, the so-called *small-small* formulation is adopted, *i.e.* the modeled terms are computed as a function of the smallest resolved scales only. Then, combining the resulting equations which govern the small and large resolved scales (for more details, see [Farhat et al., 2006]), we obtain the final equations:

$$
\begin{cases}
A(\mathcal{X}_{i_h}, \mathbf{W}_h) & = 0 \\[2mm]
\mathbf{B}(\mathcal{X}_{i_h}, \Phi_{i_h}, \mathbf{W}_h) + \displaystyle\int_\Omega \boldsymbol{\tau'}_h \nabla \Phi'_{i_h} \, d\Omega & = \mathbf{0} \\[2mm]
C(\mathcal{X}_{i_h}, \Phi_{i_h}, \mathbf{W}_h) + \displaystyle\int_\Omega \frac{C_p \mu'_t}{Pr_t} \nabla T'_h \cdot \nabla \Phi'_{i_h} \, d\Omega & = 0
\end{cases}
$$

where $\tau'_{ij} = \mu'_t(2S'_{ij} - \frac{2}{3}S'_{kk}\delta_{ij})$, $S'_{ij} = \frac{1}{2}(\frac{\partial u'_i}{\partial x_j} + \frac{\partial u'_j}{\partial x_i})$, and μ'_t is given either by the Smagorinsky, or the WALE or the Vreman models. The constants of the models are the same as those used herein in classical LES and specified in sections 8.3.1.1 and 8.3.3. In this approach, the local filter size Δ' is set in each tetrahedron T_l to its volume to the power one third. Here, C_p is the specific heat at constant pressure and Pr_t is the subgrid scale Prandtl number which is assumed to be constant.

REMARK 8.1 The laminar Navier-Stokes equations are recovered by substituting $\tau'_h = 0$ and $\mu'_t = 0$ in the above system of equations. The classical Smagorinsky LES model is recovered by substituting $\tau'_h = \tau_h$, $\mu'_t = \mu_t$, $T'_h = T_h$ and $\Phi'_{i_h} = \Phi_{i_h}$. ▯

8.4.1.1 *A priori* separation of the scales

Given a tetrahedral mesh, a corresponding dual mesh defined by cells or control volumes can always be derived. Such a dual mesh can be partitioned into macro-cells by a process known as agglomeration [Lallemand et al., 1992]. This process is graphically depicted in Figure 8.9 for the two-dimensional case.

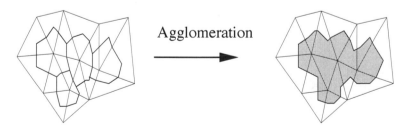

FIGURE 8.9: Unstructured mesh, dual mesh, and agglomeration of some cells of the dual mesh into a macro-cell (two-dimensional case).

Given an agglomeration into macro-cells, the idea then is to define the coarse scale component $\overline{\mathbf{W}}_h = P(\mathbf{W}_h)$, in the diffusive terms, as the average of $\mathbf{W}_h$ in the macro-cells in the following sense

$$\overline{\mathbf{W}}_h = P(\sum_k \Phi_{k_h} \mathbf{W}_{k_h}) = \sum_k \Phi_{k_h} \widetilde{\mathbf{W}}_{k_h} \tag{8.23}$$

where

$$\widetilde{\mathbf{W}}_{k_h} = \frac{\sum\limits_{j \in I_k} Vol(C_j) \mathbf{W}_{j_h}}{\sum\limits_{j \in I_k} Vol(C_j)}, \tag{8.24}$$

C_j is the cell around the vertex j, $Vol(C_j)$ denotes its volume, $I_k = \{j \ / \ C_j \subset C_{m(k)}\}$, and $C_{m(k)}$ denotes the macro-cell containing the cell C_k.

From equation (8.23) and equation (8.24) it follows that $\overline{\mathbf{W}}_h$ can also be written as

$$\overline{\mathbf{W}}_h = \sum_k \overline{\Phi}_{k_h} \mathbf{W}_{k_h} \tag{8.25}$$

where

$$\overline{\Phi}_{k_h} = \frac{Vol(C_k)}{\sum\limits_{j \in I_k} Vol(C_j)} \sum_{j \in I_k} \Phi_{j_h}. \tag{8.26}$$

Given the properties of the P1 shape functions Φ_{k_h}, it follows that the coarse scale component $\overline{\mathbf{W}}_h$ is approximated here by a continuous function which is constant in each tetrahedron contained in a macro-cell, has the same constant value in all the tetrahedrons contained in the same macro-cell, and is linear in the tetrahedrons shared by at least two macro-cells. In other words, $\overline{\mathbf{W}}_h$ can be viewed as a piecewise constant function with linear connections between its constant stages (see Figure 8.10 for a one-dimensional representation).

It also follows that

$$\mathbf{W}'_h = \mathbf{W}_h - \overline{\mathbf{W}}_h = \sum_k \Phi_{k_h}(\mathbf{W}_{k_h} - \widetilde{\mathbf{W}}_{k_h}) = \sum_k \Phi'_{k_h} \mathbf{W}_{k_h} = \sum_k (\Phi_{k_h} - \overline{\Phi}_{k_h}) \mathbf{W}_{k_h}. \tag{8.27}$$

The same averaging procedure outlined above can be applied to define the coarse scale component $\overline{\mathbf{W}} = P(\mathbf{W})$ in the convective terms, in which case Φ_k is replaced by $\mathcal{X}_k$. Hence, for the convective terms, P is defined by

$$P(\mathbf{W}_h) = \sum_k \left(\frac{Vol(C_k)}{\sum\limits_{j \subset I_k} Vol(C_j)} \sum_{j \in I_k} \mathcal{X}_{j_h} \right) \mathbf{W}_{k_h} \tag{8.28}$$

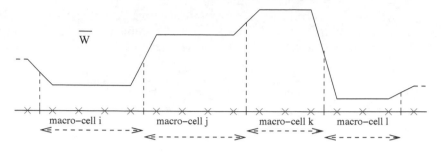

$\times$: Nodes

FIGURE 8.10: One-dimensional representation of a coarse scale component $\overline{\mathbf{W}}$.

and for the diffusive terms, P is defined by

$$P(\mathbf{W}_h) = \sum_k \left(\frac{Vol(C_k)}{\sum_{j \in I_k} Vol(C_j)} \sum_{j \in I_k} \Phi_{j_h} \right) \mathbf{W}_{k_h}. \qquad (8.29)$$

The reader can check that in both cases, P verifies $P^2 = P$ and therefore is a projector.

This completes the definition of the decompositions $\mathcal{V}_{FV_h} = \overline{\mathcal{V}}_{FV_h} \oplus \mathcal{V}'_{FV_h}$ and $\mathcal{V}_{FE_h} = \overline{\mathcal{V}}_{FE_h} \oplus \mathcal{V}'_{FE_h}$, where $\overline{\mathcal{V}}_{FV_h}$ ($\overline{\mathcal{V}}_{FE_h}$) denotes the space of approximation of the coarse scales spanned by the functions $\{\overline{\mathcal{X}}_{k_h}\}$ ($\{\overline{\Phi}_{k_h}\}$), and $\mathcal{V}'_{FV_h}$ ($\mathcal{V}'_{FE_h}$) denotes the space of approximation of the fine scales spanned by the functions $\{\mathcal{X}'_{k_h}\}$ ($\{\Phi'_{k_h}\}$). If n is the total number of nodes in the given mesh and N is the total number of macro-cells defined by the agglomeration, then the dimension of each of $\overline{\mathcal{V}}_{FV_h}$ and $\overline{\mathcal{V}}_{FE_h}$ is N, and that of each of $\mathcal{V}'_{FV_h}$ and $\mathcal{V}'_{FE_h}$ is $n - N$.

The reader can note that a dynamic version of this VMS-LES method, not described in this chapter for the sake of brevity, based on variational analogues of Germano's algebraic identity and the same agglomeration based numerical procedure has also been proposed in [Farhat et al., 2006].

8.4.2   The impact of VMS-LES vs. LES

This numerical part has many objectives: (i) to compare the relative performance of LES and VMS-LES for the simulation of bluff-body flows through our numerical framework (ii) to show the effects of SGS models, grid resolution and numerical dissipation on the quality of the obtained predictions (iii) to evaluate the performance of LES and VMS-LES for the simulation of bluff-body flows around a moving complex geometry. For tasks (i) and

(ii) we consider the flow around a circular cylinder at Reynolds number, based on the free-stream velocity and on the cylinder diameter, equal to 3900. This flow has been chosen since it is a classical and well documented benchmark (see e.g. [Parneaudeau et al., 2008] for experimental data and [Breuer, 1998], [Kravchenko and Moin, 1999], [Lee et al., 2006], [Parneaudeau et al., 2008] for numerical studies). Moreover, it contains all the features and all the difficulties encountered in the simulation of bluff-body flows also for more complex configurations and higher Reynolds numbers, at least for laminar boundary-layer separation. On the other hand the flow configuration considered for task (iii) is of direct interest in offshore engineering.

8.4.2.1 Flow around a circular cylinder (VMS-LES)

Test-case description and simulation parameters The flow over a circular cylinder at Reynolds number (based on the cylinder diameter and on the free-stream velocity) equal to 3900 is simulated. The computational domain is such that $-10 \leq x/D \leq 25$, $-20 \leq y/D \leq 20$ and $-\pi/2 \leq z/D \leq \pi/2$, where x, y and z denote the streamwise, transverse and spanwise directions respectively, the cylinder center being located at $x = y = 0$. Characteristic based conditions are used at the inflow and outflow as well as on the lateral surfaces [Steger and Warming, 1981a]. In the spanwise direction periodic boundary conditions are applied and on the cylinder surface no-slip is imposed. The free-stream Mach number is set equal to 0.1 in order to make a sensible comparison with incompressible simulations in the literature. As described in the section devoted to the numerical framework, preconditioning is used to deal with the low Mach number regime. The flow domain is discretized by two unstructured tetrahedral grids: the first one (GR1) consists of approximately 2.9×10^5 nodes. The averaged distance of the nearest point to the cylinder boundary is $0.017D$, which corresponds to $y^+ \approx 3.31$. The second grid (GR2) is obtained from GR1 by refining in a *structured* way, *i.e.* by dividing each tetrahedron in 4, resulting in approximately 1.46×10^6 nodes.

Several LES and VMS-LES simulations have been carried by varying the SGS model, the value of γ_s, and the grid resolution. All the simulations have been carried out with the same time step, such that a vortex-shedding period contains approximately 400 time steps. The time-step independence of the results has been checked.

Results for a coarse grid resolution Let us start to analyze the results obtained in the simulations carried out on the coarser grid GR1. The main parameters of these simulations are summarized in Table 8.6.

Statistics are computed by averaging in the spanwise homogeneous direction and in time on 20 vortex-shedding cycles. The main flow bulk parameters obtained in our simulations are also shown in Table 8.6, while the corresponding values obtained in experiments and LES in the literature are reported in Table 8.7.

Simulation	Turb. model	SGS model	Grid	γ_s	$\overline{C_d}$	St	l_r	$-\overline{Cp_b}$
C1	LES	Smagorinsky	GR1	0.3	1.16	0.212	0.81	1.17
C2	LES	Vreman	GR1	0.3	1.04	0.221	0.97	1.01
C3	LES	WALE	GR1	0.3	1.14	0.214	0.75	1.2
C4	VMS-LES	Smagorinsky	GR1	0.3	1.00	0.221	1.05	0.96
C5	VMS-LES	Vreman	GR1	0.3	1.00	0.22	1.07	0.97
C6	VMS-LES	WALE	GR1	0.3	1.03	0.219	0.94	1.01
C7	no model	-	GR1	0.3	0.96	0.223	1.24	0.90
C8	no model	-	GR1	0.2	0.94	0.224	1.25	0.89

Table 8.6: Flow around a circular cylinder at Rey=3900. Main simulation parameters and flow bulk coefficients for simulations on grid GR1. $\overline{C_d}$ denotes the mean drag coefficient, St the vortex-shedding Strouhal number, based on the free-stream velocity and the cylinder diameter, l_r the mean recirculation bubble length and $\overline{Cp_b}$ the base pressure.

LES data	$\overline{C_d}$	St	l_r	$-\overline{Cp_b}$
[Breuer, 1998]	[0.969,1.486]	–	[0.397,1.686]	[0.867,1.665]
[Kravchenko and Moin, 1999]	[1.04,1.38]	[0.193,0.21]	[1.,1.35]	[0.93,1.23]
[Parneaudeau et al., 2008]	–	0.208±0.001	1.56	–
[Lee et al., 2006]	[0.99,1.04]	[0.209,0.212]	[1.35,1.37]	[0.89,0.94]
Experimental data	$\overline{C_d}$	St	l_r	$-\overline{Cp_b}$
[Dong et al., 2006]	–	–	1.47	–
[Lourenco and Shih, 1993]	–	–	1.18±0.05	–
[Kravchenko and Moin, 1999]	0.99±0.05	—	–	0.88±0.05
[Ong and Wallace, 1996]	–	0.21±0.005	—	–
[Parneaudeau et al., 2008]	–	0.208±0.002	1.51	–

Table 8.7: Flow around a circular cylinder at Rey=3900. Bulk flow parameters obtained in experiments and in large eddy simulations in the literature. The data from [Dong et al., 2006] are at Re=4000 and those of Norberg (taken from [Kravchenko and Moin, 1999]) at Re=4020.

For the vortex-shedding Strouhal number, St (based on the free-stream velocity and the cylinder diameter), experimental values in the range of $[0.205, 0.215]$ are generally obtained, which well agree with those obtained in the LES in the literature (Table 8.7) and in our simulations (Table 8.6).

As for the mean drag coefficient, $\overline{C_d} = 0.99 \pm 0.05$ was obtained in the experiments by Norberg at $Re = 4020$ (data taken from [Kravchenko and Moin, 1999]), which well agrees with those computed in well resolved LES in the literature [Kravchenko and Moin, 1999], [Lee et al., 2006]. A significant overestimation of $\overline{C_d}$ is obtained in the LES simulations (C1-C3), which ranges from 4.5% to 17% with respect to the upper limit of the experimental range. Conversely, the prediction given by the VMS-LES simulations (C4-C6) and by those without any SGS model (C7-C8) are better (the maximum discrepancy from the experimental range being of 3.5%).

Clearly, the mean drag depends on the pressure distribution on the cylinder surface. Figure 8.11 shows the mean pressure coefficient distribution at the cylinder obtained in the simulations on GR1, together with experimental data. From the discrepancy between numerical results and experimental data

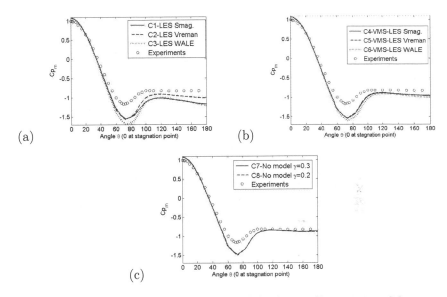

FIGURE 8.11: Flow around a circular cylinder at Rey=3900. Mean pressure coefficient distribution at the cylinder obtained in the simulations on the coarser grid GR1. (a) LES simulations on GR1 (C1-C3); (b) VMS-LES simulations (C4-C6); (c) no-model (C7-C8).

in the zone of the negative peak it is evident that in all cases the boundary layer evolution is not accurately captured in the simulations, due to the grid coarseness. Other symptoms of a too coarse grid resolution (see the discussion in [Kravchenko and Moin, 1999]) are the underestimation of the mean recirculation length l_r in all the simulations on GR1 (Table 8.6) and the V shape of the mean streamwise velocity profiles in the near wake (not shown here for sake of brevity).

Nevertheless, it is interesting to analyze the results obtained on GR1 because they show yet significant differences between LES and VMS-LES. Moreover, they are relevant to situations often encountered in the numerical simulation of complex industrial or engineering flows, in which only a rather coarse grid resolution can be achieved. First of all, in LES the discrepancy observed in the negative peak of mean C_p is larger and the differences among the different SGS models are more pronounced than in VMS-LES.

Let us analyze now the SGS viscosity introduced by the different considered SGS models. As qualitatively shown by the instantaneous isocontours of μ_{SGS}/μ on the horizontal plane at $z = 0$ reported in Figure 8.12, in LES simulations, as expected, the model is mainly acting in the wake. Generally,

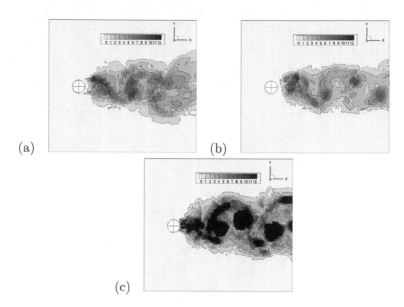

(a)   (b)

(c)

FIGURE 8.12: Flow around a circular cylinder at Rey=3900. Instantaneous isocontours of μ_{SGS}/μ obtained in the LES simulations on the coarser grid GR1. (a) Simulation C1, Smagorinsky model; (b) simulation C2, Vreman model; (c) simulation C3, WALE model. All the considered instants correspond to a maximum of the lift coefficient.

in the whole domain, the WALE model provides the highest values of SGS viscosity, while the lowest ones are given by the Vreman model. The maximum instantaneous values of μ_{SGS}/μ are of the order of 10 for the Smagorinsky and Vreman models, while local values up to 100 are given by the WALE model (the isocontour range in Figure 8.12c is saturated to 12 for sake of comparison with the other models). No adaptation of the model constants to the considered test-case has been made here. For the Vreman and WALE models we used the values recommended in the original papers ([Vreman, 2004a] and [Nicoud and Ducros, 1999] respectively), while for the Smagorinsky model the value generally indicated as *optimal* for shear flows. Thus, the Smagorinsky model in some sense benefits from the fact that it has been widely used and the knowledge on its behavior is deeper than for the other two models.

Let us analyze now the same quantities obtained in the VMS-LES simulations (Figure 8.13). Although the qualitative behavior of the SGS models

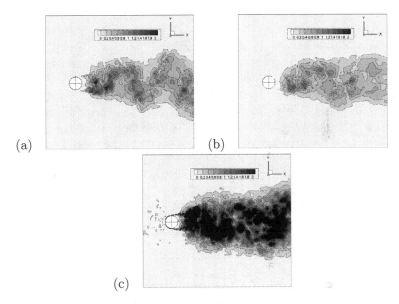

FIGURE 8.13: Flow around a circular cylinder at Rey=3900. Instantaneous isocontours of μ_{SGS}/μ obtained in the VMS-LES simulations on the coarser grid GR1. (a) Simulation C4, Smagorinsky model; (b) simulation C5, Vreman model; (c) simulation C6, WALE model. All the considered instants correspond to a maximum of the lift coefficient.

is similar to that found in LES simulation, it is evident that the SGS viscosity introduced by all the considered models is significantly lower than in the corresponding LES simulations in the whole domain and also near the cylinder. Note that the maximum isocontour value in Figure 8.13 is 2.5. This is due to the fact that in the VMS-LES case we use the so called small-small approach, *i.e.* the SGS viscosity is computed as a function of the smallest resolved scales. We recall that another fundamental difference, which can not be appreciated from the values of the *SGS* viscosity is that in the VMS-LES approach the SGS viscosity only acts on the smallest resolved scales.

The different behavior of the SGS models leads to additional inaccuracies of the LES predictions, besides those due to grid coarseness and previously discussed, which are not present in VMS-LES. For instance, Figure 8.11a shows that the base pressure is inaccurately predicted in all LES simulations except for the Vreman model (the least viscous one), leading to an inaccurate value of the mean drag coefficient (Table 8.6). Conversely, for the VMS-LES simulations the agreement with the experiments is fairly good (see also Figure 8.11b). Differences between the LES and VMS-LES results are present also in the mean velocity field (not shown here for sake of brevity). An indication is given by the different predictions of the mean recirculation length l_r given by LES simulations with respect to those obtained in VMS-LES; in particular, for a given SGS model, the VMS-LES simulations systematically predict larger values of l_r than the LES ones.

Simulations without any SGS model were also carried out. The pressure distribution obtained in the simulations without any SGS model is very similar to the one obtained in the VMS-LES ones (compare Figure 8.11c to Figure 8.11b); this results in a similar prediction of the mean drag coefficient (Table 8.6). The same is for the mean velocity field in the near-wake. This is an a-posteriori confirmation that the used MUSCL reconstruction indeed introduces a viscosity acting only on the highest resolved frequencies [Camarri et al., 2004], as the SGS viscosity in the VMS approach and that this limits its negative effects. Moreover, the results obtained with two different (low) values of the parameter γ_s are also very similar (Table 8.6 and Figure 8.11c), consistently with our previous findings [Camarri et al., 2004]. However, more downstream, when the contribution of the SGS model becomes significant, the results obtained in the no-model simulations start to significantly deviate from those obtained in LES or VMS-LES simulations. Again, an indicator of the differences in the mean velocity field is the prediction of l_r; the no-model simulations give significantly larger values than those obtained in LES and also in VMS-LES simulations (see Table 8.6). This is related to the fact that in no-model simulation the shear-layer transition and vortex formation occur more downstream than in VMS-LES simulations and even more than in LES ones. Summarizing, from this analysis it seems that the larger is the SGS viscosity introduced by the SGS models the shorter is the length of the shear layers and of the mean recirculation bubble. For the set of simulations on GR1, it seems that the no-model simulations give the best agreement with experimental data also for the prediction of l_r. However, this is an example of error compensation; indeed, the grid coarseness leads to an underestimation of l_r (see also [Kravchenko and Moin, 1999]) and the fact that in the no-model simulations the shear-layer transition and vortex formation occurs more downstream tends to compensate this underestimation. It will be shown in section 8.4.2.1 that on the finer grid GR2 the no-model simulation will give a definitely too large value of l_r.

Results for a finer grid resolution   Let us analyze in this section the results of the simulations on the finer grid GR2 in the light of these previous considerations. The main parameters of these simulations are summarized in Table 8.8.

Table 8.8: Flow around a circular cylinder at Rey=3900. Main simulation parameters and flow bulk coefficients for simulations on grid GR2. The symbols are the same as in Table 1.

Simulation	Turb. model	SGS model	Grid	γ_s	$\overline{C_d}$	St	l_r	$-\overline{C_{p_b}}$
F1	LES	Smagorinsky	GR2	0.3	0.99	0.218	1.54	0.85
F2	LES	Vreman	GR2	0.4	0.92	0.227	1.83	0.78
F3	LES	WALE	GR2	0.3	1.02	0.221	1.22	0.94
F4	VMS-LES	Smagorinsky	GR2	0.3	0.93	0.226	1.68	0.81
F5	VMS-LES	Vreman	GR2	0.4	0.90	0.228	1.92	0.76
F6	VMS-LES	WALE	GR2	0.3	0.94	0.223	1.56	0.83
F7	no model	-	GR2	0.3	0.92	0.225	1.85	0.77

Statistics are computed by averaging in the spanwise homogeneous direction and in time on 25 vortex-shedding cycles. Table 8.8 also shows the main flow bulk parameters obtained in the simulations on GR2, to be compared to those of LES in the literature and to experimental data, reported in Table 8.7. A good agreement with the reference experimental data is obtained in the prediction of the mean drag for all the simulations carried out on GR2, while on the coarser grid significant errors were observed for the LES simulations. The largest underestimation is found in the VMS-LES simulation with the Vreman model with an approximately 4% error with respect to the lower limit of the experimental range, and, thus much lower than that observed for LES simulations on the coarse grid. Figure 8.14 shows the mean pressure coefficient distribution at the cylinder obtained in the simulations on GR1, compared with experimental data.

The negative peak of mean C_p is well captured in all the simulations, except in the LES with the WALE model (F3). This indicates that the resolution near the cylinder of GR2 is adequate to well capture the boundary layer evolution,

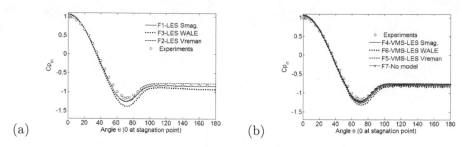

FIGURE 8.14: Flow around a circular cylinder at Rey=3900. Mean pressure coefficient distribution at the cylinder obtained in the simulations on the finer grid GR2. (a) LES simulations on GR2 (F1-F3); (b) VMS-LES (F4-F6) and no-model simulations (F7).

while the discrepancy observed in the F3 simulation is probably due to the too large SGS diffusivity introduced downstream the cylinder by the WALE model in classical LES.

The same quantities as for the coarser grid GR1 (section 8.4.2.1) are shown to analyze the behavior of the different SGS models in both classical and variational multiscale LES for the finer grid resolution. More particularly, Figures 8.15 and 8.16 show the instantaneous isocontours of μ_{SGS}/μ obtained in LES and VMS-LES simulations on the finer grid (compare with Figures 8.12 and 8.13 for GR1).

First, as expected, for all the models and in both LES and VMS-LES the SGS viscosity introduced on the finer grid is smaller that for coarser one and in a narrower region in space due to a better resolution of the wake. Nonetheless, the differences among the SGS models are the same as those discussed in section 8.4.2.1 for the coarser grid GR1, with the WALE model introducing the largest amount of SGS viscosity and the Vreman one the smallest. Finally, the effect of the small-small VMS formulation is to significantly reduce the introduced SGS viscosity for all the considered models and this is again in accordance with the observations made for the coarser grid in section 8.4.2.1. Note how on the finer grid the amount of SGS viscosity given by the Smagorinsky model, and in a larger extent by the Vreman one, in the VMS-LES simulations is very small.

As anticipated previously, for the finer grid the no-model simulation gives a value of l_r largely overestimated with respect to the reference experimental value. For the WALE model, the best agreement with the experiments is found in the VMS-LES formulation, as for the mean velocity field (not shown here) and the value of l_r (see Table 8.8). However, this is not the case for the other two considered SGS models. Indeed, for the Smagorinsky model, the best results are obtained in the LES simulation (F1), while it seems that a too low SGS viscosity is provided by this model in the VMS-LES formulation (F4), as indicated by the overestimation of l_r in Table 8.8 and by the shape of

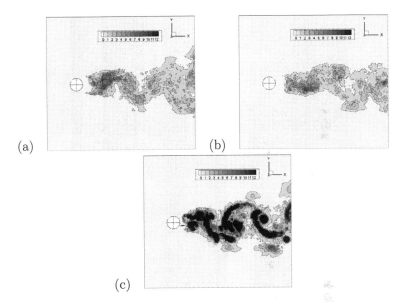

FIGURE 8.15: Flow around a circular cylinder at Rey=3900. Instantaneous isocontours of μ_{SGS}/μ obtained in the LES simulations on the finer grid GR2. (a) Simulation F1, Smagorinsky model; (b) simulation F2, Vreman model; (c) simulation F3, WALE model. All the considered instants correspond to a maxim of the lift coefficient.

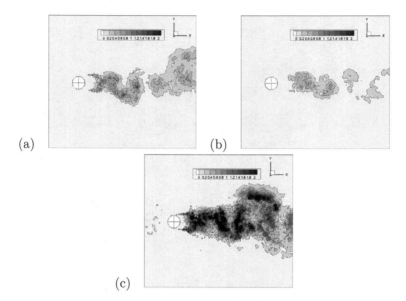

FIGURE 8.16: Flow around a circular cylinder at Rey=3900. Instantaneous isocontours of μ_{SGS}/μ obtained in the VMS-LES simulations on the finer grid GR2. (a) Simulation F4, Smagorinsky model; (b) simulation F5, Vreman model; (c) simulation F6, WALE model. All the considered instants correspond to a maximum of the lift coefficient.

the velocity profiles (not shown here for sake of brevity). Finally, the Vreman model is found to be not dissipative enough even in classical LES (F3) and the results further deteriorate in VMS-LES (F5).

Thus, it seems that for the considered flow a significant sensitivity to the SGS model is present also in VMS-LES and that the introduction of a proper amount of SGS viscosity remains a crucial issue also in this approach.

8.4.2.2 Vortex-induced motion of a complex geometry

FIGURE 8.17: Spar geometry.

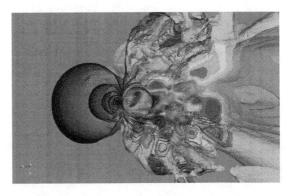

FIGURE 8.18: Flow past a spar at Re=3×10^5. VMS-LES with Smagorinsky model, velocity module.

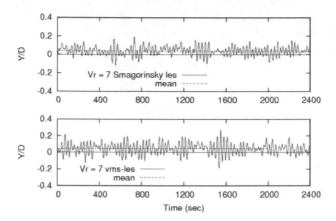

FIGURE 8.19: Flow past a spar at Re=3×10^5. Time variation of the transverse position of the spar at a reduced velocity of 7 m/s. for LES-Smagorinsky (upper curve) and VMS-Smagorinsky (lower curve).

The prediction of vortex-induced motion of a complex spar geometry is an important motivation for VMS modeling [Sirnivas et al., 2006]. The spar geometry consists of a cylinder equipped with helicoidal strakes, see Figure 8.17. Each strake produces in the flow a shear layer that interacts with the large flow structures and inhibits to a significant extent the von Karman vortex street, see Figure 8.18. This is a typical backscatter effect and we investigate the impact of the choice of a VMS model on the quality of a LES prediction. In our computations, the obstacle is maintained by elastic moorings and moves under the effect of the vortex shedding. The fluid-structure coupling is computed at several reduced velocity between 4 and 9 (m/s) and Reynolds numbers between 2×10^5 and 4×10^5. The mesh involves 5×10^5 vertices and the computation is performed during 40 periods before statistics are computed.

The behavior of the transverse position of the spar is a key output to be accurately predicted. Figure 8.19 shows the time variation of the transverse position of the spar at a reduced velocity of 7 m/s for LES and VMS models equipped with Smagorinsky's subgrid scale model. The r.m.s. computed in LES with the Smagorinsky model is 0.048, and the r.m.s. in VMS-LES is 0.070, which compares better with the experimental data of 0.077. The agreement of the VMS calculations with experiments for the different velocities is demonstrated in Table 8.9.

Reduced velocity	4	5	6	7	8	9
LES	-	-	-	.048	-	-
VMS-LES	.0018	.020	.04	.070	.12	.118
Experiments	.0018	.025	.05	.077	.13	.125

Table 8.9: Flow past a spar at Re=3×10^5. Vortex-induced motion: RMS transverse deviation.

8.5 Hybrid RANS/LES

8.5.1 Model features and description

A major limitation of LES for the simulation of complex flows is the fact that its cost increases as the flow Reynolds number rises. Indeed, the grid has to be fine enough to resolve a significant part of the turbulent scales, and this becomes particularly critical in the near-wall regions where small vortical structures play a key role. One way to overcome this limitation is to introduce Reynolds Average Navier-Stokes (RANS) modeling in these regions.

For this purpose, hybrid models have recently been proposed in the literature (see for example [Travin et al., 1999], [Spalart et al., 1997], [Labourasse and Sagaut, 2002], [Frohlich and von Terzi, 2008], [Vengadesan and Nithiarasu, 2007], [Camarri et al., 2005]) in which RANS and LES approaches are combined together in order to obtain simulations as accurate as in the LES case but at reasonable computational costs. These hybrid methods can be divided in *zonal* approaches, in which RANS and LES are used in a-priori defined regions, and the so called *universal* models, which should be able to automatically switch from RANS to LES throughout the computational domain. In the perspective of the simulation of massively separated unsteady flows in complex geometry, as occur in many cases of engineering or industrial interest, we are primarily interested in universal hybrid models. In this context, we have proposed a new strategy for blending RANS and LES approaches in a hybrid model.

We begin this section by recalling two different hybrid strategies that are representative examples of universal RANS/LES hybridization before to present the new hybrid model that we have proposed.

8.5.2 Detached eddy simulation

The original DES model [Spalart et al., 1997] is an extension of the Spalart-Allmaras (**S-A**) RANS model, which we recall in short. In the S-A model, the turbulent eddy-viscosity is given by

$$\nu_t = \tilde{\nu} f_{v1}, \quad f_{v1} = \frac{\chi^3}{\chi^3 + C_{v1}^3}, \quad \chi := \frac{\tilde{\nu}}{\nu} \tag{8.30}$$

$$\frac{\partial \tilde{\nu}}{\partial t} + u_j \frac{\partial \tilde{\nu}}{\partial x_j} = C_{b1}[1 - f_{t2}]\tilde{S}\tilde{\nu} + \frac{1}{\sigma}\{\nabla \cdot [(\nu + \tilde{\nu})\nabla\tilde{\nu}] + C_{b2}|\nabla\nu|^2\} - \tag{8.31}$$

$$\left[C_{w1}f_w - \frac{C_{b1}}{\kappa^2}f_{t2}\right]\left(\frac{\tilde{\nu}}{d}\right)^2 + f_{t1}\Delta U^2 \tag{8.32}$$

$$\tag{8.33}$$

$$\tilde{S} \equiv S + \frac{\tilde{\nu}}{\kappa^2 d^2}f_{v2}, \quad f_{v2} = 1 - \frac{\chi}{1 + \chi f_{v1}} \tag{8.34}$$

where

$$S = \equiv \sqrt{2\Omega_{ij}\Omega_{ij}}$$

$$\Omega_{ij} \equiv \frac{1}{2}\left(\frac{\partial u_i}{\partial x_j} - \frac{\partial u_j}{\partial x_i}\right)$$

$$f_w = g\left[\frac{1 + C_{w3}^6}{g^6 + C_{w3}^6}\right]^{1/6}, \quad g = r + C_{w2}(r^6 - r), \quad r \equiv \frac{\tilde{\nu}}{\tilde{S}\kappa^2 d^2}$$

$$f_{t1} = C_{t1}g_t \exp\left(-C_{t2}\frac{\omega_t^2}{\Delta U^2}[d^2 + g_t^2 d_t^2]\right)$$

$$f_{t2} = C_{t3}\exp(-C_{t4}\chi^2).$$

Symbols $\sigma, C_{b1}, C_{b2}, \kappa, C_{w1}, C_{w2}, C_{w3}, C_{v1}, C_{t1}, C_{t2}, C_{t3}, C_{t4}$ hold for constants. Symbol d holds for the distance to the closest surface.

The idea is to keep the S-A model near the wall and for the rest of computational domain, to transform the S-A model into a kind of Smagorinsky LES model. To accomplish this, a modified distance function is introduced:

$$\tilde{d} = min[d, C_{DES}\Delta], \tag{8.35}$$

where C_{DES} is a constant and Δ is the largest dimension of the grid cell in question.

8.5.3   Limited numerical scales (LNS) approach

The basic idea of the LNS model [Batten et al., 2004] is to multiply the Reynolds stress tensor, given by the RANS closure, by a blending function, which permits to switch from the RANS to the LES approach.

Validation of this hybrid approach in a finite volume/finite element framework on unstructured meshes can be found in [Camarri et al., 2005], [Koobus et al., 2007]. In these works, the standard $k - \varepsilon$ model [Launder and Spalding, 1979] is used for the RANS closure, in which the Reynolds stress tensor is modeled as follows, by introducing a turbulent eddy-viscosity, μ_t:

$$R_{ij} \simeq \mu_t \underbrace{\left[\frac{\partial \widetilde{u}_i}{\partial x_j} + \frac{\partial \widetilde{u}_j}{\partial x_i} - \frac{2}{3} \frac{\partial \widetilde{u}_l}{\partial x_l} \delta_{ij} \right]}_{\widetilde{P}_{ij}} - \frac{2}{3} \langle \rho \rangle k \delta_{ij} \ , \tag{8.36}$$

where the tilde denotes the Favre average, the overbar time averaging, δ_{ij} is the Krönecker symbol and k is the turbulent kinetic energy. The turbulent eddy-viscosity μ_t is defined as a function of k and of the turbulent dissipation rate of energy, ε, as follows:

$$\mu_t = C_\mu \frac{k^2}{\varepsilon} \ , \tag{8.37}$$

in which C_μ is a model parameter, set equal to the classical value of 0.09 and k and ε are obtained from the corresponding modeled transport equations [Launder and Spalding, 1979].

The LNS equations are then obtained from the RANS equations by replacing the Reynolds stress tensor R_{ij}, given by equation (8.36), with the tensor L_{ij}:

$$L_{ij} = \alpha R_{ij} = \alpha \mu_t \widetilde{P}_{ij} - \frac{2}{3} \langle \rho \rangle \, (\alpha k) \, \delta_{ij} \ , \tag{8.38}$$

where α is the damping function ($0 \leq \alpha \leq 1$), varying in space and time.

In the LNS model, the damping function is defined as follows:

$$\alpha = \min \left\{ \frac{\mu_{sgs}}{\mu_t}, 1 \right\} \tag{8.39}$$

in which μ_{sgs} is the SGS viscosity obtained from a LES closure model.

As discussed in [Batten et al., 2004], the model should work in the LES mode where the grid is fine enough to resolve a significant part of the turbulence scales, as in LES; elsewhere ($\alpha = 1$), the $k - \varepsilon$ RANS closure should be recovered.

8.5.4 A second-generation hybrid model

In this subsection, we present an hybrid RANS/LES model that is designed to combine two independent LES and RANS models.

As in Labourasse and Sagaut [Labourasse and Sagaut, 2002], the following decomposition of the flow variables is adopted:

$$W = \underbrace{< W >}_{RANS} + \underbrace{W^c}_{correction} + W^{SGS}$$

where $< W >$ are the RANS flow variables, obtained by applying an averaging operator to the Navier-Stokes equations, W^c are the remaining resolved fluctuations (*i.e.* $< W > +W^c$ are the flow variables in LES) and W^{SGS} are the unresolved or SGS fluctuations.

Writing the Navier-Stokes equations for the averaged flow $\langle W \rangle$ and applying a filtering operator, the LES equations are obtained and we get first a closure term given by a RANS turbulence model and then a SGS term. An equation for the resolved fluctuations W^c can thus be derived (see also [Labourasse and Sagaut, 2002]).

The basic idea of the proposed hybrid model is to solve the equation for the averaged flow in the whole domain and to correct the obtained averaged flow by adding the remaining resolved fluctuations (computed through the equation of the resolved fluctuations), wherever the grid resolution is adequate for a LES. To identify the regions where the additional fluctuations must be computed, we introduce a *blending function*, θ, smoothly varying between 0 and 1. When $\theta = 1$, no correction to $\langle W \rangle$ is computed and, thus, the RANS approach is recovered. Conversely, wherever $\theta < 1$, additional resolved fluctuations are computed; in the limit of $\theta \to 0$ the full LES approach is recovered. For θ going from 1 to 0, *i.e.* when, following the definition of the blending function, the grid resolution is intermediate between one adequate for RANS and one adequate for LES, the term containing the LES contribution in the equation of the resolved fluctuations is damped through multiplication by $1 - \theta$. Although it could seem rather arbitrary from a physical point of view, this is aimed to obtain a smooth transition between RANS and LES. More specifically, we wish to obtain a progressive addition of fluctuations when the grid resolution increases and the model switches from the RANS to the LES mode.

Summarizing, the ingredients of the proposed approach are: a RANS closure model, a SGS model for LES and the definition of the blending function.

8.5.4.1 RANS closure

As far the closure of the RANS equations is concerned, the standard $k - \varepsilon$ model is a good basis.

Standard $k - \varepsilon$ model

This is an eddy-viscosity model with a turbulent viscosity μ_t defined from two extra variables (two-equation model) the turbulent kinetic energy k and its dissipation rate ε, and given by expression (8.37). The Reynolds tensor, the main term to model, is then given by equation (8.36).

The k and ε fields are evaluated by solving the two extra transport equations:

$$\frac{\partial \langle \rho \rangle k}{\partial t} + (\langle \rho \rangle \widetilde{u}_j k)_{,j} = \left[\left(\mu + \frac{\mu_t}{\sigma_k} \right) \frac{\partial k}{\partial x_j} \right]_{,j} + R_{ij} \frac{\partial \widetilde{u}_i}{\partial x_j} - \langle \rho \rangle \varepsilon \quad , \tag{8.40}$$

$$\frac{\partial \langle \rho \rangle \varepsilon}{\partial t} + (\langle \rho \rangle \varepsilon \widetilde{u}_j)_{,j} = \left[\left(\mu + \frac{\mu_t}{\sigma_\epsilon} \right) \frac{\partial \varepsilon}{\partial x_j} \right]_{,j} + C_{\epsilon 1} \left(\frac{\varepsilon}{k} \right) R_{ij} \frac{\partial \widetilde{u}_i}{\partial x_j} - C_{\epsilon 2} \langle \rho \rangle \frac{\varepsilon^2}{k} \quad , \tag{8.41}$$

where constants $C_{\epsilon 1}$, $C_{\epsilon 2}$, σ_k and σ_ϵ are defined by:

$$C_{\epsilon 1} = 1.44 \qquad C_{\epsilon 2} = 1.92 \qquad \sigma_k = 1.0 \qquad \sigma_\epsilon = 1.3$$

Historically these constants are deduced from the application of the model to simple turbulent flows. They also can be mathematically derived by the Renormalization Group method, see for example [Yakhot and Orszag, 1986].

Wall treatment by wall law The k,ε variables, the velocity and the temperature can have stiff behavior and exhibit small scales near a wall. However, their behavior presents in many case some common features, such as obeying the logarithmic law for the tangent velocity component. In the theory and in the $k - \varepsilon$ formulation, this is a consequence of the equilibrium between turbulent kinetic energy production and dissipation, which arises in a large enough region close to wall. Wall law methods use this feature in order to avoid the costly discretization of the wall behavior of these variables. Closer to the wall, a different behavior needs be taken into account. This is well modeled in Reichardt's law (see for example [Hinze, 1959]) which writes in terms of U^+ and y^+:

$$U^+ = \frac{\overline{U}}{U_\tau} \quad , \tag{8.42}$$

$$y^+ = \frac{\rho U_\tau}{\mu} y \quad , \tag{8.43}$$

where $\overline{U}$ is the statistical average of velocity. The friction velocity U_τ is given by:

$$U_\tau = \sqrt{\frac{\tau_p}{\rho}} \quad . \tag{8.44}$$

Then we set:

$$U^+ = \frac{1}{\kappa} ln \left(1 + \kappa y^+ \right) + 7.8 \left(1 - e^{-\frac{y^+}{11}} - \frac{y^+}{11} e^{-0.33 y^+} \right) \quad . \tag{8.45}$$

where $\kappa = 0.41$ si the von Kàrmàn constant.

A low Reynolds version The low Reynolds $k - \varepsilon$ model proposed in [Goldberg et al., 1998], [Goldberg and Ota, 1990] is presented now. Let:

$$\mu_t = C_\mu f_\mu \rho \frac{k^2}{\epsilon} \tag{8.46}$$

where $C_\mu = 0.09$ is the usual coefficient and f_μ a so-called damping function (for damping the effect of model when closer to wall):

$$f_\mu = \frac{1 - e^{-A_\mu R_t}}{1 - e^{-R_t^{1/2}}} max(1, \psi^{-1}) \tag{8.47}$$

where $\psi = R_t^{1/2}/C_\tau$. The turbulent Reynolds number $R_t = k^2/(\nu\varepsilon)(\nu = \mu/\rho)$ allows to avoid evaluating the distance to wall. $A_\mu = 0.01$ is a constant. k et ε are solution of:

$$\frac{\partial \langle\rho\rangle k}{\partial t} + (\langle\rho\rangle \tilde{u}_j k)_{,j} = \left[\left(\mu + \frac{\mu_t}{\sigma_k}\right)\frac{\partial k}{\partial x_j}\right]_{,j} + R_{ij}\frac{\partial \tilde{u}_i}{\partial x_j} - \langle\rho\rangle\varepsilon \ , \quad (8.48)$$

$$\frac{\partial \langle\rho\rangle\varepsilon}{\partial t} + (\langle\rho\rangle\varepsilon\tilde{u}_j)_{,j} = \left[\left(\mu + \frac{\mu_t}{\sigma_\epsilon}\right)\frac{\partial \varepsilon}{\partial x_j}\right]_{,j} +$$

$$\left(C_{\epsilon 1} R_{ij}\frac{\partial \tilde{u}_i}{\partial x_j} - C_{\epsilon 2}\langle\rho\rangle\varepsilon + E\right) T_\tau^{-1} \ ,$$

$$\tag{8.49}$$

where T_τ is the realizable time scale:

$$T_\tau = \frac{k}{\epsilon} max(1, \psi^{-1}), \tag{8.50}$$

this time scale is k/ε for large values of R_t and thus of ψ, but tends to be equal to the Kolmogorov scale $C_\tau(\nu/\varepsilon)^{1/2}$ for $R_t << 1$. Constants are defined as $C_\tau = 1.41$, $C_{\epsilon 1} = 1.42$, $C_{\epsilon 2} = 1.83$, lastly:

$$E = \rho A_E V(\varepsilon T_\tau)^{0.5}\xi \tag{8.51}$$

where $A_E = 0.3$, $V = max(\sqrt{k}, (\nu\varepsilon)^{0.25})$ and $\xi = max(\frac{\partial k}{\partial x_i}\frac{\partial \tau}{\partial x_i}, 0)$, with $\tau = k/\varepsilon$.

8.5.4.2   RANS and LES combination

For the LES mode, we wish to recover the variational multiscale approach described in section 8.4. Thus, the Galerkin projection of the equations for the averaged flow and for the correction term in the proposed hybrid model become respectively:

$$\left(\frac{\partial \langle W\rangle}{\partial t}, \psi_l\right) + (\nabla \cdot F_c(\langle W\rangle), \psi_l) + (\nabla \cdot F_v(\langle W\rangle), \phi_l) =$$
$$- (\tau^{RANS}(\langle W\rangle), \phi_l) \quad l = 1, N \tag{8.52}$$

$$\left(\frac{\partial W^c}{\partial t}, \psi_l\right) + (\nabla \cdot F_c(\langle W\rangle + W^c), \psi_l) - (\nabla \cdot F_c(\langle W\rangle), \psi_l) +$$
$$(\nabla \cdot F_v(W^c), \phi_l) = (1 - \theta)\left[\left(\tau^{RANS}(\langle W\rangle), \phi_l\right) - \left(\tau^{LES}(W'), \phi_l'\right)\right] \quad l = 1, N \tag{8.53}$$

where $\tau^{RANS}(\langle W \rangle$ is the closure term given by a RANS turbulence model, W' and ϕ'_l denote the small resolved component of $\langle W \rangle + W^c$ and ϕ_l as defined in section 8.4, and $\tau^{LES}(W')$ is given by one of the SGS closures described in section 8.3.

As a possible choice for θ, the following function is used in the present study:

$$\theta = F(\xi) = tanh(\xi^2) \tag{8.54}$$

where ξ is the *blending parameter*, which should indicate whether the grid resolution is fine enough to resolve a significant part of the turbulence fluctuations, *i.e.* to obtain a LES-like simulation. The choice of the *blending parameter* is clearly a key point for the definition of the present hybrid model. In the present study, different options are proposed and investigated, namely: the ratio between the eddy-viscosities given by the LES and the RANS closures, $\xi_{VR} = \mu_s/\mu_t$, which is also used as a blending parameter in LNS [Batten et al., 2004] and $\xi_{LR} = \Delta/l_{RANS}$, l_{RANS} being a typical length in the RANS approach, *i.e.* $l_{RANS} = k^{3/2}\epsilon^{-1}$ and, Δ measures the local mesh size.

To avoid the solution of two different systems of PDEs and the consequent increase of required computational resources, equations (8.52) and (8.53) can be recast together as:

$$\left(\frac{\partial W}{\partial t}, \psi_l\right) + (\nabla \cdot F_c(W), \psi_l) + (\nabla \cdot F_v(W), \phi_l) =$$
$$-\theta\left(\tau^{RANS}(\langle W \rangle), \phi_l\right) - (1-\theta)\left(\tau^{LES}(W'), \phi'_l\right) \quad l = 1, N \tag{8.55}$$

where W stands now for $\langle W \rangle + W^c$.

Clearly, if only equation (8.55) is solved, $\langle W \rangle$ is not available at each time step. Two different options are possible: either to use an approximation of $\langle W \rangle$ obtained by averaging and smoothing of W, in the spirit of VMS, or to simply use in equation (8.55) $\tau^{RANS}(W)$. The second option is adopted here as a first approximation.

8.5.5   The interest in hybridizing RANS and VMS-LES

In this numerical part, we evaluate the performance of our new hybrid model for the simulation on unstructured grids of the flow around a circular cylinder. The obtained numerical results are contrasted with those predicted by RANS and various hybrid models, and compared with experimental data.

Flow around a circular cylinder (Hybrid RANS/LES) The new proposed hybrid model (*Fluctuation Correction Model*, FCM) has been applied to the simulation of the flow around a circular cylinder at $Re = 140000$ (based on the far-field velocity and the cylinder diameter). The domain dimensions are: $-5 \leq x/D \leq 15$, $-7 \leq y/D \leq 7$ and $0 \leq z/D \leq 2$ where x, y and z denote the streamwise, transverse and spanwise direction, respectively. The cylinder of unit diameter is centered on $(x, y) = (0, 0)$. Two grids have been used, the first one (GR1) has 4.6×10^5 nodes, while the second one has (GR2) 1.4×10^6 nodes. Both grids are composed of a structured part around the cylinder boundary and a unstructured part in the rest of the domain. The inflow conditions are the same as in the DES simulations of Travin et al. [Travin et al., 1999]. In particular, the flow is assumed to be highly turbulent by setting the inflow value of eddy-viscosity to about 5 times the molecular viscosity as in the DES simulation of Travin et al. [Travin et al., 1999]. This setting corresponds to a free-stream turbulence level $\overline{u'^2}/U_0$ (where u' is the inlet velocity fluctuation and U_0 is the free-stream mean velocity) of the order of 4%. As discussed also by Travin et al. [Travin et al., 1999], the effect of such a high level of free-stream turbulence is to make the boundary layer almost entirely turbulent also at the relatively moderate considered Reynolds number. For the purpose of these simulations, the Steger-Warming conditions [Steger and Warming, 1981a] are imposed at the inflow and outflow as well as on the upper and lower surface $(y = \pm H_y)$. In the spanwise direction periodic boundary conditions are applied and on the cylinder surface no-slip boundary conditions are set.

The RANS model is that based on the Low-Reynolds approach [Goldberg et al., 1998]. The LES closure is based on the VMS approach (see section 8.4). The SGS models used in the simulations are those described in section 8.3. The V6 scheme has been used and the numerical parameter γ_s, which controls the amount of numerical viscosity introduced in the simulation, has been set equal to 0.2. The main parameters characterizing the simulations carried out with the FCM are summarized in Table 8.10. The main flow bulk pa-

Simulation	Blending parameter	Grid	LES-SGS model
FCM1	VR	GR1	Smagorinsky
FCM2	LR	GR1	Smagorinsky
FCM3	LR	GR2	Smagorinsky
FCM4	LR	GR1	Vreman
FCM5	LR	GR1	Wale

Table 8.10: Supercritical flow past a cylinder. Simulation name and their main characteristics. VR stands for viscosity ratio and LR for length ratio.

rameters obtained in the present simulations are summarized in Table 8.11, together with the results of DES simulations in the literature and some experimental data. Let us analyze, first, the sensitivity to the blending parameter, by comparing the results of the simulation FCM1 and FCM2. The results are practically insensitive to the definition of the blending parameter. Conversely, the grid refinement produced a decrease of $\bar{C}_d$ and a delay in the boundary layer separation (compare FCM2 and FCM3). However, note that,

Data from	Re	$\overline{C_d}$	C_l'	St	l_r	θ_{sep}
FCM1	$1.4 \ 10^5$	0.62	0.083	0.30	1.20	108
FCM2	$1.4 \ 10^5$	0.62	0.083	0.30	1.19	108
FCM3	$1.4 \ 10^5$	0.54	0.065	0.33	1.13	115
FCM4	$1.4 \ 10^5$	0.65	0.077	0.28	1.14	109 (99)
FCM5	$1.4 \ 10^5$	0.66	0.094	0.28	1.24	109 (100)
Numerical data (DES)						
[Travin et al., 1999]	$1.4 \ 10^5$	0.57-0.65	0.08-0.1	0.28-0.31	1.1 -1.4	93-99
[Lo et al., 2005]	$1.4 \ 10^5$	0.6-0.81	–	0.29-0.3	0.6-0.81	101-105
Experiments						
[James et al., 1980]	$3.8 \ 10^6$	0.58	–	0.25	–	110
[Achenbach, 1968]	$5 \ 10^6$	0.7	–	–	–	112
[Schewe, 1983]	$8 \ 10^6$	0.52	0.06	0.28	–	–

Table 8.11: Supercritical flow past a cylinder. Main bulk flow quantities. $\overline{C_d}$ denotes the mean drag coefficient, C_l' the r.m.s. of the lift coefficient, St the Strouhal number, l_r the mean recirculation length: the distance on the centerline direction from the surface of the cylinder to the point where the time-averaged streamwise velocity is zero, and θ_{sep} the separation angle.

for unstructured grids, the refinement changes the local quality of the grid (in terms of homogeneity and regularity of the elements) and this may enhance the sensitivity of the results.

The agreement with the DES results is fairly good. As for the comparison with the experiments, as also stated in Travin et al. [Travin et al., 1999], since our simulations are characterized by a high level of turbulence intensity at the inflow, it makes sense to compare the results with experiments at higher Reynolds number, in which, although the level of turbulence intensity of the incoming flow is very low, the transition to turbulence of the boundary layer occurs upstream separation. The agreement with these high Re experiments is indeed fairly good, as shown in Table 8.11 and in Figure 8.20. The behavior of the separation angle requires a brief discussion. There is a significant discrepancy between the values obtained in DES and the experimental ones. For our simulations, the values of θ_{sep} shown in Table 8.11 are estimated by considering the point at which the C_p distribution over the cylinder becomes nearly constant (see e.g. Figure 8.20), as usually done in experimental studies. Indeed, the reported values are generally in better agreement with the experiments than those obtained by DES. However, if we estimate the separation angle from the streamlines of the average or instantaneous velocity fields, significantly lower values are found (reported in parentheses in Table 8.11 for the simulations FCM4 and FCM5); these values are closer to those obtained by DES. Finally, the model works in RANS mode in the boundary layer and in the shear-layers detaching from the cylinder, while in the wake a full LES correction is recovered. This is shown, for instance, in Figure 8.21, in which the instantaneous isocontours of spanwise vorticity obtained in the simulation

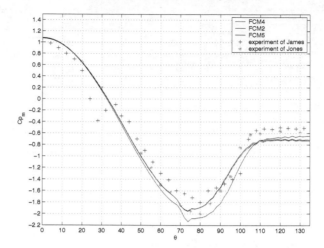

FIGURE 8.20: Supercritical flow past a cylinder. $\bar{C}_p$ on the cylinder surface compared to numerical and experimental results.

FCM2 are reported, to which the isolines of the blending function $\theta = 0.1$ and $\theta = 0.9$ are superimposed.

As for the behavior with grid refinement, it is, at least qualitatively, correct. Indeed, the extension of the zone in the detaching shear-layers in which the model works in RANS mode decreases with grid refinement, as shown for instance in Figure 8.22, reporting a zoom near the cylinder of the instantaneous isocontours of the blending function θ, obtained in the simulations FCM2 and FCM3.

8.6   Concluding remarks

Some modern methods for the prediction of separated flows around blunt (possibly complex) geometries have been presented in this chapter. The chapter recall the main steps followed in our research activity to integrate a turbulence modeling relying entirely or partially on the large-eddy simulation approach into an *industrial*-type numerical solver.

Our starting point was indeed a numerical solver for compressible flows on unstructured grids, through a mixed finite-element/finite-volume formulation, designed and validated for RANS simulations. The first main action was to move to a LES approach in order to deal with massively separated and highly unsteady 3D flows for which the RANS approach encounters accuracy problems. This carries some issues to be tackled. First of all, the

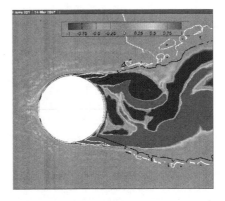

FIGURE 8.21: Supercritical flow past a cylinder. Instantaneous isocontours of spanwise vorticity (simulation FCM2). The black and white lines are the isolines of the blending function $\theta = 0.1$ and $\theta = 0.9$.

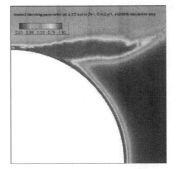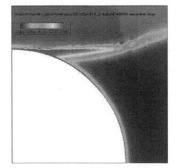

FIGURE 8.22: Supercritical flow past a cylinder. Instantaneous isocontours of the blending function θ. Simulations FCM2 (a) and FCM3 (b).

numerical method has to be adapted to the new turbulence treatment. The most critical point is the possible negative interaction between the LES closure model and the numerical viscosity, required for numerical stability of our co-located discretization. Our recommendation is to keep a physically-based closure model and to modify the numerical MUSCL reconstruction in order to obtain a numerical viscosity proportional to high-order (preferably 6th order) spatial derivatives. In this way, the numerical dissipation is concentrated on a narrow band of the smallest resolved scales and thus its interaction with the SGS dissipation, usually proportional to second-order spatial derivatives, is reduced. Moreover, a coefficient controls the numerical dissipation to the smallest amount required to stabilize the simulation.

As for SGS models, some difficulties due to the use of *classical* models on unstructured grids appear. For instance, the dynamic version of the Smagorinsky model gives, as generally found in the literature, more accurate results than the Smagorinsky model, but results in a dramatic increase in the computational cost, much larger than that observed for structured grids. This leads to prefer the variational multiscale formulation, which was adapted to the present mixed finite-element/finite-volume discretization on unstructured grids. As shown in the present chapter, this approach gives results as accurate as the dynamic Smagorinsky LES model but at computational costs comparable with those of the classical Smagorinsky LES model. Further, a paramount interest of VMS-LES is its lower damping of the largest resolved flow eddies.

Another critical issue for the use of LES for the simulation of flows at high Reynolds numbers is the near wall resolution. Indeed, the grid has to be fine enough to resolve a significant part of the turbulent scales, and this becomes particularly critical in the near-wall regions. This motivates to design new hybrid models, in which RANS and LES approaches are combined.

A new strategy for blending RANS and LES has been described here. It is based on a decomposition of the flow variables in a RANS part and a correction part, which takes into account the resolved fluctuations. To identify the zones in which the correction must be computed and added to the RANS part, a blending function is introduced so that the model works in RANS mode where the grid is coarse and tends with continuity to LES as the grid refinement becomes adequate. For the closure of the LES part, the VMS approach has been integrated in the proposed hybridization strategy. As a first choice, we use a simplified version of the model in which only one set of unknowns is computed. The proposed method has been applied to the hybrid simulation of a flow around a circular cylinder.

Through this chapter we have tried to show how the result of a computation can be assessed. The predictivity of LES methods remains a difficult issue, addressed in regular conferences, see [Meyers et al., 2008]. Modeling error remains relatively large. We have shown that mesh convergence is mandatory, while needing a careful analysis. However, through the combined use of a robust, but yet accurate enough, numerics, of unstructured grids and of VMS-

LES or hybrid RANS/LES models, the accurate prediction of a large class of industrial turbulent flows, including those characterized by massive 3D separation, is certainly within the reach of present numerical tools.

Chapter 9

Numerical algorithms for free surface flow

Alexandre Caboussat

University of Houston, Department of Mathematics
4800 Calhoun Rd,
Houston, Texas 77204-3008, USA

Guillaume Jouvet

Ecole Polytechnique Fédérale de Lausanne, Institute of Analysis and Scientific Computing
1015 Lausanne, Switzerland

Marco Picasso

Ecole Polytechnique Fédérale de Lausanne, Institute of Analysis and Scientific Computing
1015 Lausanne, Switzerland

Jacques Rappaz

Ecole Polytechnique Fédérale de Lausanne, Institute of Analysis and Scientific Computing
1015 Lausanne, Switzerland

9.1 Introduction

Free surface flows arise in numerous applications of many fields of physics or engineering. A list of applications includes (but is not limited to) fluid-structure interactions for blow flow modeling [Guidoboni et al., 2009], [Quarteroni and Formaggia, 2003], bubbles flow [Bunner and Tryggvason, 2002a], [Kuzmin and Turek, 2004], [Sussman et al., 2007], mold casting [Caboussat et al., 2005], [Cummins et al., 2005], [François et al., 2006], [Maronnier et al., 1999], glaciers [Jouvet et al., 2009], [Jouvet et al., 2008], [Rappaz and Reist, 2005], [Picasso et al., 2004], visco-elastic material such as glue or honey [Bonito et al., 2006], [Shelley et al., 1997], or aluminum processes [Flück et al., 2009]. Among all these applications, the common denominators are certainly the complicated behavior of the interfaces between the different phases, and the non-trivial coupling between various physical models on each side of those interfaces.

One closely related topic is the modeling of particle flow, that also involves the interaction between two phases [Glowinski et al., 2001], [Li and Ito, 2006], [Peskin, 1980]. One particular field of applications is again the numerical simulation of biological systems, such as red blood cells [Pan and Wang, 2009] or valves [de Hart, 2002].

For these various reasons, the modeling of multi-phases flow has been a very active field of research for the last decades, with an impressive amount of simulations and publications. We focus in this chapter on two particular applications: the modeling of liquid flows in the field of mold casting on the one hand, and the modeling of ice flows such as glaciers on the other hand. At first glance, these two applications have nothing in common. Mold casting usually involves liquid flows that propagate in complex geometries, with high Reynolds numbers, and many topological changes. Glaciers are modeled by almost stationary models, with very low Reynolds numbers and high viscosities. In both cases, the position of the interface is a crucial point to guarantee accurate results.

The literature contains numerous models for complex liquid-gas free surfaces problems, see e.g. [François et al., 2006], [Scardovelli and Zaleski, 1999]. In most of the numerical models [Bunner and Tryggvason, 2002a], [Codina and Soto, 2002], [Li and Renardy, 2000], [Sussman et al., 1998], [Sussman and Puckett, 2000], [Unverdi and Tryggvason, 1992], it has been assumed that the behavior of the liquid-gas mixture is that of an incompressible two-phase flow. Compressibility effects in two-phase flows have been considered e.g. in [Abgrall et al., 2003], [Abgrall and Saurel, 2003], [Shyue, 1999b], while methods mixing an incompressible liquid and a compressible gas have been proposed in [Caiden et al., 2001], [Fedkiw et al., 1998]. For ice flows, involving large viscosity, the influence of the surrounding medium is less important and usually considered as vacuum [Jouvet et al., 2008], [Maronnier et al., 1999], [Maronnier et al., 2003].

When dealing with changes of topologies of the fluid domain (formation of bubbles, breakage of ice), Eulerian methods are usually adopted. Such two-phases flows are computationally expensive in three space dimensions since (at least) both the velocity and pressure must be computed at each grid point of the whole multi-phases domain.

The purpose of this chapter is to present a mathematical and numerical framework that is able to include both the fast filling of liquid flow, and the slow evolution of ice flow. The features of the model include a *volume-of-fluid method* to track a liquid domain that can exhibit complex topology changes, and conserve the mass of ice/liquid. The incompressible liquid can be modeled either as a Newtonian or as a non-Newtonian fluid, by introducing nonlinear implicit laws for the fluid viscosity. In both cases, the liquid phase is initially surrounded by vacuum. Interfacial effects, such as the external pressure in the surrounding gas, or the surface tension effects, or both, can be taken into account on the liquid-gas free surface in a second step.

The novel numerical method is based on a time splitting algorithm [Glowinski, 2003] and a two-grids method [Hackbusch, 1985]. This original approach allows the various phenomena to be decoupled. Stabilized finite element techniques [Franca and Frey, 1992] are used to solve the diffusion phenomena using an unstructured mesh of the cavity containing the liquid. A forward characteristics method [Pironneau, 1989] on a structured grid allows advection phenomena to be solved efficiently. The complete description of the model can be found in [Bonito et al., 2006], [Caboussat, 2006], [Caboussat et al., 2005], [Jouvet et al., 2009], [Jouvet et al., 2008], [Maronnier et al., 1999], [Maronnier et al., 2003].

The chapter is organized as follows. Section 9.2 is a brief review of numerical methods for two-phases flow with free boundaries. An introduction to ice dynamics is given in Section 9.3. Section 9.4 introduces the two models, namely the model for incompressible liquid flows surrounded with vacuum and the model of ice flows for glacier simulations. Despite having very different behaviors, both models are incorporated into the same numerical framework, that allows to decouple advection and diffusion operators. A time splitting scheme is described in Section 9.5, and a multi-grids method is presented in Section 9.6. Approximation of interfacial effects, i.e. the effects of the surrounding gas and the surface tension effects, are discussed in Section 9.7. Finally, a wide range of numerical results are presented. Results of simulations of liquid fluids are presented in Section 9.8, while simulations and predictions of various glaciers based on real data are presented in Section 9.9.

9.2 A short review on two-phases flow with free surfaces

Free boundary problems are situations when the boundary of the domain of interest is actually unknown, or when there exists an interface between two (liquid) phases. Numerical methods for solving free surface problems have numerous engineering applications. Indeed, problems with free surfaces appear e.g. in fluid-structure interactions [Grandmont and Maday, 2000], [Quaini and Quarteroni, 2007], blood flows in moving arteries [Guidoboni et al., 2009], [Quarteroni and Formaggia, 2003], heart simulations [Peskin, 1977], [Peskin, 1980], motion of glaciers [Jouvet et al., 2009], [Jouvet et al., 2008], [Picasso et al., 2004], visco-elastic flows [Bonito et al., 2006], [Renardy et al., 2004], [Shelley et al., 1997], mold filling [Codina and Soto, 2002], [François et al., 2006], [Kothe et al., 1998], bubbles and droplets simulations [Kuzmin and Turek, 2004], [Popinet and Zaleski, 1999], [Renardy et al., 2003], naval engineering [Parolini and Quarteroni, 2004], or particle flows [Glowinski et al., 2001] to cite a few.

In order to account for the various characteristics of each of these problems, different numerical techniques have been introduced in the past decades. A short survey of some of them is the purpose of the following. From the theoretical point of view, the analysis of moving and free boundary problems remains complicated, due to the deeply nonlinear nature of these problems. A discussion of the state-of-the-art of the literature in that direction is not included here. Some partially incomplete information can be found in [Caboussat, 2005] and references therein. From the numerical point of view, the proposed methods have to balance between accuracy and computational efficiency, the latter being particularly important when dealing with industrial applications.

In the following, we focus mainly on two-phases free surface flows, namely a liquid and a gas enclosed in a bounded cavity denoted by Λ. In mold filling problems or metallurgy problems, large modifications in the topology of the liquid domain, complex geometries, large Reynolds numbers and turbulent flow can be expected. For ice flow in glaciers, the liquid domain does not suffer strong deformations, and the viscosity of the flow is higher. Despite these very different behaviors, an accurate approximation of the position of the surface of the liquid is required in both cases. In the following, a short review on the techniques for the simulation of free surface flow is given. This review is not exhaustive, but focuses on the topics developed more precisely in this chapter.

9.2.1 Incompressible and compressible media

In a two-phases flow, we can usually assume that both media are either incompressible, see e.g. [Bänsch, 2001], [Boffi and Gastaldi, 2003], [Sussman et al., 1998], or compressible, see e.g. [Abgrall et al., 2003], [Abgrall and Saurel, 2003], [Shyue, 1999a], [Shyue, 1999b]. In the incompressible case, the velocity **v** and pressure p usually satisfy the *incompressible Navier-Stokes equations* in the whole cavity and in a given time interval:

$$\rho \frac{\partial \mathbf{v}}{\partial t} + \rho(\mathbf{v} \cdot \nabla)\mathbf{v} - \nabla \cdot \boldsymbol{\sigma} = \mathbf{f},$$
$$\nabla \cdot \mathbf{v} = 0,$$

where $\boldsymbol{\sigma}$ is the stress tensor, ρ the density (constant by material) and **f** the external forces. The viscosity is denoted by μ. If the fluid is a Newtonian fluid,

$\sigma = 2\mu\mathbf{D}(\mathbf{v}) + p\mathbf{I}$, where $\mathbf{D}(\mathbf{v}) = \dfrac{1}{2}(\nabla\mathbf{v} + \nabla\mathbf{v}^T)$ is the rate of deformation tensor.

In the compressible case, the unknowns are the velocity $\mathbf{v}$, the pressure p and the total energy per unit volume E. They usually satisfy the *compressible Euler equations* in the whole liquid-gas domain. In both approaches, the same equations are solved in the whole domain but with different physical quantities in each phase.

One can also assume that an incompressible liquid is interacting with a compressible gas, see e.g. [Caboussat et al., 2005], [Caiden et al., 2001], [Fedkiw et al., 1998]. One typical example is to consider the (incompressible) Navier-Stokes equations in the liquid domain, and the Euler equations in the gas domain. In all cases, interfacial effects can be added on the free surface between media, and initial and boundary conditions are added to make the mathematical problem well-posed. Such models are expensive from a computational point of view since the conservation equations are solved in the whole two-phases domain. In contrast with these approaches, our model allows the computation to be performed only in the liquid region, thus reducing the computational complexity.

9.2.2 Eulerian vs. Lagrangian techniques

Independently of the nature of the flow (incompressible vs. compressible), a numerical procedure must be added to compute the motion of the interfaces between phases. Two main classes of methodologies can be distinguished: the *Lagrangian* methods and the *Eulerian* methods. The Lagrangian methods (including the *ALE methods*) are based on the displacement of a system of coordinates at each point of the free surface to track the displacement of the interface between the two phases. Eulerian methods introduce a new variable in the model, denoted here by φ, to track the presence or absence of one of the two phases in the whole domain, see Figure 9.1. Some reviews can be found e.g. in [Hou, 1995], [Kothe et al., 1998], [Scardovelli and Zaleski, 1999].

9.2.3 Lagrangian methods

Lagrangian methods, and Arbitrary Lagrangian Eulerian (ALE) methods [Maury, 1999], [Picasso et al., 2004], [Flück et al., 2009], are mainly used when the displacement of the liquid domain is small, and when large distortions or changes of topology of the domain are not expected. A review of Lagrangian techniques may be found for instance in [Kothe, 1998].

In Lagrangian methods, every point of the liquid domain (i.e. every particle of liquid) is moved with the liquid velocity. The interface is stretched to describe the position of the liquid domain. From the discrete viewpoint, the vertices of the mesh are moving at each time step, as illustrated in Figure 9.2. If the deformation of the liquid domain is large, for instance for high Reynolds

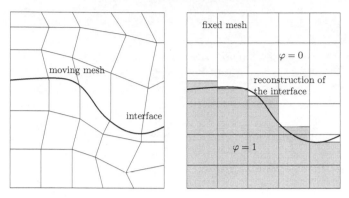

FIGURE 9.1: Two categories of methods for the tracking of free surfaces. Left: Lagrangian methods, right: Eulerian methods (volume-of-fluid formulation).

numbers or for flows with complex topological shapes, the stretching of the domains can lead to degenerate elements of the mesh. This situation is illustrated in Figure 9.2 in the case of a breaking wave: when the wave breaks, the topology of the liquid domain is changed. The mesh is distorted, its connectivity rules are no longer valid, the computation is impossible without a complete remeshing of the domain. This requires a total or partial remeshing of the domain (*rezoning*), which can be computationally very expensive.

On the other hand, Lagrangian methods offer a very good approximation of the interface, since the mesh always matches with the interface between the two media. Indeed, the interface is approximated by a piecewise linear curve and the accuracy on the numerical approximation of the boundary only depends on the mesh size. Boundary conditions, such as interfacial effects, are enforced easily at the grid points lying on the boundary.

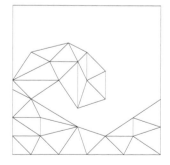

FIGURE 9.2: Lagrangian methods and moving meshes: between two time steps, the mesh in the liquid domain may be stretched until degeneracy.

9.2.4 Arbitrary Lagrangian Eulerian (ALE) methods

The *Arbitrary Lagrangian Eulerian* method has been introduced in [Hirt et al., 1974] to deal with small deformations. One important field of investigations for these methods is fluid-structure interactions [Gerbeau and Lelièvre, 2009]: structure dynamics are typically described in a Lagrangian frame of reference, while fluid equations can be written in Eulerian coordinates. The moving domain is denoted by Ω_t and given by the structure dynamics. It is mapped at each time t into a reference domain, denoted by Ω_0, by an *arbitrary* mapping $A_t : \Omega_0 \to \Omega_t$, where $\mathbf{x} = A_t(\xi)$, $\xi \in \Omega_0$. Introducing the notion of *domain velocity* $\mathbf{w} = \dfrac{\partial A_t}{\partial t}$ (also called *mesh velocity*), a generic advection equation or conservation law in the moving frame of reference, for instance

$$\frac{\partial \mathbf{v}}{\partial t} + \nabla F(\mathbf{v}) = 0, \tag{9.1}$$

defined on Ω_t, can be written under an ALE form, that is

$$\left.\frac{\partial \mathbf{v}}{\partial t}\right|_\xi - \mathbf{w}\nabla\mathbf{v} + \nabla F(\mathbf{v}) = 0, \tag{9.2}$$

thanks to the relation

$$\left.\frac{\partial \mathbf{v}}{\partial t}\right|_\xi = \left.\frac{\partial \mathbf{v}}{\partial t}\right|_\mathbf{x} + \mathbf{w}\nabla\mathbf{v}. \tag{9.3}$$

The mesh velocity $\mathbf{w}$ is distinguished from the motion of the liquid particles. The governing equations are modified and written on a fixed reference domain by adding the advection term (9.3). A careful choice of this velocity may prevent the elements of the mesh to become singular. Figure 9.3 illustrates the ALE method in the case of fluid-structure problems; the moving domain allows to take into account the flexibility of the structure of the artery without following every particle of liquid going through the section of it. The radius of the artery varies but the inflow and outflow sections are fixed.

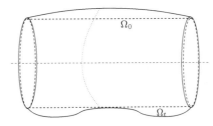

FIGURE 9.3: Arbitrary Lagrangian Eulerian methods. In full line: deformed domain, in dashed line: reference domain.

ALE methods have been widely used in the modeling of flow with small or slow domain deformation, i.e. blood flow [Boffi and Gastaldi, 2004], [Guidoboni et al., 2009], motions of glacier [Kirner, 2007], [Picasso et al., 2004], fluid dynamics in aluminum cells [Gerbeau et al., 2003], [Flück et al., 2009] and elasticity. For instance, in [Picasso et al., 2004], the mesh is moved at each time step to be adjusted to the height of the glacier, but the projection of the mesh on the bedrock surface on which the ice is lying is unchanged. Again changes of topology of the liquid domain cannot be dealt with. In aluminum cells, the magneto-hydrodynamic problem provides numerous difficulties such as the computation of the surface tension effects [Gerbeau and Lelièvre, 2009], [Flück et al., 2009], or the formation of bubbles of gas in the aluminum bath, [Romerio et al., 2005].

9.2.5   Particles methods

The particles methods are *mesh-free* methods that allow to track the motion of the liquid domain. The initial *particle in cell* method (PIC) has been introduced in the sixties [Harlow, 1964]. Then front-tracking methods (or surface tracking methods in two dimensions, see for instance [Torres and Brackbill, 2000]) have been derived from the original PIC method. The common goal is the capture the interface between the two media by using mass-free particles which are moving with the liquid velocity, independently of any mesh. This means that, for each marker j, its position $\mathbf{x}_j$ satisfies a Lagrangian equation $d\mathbf{x}_j/dt = \mathbf{v}(\mathbf{x}_j)$, where $\mathbf{v}$ is the liquid velocity.

Among them, the volume markers method introduces markers in the whole liquid domain, while the surface markers method introduces markers only on the interface between the liquid and the gas, as illustrated in Figure 9.4. The most famous volume markers method is the *markers and cells* (MAC) method [Harlow and Welch, 1965]. Several other methods have been inspired from the MAC method, see e.g. [McKee et al., 2004] for a review. A cell filled with volume markers is a liquid cell, while an cell empty of markers belongs to the gas domain. Surface markers methods have been introduced for instance in [Kothe, 1998], [Scardovelli and Zaleski, 1999]. The interface is defined by a set of particles, see Figure 9.4 (right). Examples may be found in [Aulisa et al., 2004], [Popinet and Zaleski, 1999].

These methods to track the interface are usually mixed with Eulerian methods [Bunner and Tryggvason, 2002a], [Bunner and Tryggvason, 2002b], [Kim and Lee, 2003], [Kim et al., 2003], [Shin and Juric, 2002], to capture details of the liquid front on finer scales (compared to the size of the fixed mesh) while keeping the advantages of the Eulerian approaches (see below) [Tryggvason et al., 2001]. Generally speaking, surface markers are more used than volume markers since they allow to track exactly the location of the interface with a smaller computational cost.

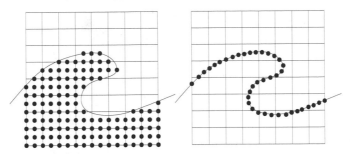

FIGURE 9.4: Markers methods. Left: volume markers method, the liquid domain is given by massless particles; right: surface markers: massless particles are located only on the free surface.

9.2.6 Immersed boundary methods

When dealing with a structure embedded within the flow, Lagrangian-type methods can be used to track the position of the structure. The *immersed boundary* methods, originally developed in [Peskin, 1977], [Peskin, 1980] for cardiovascular flows, are examples of such methods. While the incompressible (or compressible) equations are solved in the whole domain in Eulerian variables, a Lagrangian point of view is used for the simulation of an embedded flexible structure. The immersed structure is discretized as a sequence of points $\mathbf{X}_j$. The situation is illustrated in Figure 9.5.

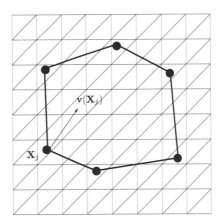

FIGURE 9.5: Immersed boundary methods. The structure immersed in the fluid is discretized by a finite number of points $\mathbf{X}_j$. The structure exerts a force on the surrounding fluid; conversely the structure is advected with the fluid velocity.

The structure is exerting a force on the fluid. Each of the vertices $\mathbf{X}_j$ of this discretization therefore influences the flow in a neighborhood of its position via a *local force*. The force exerted at the point $\mathbf{x}$ of the fluid is typically characterized by the sum of all local contributions:

$$\mathbf{F}(\mathbf{x}) = \sum_{|\mathbf{X}_j - \mathbf{x}| < \varepsilon} \mathbf{F}(\mathbf{X}_j) D_\varepsilon(\mathbf{x} - \mathbf{X}_j),$$

where $D_\varepsilon(\cdot)$ is a *numerical approximation* of the Dirac δ-function and ε is a neighborhood size, around the embedded interface [Pan and Wang, 2009]. This expression implies that all the vertices lying in a neighborhood of size ε of the point $\mathbf{x}$ in the fluid contribute to the force on the fluid. Conversely, those points $\mathbf{X}_j$ describing the flexible structure are moving in the fluid along the characteristics. The velocity of those points is obtained by interpolation of the velocity of the fluid in a neighborhood of the point $\mathbf{X}_j$. For instance, one way to define the interpolated velocities [Pan and Wang, 2009] can be as a weighted average:

$$\mathbf{V}(\mathbf{X}_j) = \sum_{\mathbf{x}:|\mathbf{X}_j - \mathbf{x}| < \varepsilon} \varepsilon^2 \mathbf{v}(\mathbf{x}) D_\varepsilon(\mathbf{X}_j - \mathbf{x}).$$

Applications to the modeling of (flexible) red blood cells have been presented in [Pan and Wang, 2009], [Tsubota et al., 2006], [Wang et al., 2009].

Eulerian methods introduce an additional unknown function in the whole cavity in order to track the presence of liquid or gas, together with a corresponding additional equation. The *level sets* methods, see [Codina and Soto, 2002], [Osher and Fedkiw, 2001], [Osher and Fedkiw, 2003], [Sussman et al., 1999], the *volume-of-fluid* (VOF) methods, see [Hirt and Nichols, 1981], [Rider and Kothe, 1998], [Xiao and Ikebata, 2003], or the *pseudo-concentration* methods, see e.g. [Thompson, 1986], [Unverdi and Tryggvason, 1992], are the most important examples.

Let us introduce here some notations. Let $T > 0$ denote a finite time and Λ a bounded cavity. Let $\Omega_t \subset \Lambda$, $t \in (0, T)$ denote the liquid domain. The gas domain is defined by $\Lambda \backslash \overline{\Omega}_t$ and the liquid-gas interface is denoted by Γ_t. The velocity and pressure are denoted by $\mathbf{v}$ and p. The additional unknown used to track the presence or not of liquid is denoted by $\varphi : \Lambda \times (0, T) \to \mathbb{R}$.

9.2.7　Level sets methods

In the level sets method, the free boundary is defined by the level line of a smooth function φ:

$$\Gamma_t = \left\{ \mathbf{x} \in \Lambda : \varphi(\mathbf{x}, t) = 0 \right\}, \quad t \in (0, T).$$

The level set function is assumed to be positive in the liquid, and negative in the gas:

$$\varphi(\mathbf{x},t) = \begin{cases} > 0, & \text{if } \mathbf{x} \in \Omega_t, \\ < 0, & \text{if } \mathbf{x} \in \Lambda \backslash \overline{\Omega}_t, \\ = 0, & \text{if } \mathbf{x} \in \Gamma_t. \end{cases}$$

The main advantage of the level sets approach is the regularity of the function $\varphi(\cdot,t)$ (typically $\varphi(\cdot,t)$ is $C^2(\Lambda)$ to approximate surface tension effects). It allows to accurately approximate interfacial effects on Γ_t, and capture motions of smooth surfaces. Figure 9.6 visualizes such a situation in two dimensions.

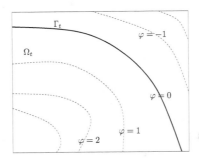

FIGURE 9.6: Level sets methods. The interface between the liquid and the gas is the zero level line of the function φ. The level lines corresponding to the negative values of φ are in the gas domain.

Without any addition of mass, the motion of the interface is obtained by advecting the values of φ with velocity field $\mathbf{v}$. Under the assumption that each particle of liquid moves with the liquid velocity along the characteristic curves and that each particle on the interface Γ_t remains on Γ_t, φ satisfies (in a weak sense):

$$\frac{\partial \varphi}{\partial t} + \mathbf{v} \cdot \nabla\varphi = 0. \tag{9.4}$$

More precisely, $\mathbf{v}(\mathbf{X}(t),t) = \mathbf{v}(\mathbf{X}(0),0)$, where $\mathbf{X}(t)$ is the trajectory of a fluid particle, thus $\mathbf{X}'(t) = \mathbf{v}(\mathbf{X}(t),t)$. Note that this relation requires to define a velocity $\mathbf{v}$ in both fluids, i.e. on both sides of the interface. When only the normal velocity to the interface is known (as it is the case in dendritic growth for instance [Burman et al., 2004]), (9.4) is transformed by setting $v_N = \mathbf{v} \cdot \dfrac{\nabla\varphi}{||\nabla\varphi||}$ to obtain an *Hamilton-Jacobi* equation:

$$\frac{\partial \varphi}{\partial t} + v_N ||\nabla\varphi|| = 0. \tag{9.5}$$

The quantity v_N denotes the normal velocity along the gradient of φ. The numerical solution of the Hamilton-Jacobi equation (9.5) and the computation of stable solutions is a difficult task, and usually requires high order algorithms. The ENO (essentially non-oscillatory) initially introduced by [Osher and Sethian, 1988], and the WENO (weighted essentially non-oscillatory) schemes, see for instance [Jiang and Peng, 2000], are based on high order finite differences approximations of each of the derivatives appearing in (9.5). Solution methods for the stationary Hamilton-Jacobi equation are based on *fast marching methods* [Sethian, 1999]; they are closely related to numerical methods for the Eikonal equation [Dacorogna et al., 2003].

One well-known drawback of the level sets approach is the *deterioration* of the function φ. When the function φ becomes flat in the neighborhood of the free surface, the accuracy of the interface decreases dramatically. One remedy consists in rescaling the function, so that it remains a *distance function* (the function $\varphi(\mathbf{x}, t)$ represents the signed distance between a point $\mathbf{x}$ of the domain and the interface at each time step).

Several techniques to re-build a function with such properties may be found in the literature, see e.g. [Cummins et al., 2005], [Gomes and Faugeras, 2000], but none of them can guarantee automatically the conservation of the mass of liquid. Among all these techniques, let us mention a standard technique for regularizing the level set function φ, see for example [Burger and Osher, 2005]. It consists in reinitializing it periodically by solving

$$\frac{\partial \varphi}{\partial t} + \text{sign}(\varphi_0) \left(\|\nabla \varphi\| - 1 \right) = 0, \tag{9.6}$$

with initial condition $\varphi(\mathbf{x}, 0) = \varphi_0(\mathbf{x})$, which admits as a stationary solution the signed distance to the initial interface $\{\mathbf{x} \in \Lambda : \varphi_0(\mathbf{x}) = 0\}$. Here, φ_0 plays the role of the level set function before regularization. The equation (9.6) is again a first-order Hamilton-Jacobi equation and can be solved numerically using the same methods as the ones discussed above.

REMARK 9.1 Pseudo-concentration methods   One variation of the level set approach is the pseudo-concentration method [Ceniceros and Roma, 2004], [Thompson, 1986], [Unverdi and Tryggvason, 1992]. The free boundary Γ_t is also defined by a level set of a smooth function in the neighborhood of the interface; this function has one fixed value in one media and another in the other media and is smoothed in a neighborhood of the interface. For instance

$$\varphi(\mathbf{x}, t) = \begin{cases} \gamma, & \text{if } d(\mathbf{x}) > \gamma, \\ -\gamma, & \text{if } d(\mathbf{x}) < -\gamma, \\ d(\mathbf{x}) & \text{if } |d(\mathbf{x})| < \gamma, \end{cases} \tag{9.7}$$

where $d(\mathbf{x})$ is the signed distance between $\mathbf{x}$ and the interface and γ is a fixed threshold. $\square$

REMARK 9.2   In both the level sets methods and the pseudo-concentration methods, φ is smooth around the interface. This property facilitates the numerical approximation of interfacial effects, since the normal vector $\mathbf{n}$ and the curvature κ of the interface may be expressed by

$$\mathbf{n}(\mathbf{x}, t) = -\frac{\nabla \varphi}{||\nabla \varphi||}, \qquad \kappa(\mathbf{x}, t) = -\nabla \cdot \frac{\nabla \varphi}{||\nabla \varphi||}. \qquad (9.8)$$

Such interfacial effects are crucial in slow moving processes, such as bubbly flows [Tryggvason et al., 2001], [Josserand and Zaleski, 2003] or some viscoelastic flows [Renardy et al., 2004], [Shelley et al., 1997]. ⬚

9.2.8 Volume-of-fluid methods

In the *volume-of-fluid* method (VOF), the fluid domain is tracked by its characteristic function (or *volume fraction of fluid*), that is

$$\varphi(\mathbf{x}, t) = \begin{cases} 1, & \text{if } \mathbf{x} \in \Omega_t, \\ 0, & \text{otherwise.} \end{cases} \qquad (9.9)$$

This function jumps over the interface. In most of the VOF methods, it satisfies the advection equation (9.4) when the velocity is continuous across the interface, see for instance [Kothe et al., 1999], [Rider and Kothe, 1998], since, from a Lagrangian point of view, the function φ is constant along the trajectories of the fluid particles. The mass of fluid is conserved as long as the numerical scheme is a discrete form of a conservative advection equation. The volume-of-fluid method is also called *volume tracking* method, since it is able to capture rather than follow the interface. The VOF method implicitly takes into account the possible changes of topology of the liquid domain.

The drawback of the volume-of-fluid methods is for the approximation of the interfacial effects; for instance, the computation of the curvature of the interface is difficult since it involves the derivatives of the non-smooth function φ at the interface. Therefore some regularization is needed. Moreover, since φ is discontinuous at the interface, numerical diffusion is added. One category of techniques tends to regularize the volume fraction of fluid in order to estimate its derivatives: in [Brackbill et al., 1992], [Caboussat, 2006], the smoothing of the volume fraction of fluid is done by convolution; in [Cummins et al., 2005], the smoothing with kernel functions is compared with the interpolation of the interface with a *height* function and an estimation via the reconstruction of a distance function.

The numerical diffusion introduced by the advection of the characteristic function implies that the reconstruction of the interface from the values of the volume fraction of fluid is difficult. Figure 9.7 visualizes one simple example, when $||\mathbf{v}|| \, \Delta t = 1.5 \Delta x$, where Δt denotes the time step, and Δx denotes the step size. The first row illustrates the effects of the numerical diffusion after two time steps, while the second row shows a corrected solution.

As illustrated in Figure 9.7 (top), the numerical diffusion induces a significant loss of accuracy around the interface. The most simple algorithm to

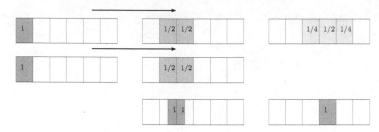

FIGURE 9.7: Volume-of-fluid method: Numerical diffusion of the volume fraction of fluid for a simple case of the advection of φ; Top: without any reconstruction of the interface; Bottom: with the SLIC method, in which the interface is defined as a straight line before the advection.

reduce the numerical diffusion is the SLIC algorithm (*Simple Line Interface Calculation*), developed first in [Noh and Woodward, 1976]. The SLIC algorithm reconstructs the liquid front by defining simple lines inside the cells, as illustrated in Figure 9.7 (bottom). These straight lines are parallel to one of the coordinate directions and their direction and position are deduced from the values of the volume fraction of fluid in the cells in the neighborhood of the considered cell, see e.g. [Hirt and Nichols, 1981], [Maronnier et al., 1999], [Maronnier et al., 2003].

The SLIC algorithm is only a first order algorithm with respect to the mesh size [Rider and Kothe, 1998]. The PLIC algorithm (*Piecewise Linear Interface Calculation*) is also geometric in nature and has been introduced to increase the order of convergence of SLIC for the reconstruction algorithm of the interface, see [Aulisa et al., 2003], [Rider and Kothe, 1998], [Scardovelli and Zaleski, 1999] and references therein. The PLIC methods are second order algorithms [Rider and Kothe, 1998]. Instead of constructing the interface by simple lines only along the coordinates directions, all directions are allowed for the interface line inside one cell, defining the interface as a discontinuous chain of segments, with asymptotic small discontinuities, see Figure 9.8 (right). The key point in the algorithm is the determination of the direction of each segment of the reconstructed interface, which corresponds basically to the characterization of its normal vector [Rider and Kothe, 1998].

Generalizations for multi-phases flow have been proposed for instance in [Caboussat et al., 2008], [Choi and Bussman, 2006], [Schofield et al., 2009].

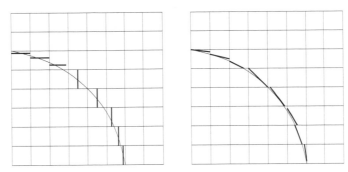

FIGURE 9.8: Volume-of-fluid method: Reconstruction of the interface with SLIC and PLIC algorithms. Left: the SLIC algorithm reconstructs the interface as a set of horizontal or vertical segments. Right: the PLIC algorithm allows the segments to be oriented arbitrarily.

9.3 Some preliminary remarks on ice and glacier modeling

Ice flow is another example of free surface flows, but with drastically different characteristics than those involved in mold filling applications. Glacier modeling is relevant for future management of natural risks, hydroelectric plants, water supply for agriculture, and tourism. Glaciers are strongly suffering from the current global warming, implying a retreat of most of them, as emphasized in the IPCC 2007 report [IPCC, 2007]. Actually, such a retreat has been documented by glaciologists in the Swiss Alps for more than a century, see e.g. [Mercanton, 1916].

The dynamics of a glacier are driven by different phenomena. Due to gravity, ice is flowing down to the valley. According to empirical experiences, ice can be considered as an *incompressible non-Newtonian fluid* and the inertial force can be disregarded. The mechanical law of ice deformation (so called "Glen's flow law") has been formalized in the 50's by Glen [Glen, 1958]. In the upper part of the glacier, snow fall exceeds snow melt, resulting in an addition of mass (*accumulation*). The melting of ice (due to higher temperatures) is more important in the lower part of the glacier (*ablation*). Therefore, given a glacier shape of non-Newtonian ice, the mass and momentum equations reduce to a nonlinear stationary Stokes problem under external weather conditions. A sliding law is added on the bedrock to account for the sliding of the ice on the rock. Note that the sliding of the ice is the most difficult quantity to measure in practice.

Due to the thin shape of glaciers, the first numerical models appearing in the literature (in the 80s) have been based on the assumption of a shallow flow. The most popular one, called *Shallow Ice Approximation*, has been introduced by Hutter, see [Hutter, 1983], [Paterson, 1994]. Later on, several higher order shallow ice models have been proposed, among them, the *First Order Approximation* [Blatter, 1995]. In any simplified model, the assumption of a shallow flow leads to the elimination of the pressure (usually assumed to be hydrostatic) and of some component of the velocity. The decrease of the number of unknowns reduce the computational cost of the numerical resolution, that can be done by using either finite differences [Blatter, 1995], [Bueler et al., 2007], [Pattyn, 2003], finite volumes [Deponti et al., 2006] or finite elements [Picasso et al., 2004]. More recently, the full Stokes model has been used for computing the ice flow [Gagliardini et al., 2007], [Jouvet et al., 2008] in order to avoid the drawbacks of shallow models, such as the inaccuracy of the solution at the glacier margins [le Meur et al., 2004].

When modeling the position and displacement of the ice, one can choose a Lagrangian (or Arbitrary Lagrangian Eulerian) approach, this being justified that the ice motion is slow and the deformation is small [Deponti et al., 2006], [Picasso et al., 2004]. These methods are adapted to a coupling with simplified models, such as the 'Shallow Ice Approximation' or 'First Order Approximation' models. Indeed, Lagrangian or ALE methods can estimate accurately the ice surface slope that appears in the equations of any shallow model. However this type of approach does not allow to take into account changes of topology (typically when the glacier breaks into several pieces over an accidented bedrock) that happen over long periods of time. Following section 9.2, we focus on Eulerian methods which overcome this drawback.

Level set methods in glaciology have been considered in [Pralong and Funk, 2004] to compute the onset of crevasse formation in 2D, but conservation of mass is hard to obtain. Therefore a volume-of-fluid approach is favored here to conserve the mass of ice. Mass conservation along the ice-air interface yields a transport equation, similar to (9.4) which can be used to determine the evolution of the glacier shape. The major difference with the above mentioned model, resides in the changes in the ice mass due to accumulation/ablation. A source term – the so-called mass balance – is added to the right-hand side of this transport equation to account for those terms containing the climatic input of the model. The determination of the glacier surface mass balance results from a combination of long-term measurements, performed by glaciologists, and parameter identifications, based on climate data and observations of surface elevation change [Huss et al., 2008a]. More details about the measured bedrock elevation can be found in [Farinotti et al., 2009].

Climatic inputs are not entirely predicable for future times, and several scenarios can be explored in order to predict the retreat of glaciers. Ultimately, based on climate models in seasonal resolution [Frei, 2007], [Huss et al., 2008b], climatic scenarios can be defined allowing the computation of future mass balances, and predictions can be made.

9.4   Modeling

9.4.1   Modeling of liquid flow

Inspired from [Maronnier et al., 1999], [Maronnier et al., 2003], we consider an incompressible liquid flow surrounded by vacuum. Let Λ, with boundary $\partial\Lambda$, be a cavity of $\mathbb{R}^d$, $d = 2, 3$ in which a liquid must be confined, and let $T > 0$ be the final time of simulation. For any given time $t \in (0, T)$, let $\Omega_t \subset \Lambda$, with boundary $\partial\Omega_t$, be the domain occupied by the liquid, let $\Gamma_t = \partial\Omega_t \backslash \partial\Lambda$ be the free surface between the liquid and the surrounding medium, and let Q_T be the space-time domain containing the liquid, i.e. $Q_T = \{(\mathbf{x}, t) : \mathbf{x} \in \Omega_t, 0 < t < T\}$.

In the liquid region, the velocity field $\mathbf{v} : Q_T \to \mathbb{R}^d$ and the pressure field $p : Q_T \to \mathbb{R}$ are assumed to satisfy the time-dependent, incompressible Navier-Stokes equations, that is

$$\rho\frac{\partial\mathbf{v}}{\partial t} + \rho(\mathbf{v}\cdot\nabla)\mathbf{v} - 2\nabla\cdot(\mu\mathbf{D}(\mathbf{v})) + \nabla p = \mathbf{f} \qquad \text{in } Q_T, \qquad (9.10)$$

$$\nabla\cdot\mathbf{v} = 0 \qquad \text{in } Q_T. \qquad (9.11)$$

Here $\mathbf{D}(\mathbf{v}) = 1/2(\nabla\mathbf{v} + \nabla\mathbf{v}^T)$ denotes the rate of deformation tensor, ρ the constant density, μ is the dynamic viscosity, and $\mathbf{f}$ the external forces.

REMARK 9.3 Various choices of viscosity   For laminar isothermal flows, the dynamic viscosity μ is constant $\mu = \mu_L$ (Newtonian fluid). When considering turbulent flows, a simple turbulent viscosity can be added $\mu = \mu_L + \mu_T$, where:

$$\mu_T = \mu_T(\mathbf{v}) = \alpha_T\rho\sqrt{2\mathbf{D}(\mathbf{v}) : \mathbf{D}(\mathbf{v})},$$

where α_T is a parameter to be chosen. The use of a turbulent viscosity is required when large Reynolds numbers and thin boundary layers are involved (Non-Newtonian fluid). When considering Bingham flows (in mud flows or avalanches for instance), a plastic viscosity can be added:

$$\mu_B = \frac{\alpha_0\rho}{\sqrt{2\mathbf{D}(\mathbf{v}) : \mathbf{D}(\mathbf{v}) + \varepsilon}},$$

where α_0 and ε are two parameters to be chosen. Another nonlinear model for the viscosity is presented in section 9.4.2 when modeling ice flow in glaciers.
☐

We use a volume-of-fluid formulation, and let $\varphi : \Lambda \times (0, T) \rightarrow \mathbb{R}$ be the characteristic function of the liquid domain Q_T. The function φ equals one if the liquid is present, zero if it is not, thus $\Omega_t = \{\mathbf{x} \in \Lambda : \varphi(\mathbf{x}, t) = 1\}$. In order to describe the kinematics of the free surface, φ must satisfy (in a weak sense):

$$\frac{\partial \varphi}{\partial t} + \mathbf{v} \cdot \nabla \varphi = 0 \qquad \text{in } \Lambda \times (0, T), \tag{9.12}$$

where the velocity $\mathbf{v}$ is extended continuously in the neighborhood of Q_T such that $\nabla \cdot \mathbf{v} = 0$ [Cattabriga, 1961]. Relationship (9.12) is actually defined in a neighborhood of the interface Γ_t. At initial time, the characteristic function of the liquid domain φ is given, which defines the initial liquid region $\Omega_0 = \{\mathbf{x} \in \Lambda : \varphi(\mathbf{x}, 0) = 1\}$. The initial velocity field $\mathbf{v}(0) = \mathbf{v}_0$ is prescribed in Ω_0. A summary of the notations is illustrated in Figure 9.9, for the situation of the filling of a cavity.

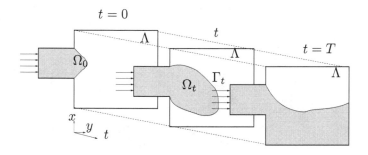

FIGURE 9.9: Filling of a cavity with a liquid. The fluid is injected from the left side and fills the cavity. Notations in the space-time domain.

The boundary conditions for the velocity field are the following. On the boundary of the liquid region being in contact with the walls (i.e. $\partial \Omega_t \cap \partial \Lambda$), inflow, slip or Signorini boundary conditions are enforced, see [Maronnier et al., 1999], [Maronnier et al., 2003]. On the free surface Γ_t, we establish the equilibrium of forces on the free surface. Tangential forces are neglected on the free boundary, as the two media are assumed not to 'slide' on each other. When we have two fluids with both velocity $\mathbf{v}$ and pressure p defined on both sides of the interface, the equilibrium balance at the interface reads

$$[-p\mathbf{n} + 2\mu \mathbf{D}(\mathbf{v}) \cdot \mathbf{n}] = \sigma \kappa \mathbf{n} \qquad \text{on } \Gamma_t, \quad t \in (0, T), \tag{9.13}$$

where $[\cdot]$ denotes the jump across the interface, σ is the surface tension coefficient (a physical constant that depends on the two media), κ is the curvature of the interface, and $\mathbf{n}$ is the unit normal vector of the liquid-gas free surface oriented outside the liquid.

In a first step we consider the interaction between a liquid and a vacuum; this implies that both velocities and pressure in the vacuum are vanishing, and the vacuum does not exert any force on the fluid. If, in addition, we neglect surface tension effects, the boundary condition (9.13) on the free surface becomes

$$-p\mathbf{n} + 2\mu\mathbf{D}(\mathbf{v}) \cdot \mathbf{n} = 0 \qquad \text{on } \Gamma_t, \quad t \in (0, T). \tag{9.14}$$

The mathematical model is thus well-posed (i.e. with an adequate number of boundary and initial conditions). The addition of some terms in (9.13) are discussed in section 9.7.

9.4.2 Modeling of ice flow

Let us extend the model presented in section 9.4.1 to glacier dynamics, and highlight the similarities and differences. Unless specified otherwise, the notations are similar.

At time t, the ice domain is denoted by Ω_t, the bedrock-ice interface is $\Gamma_{B,t}$ and the ice-air interface is Γ_t. The ice region in the space-time domain is denoted by Q_T, while $\mathbf{v}$ and p still denote the ice velocity and pressure, respectively. When considering the motion of a glacier during years or centuries, ice can be considered as an incompressible non-Newtonian fluid. Moreover, a dimensionless scaling shows that inertial terms can be disregarded. Therefore, the mass and momentum equations reduce at time t to a stationary nonlinear Stokes problem in the ice domain: Find $\mathbf{v} : \Omega_t \to \mathbb{R}^d$ and $p : \Omega_t \to \mathbb{R}$ such that

$$- 2\nabla \cdot (\mu(\mathbf{v})\mathbf{D}(\mathbf{v})) + \nabla p = \rho\mathbf{g}, \tag{9.15}$$

$$\nabla \cdot \mathbf{v} = 0. \tag{9.16}$$

Here the right-hand side force $\mathbf{f} = \rho\mathbf{g}$ only incorporates gravity effects in that particular case. Glen's flow law [Glen, 1958], [Hutter, 1983] holds for the viscosity $\mu = \mu(\mathbf{v})$. More precisely, for a given velocity field $\mathbf{v}$, the viscosity μ satisfies the following implicit nonlinear equation:

$$\frac{1}{2\mu} = A \left(\sigma_0^{m-1} + \left(2\mu \sqrt{\frac{1}{2}\left(\mathbf{D}(\mathbf{v}) : \mathbf{D}(\mathbf{v})\right)} \right)^{m-1} \right), \tag{9.17}$$

where A is a positive number known as the rate factor, $m \geq 1$ is Glen's exponent, and $\sigma_0 > 0$ is a regularization parameter which prevents infinite viscosity for zero strain. It should be noted that A depends on ice temperature but, since temperature variations are very small in most Alpine glaciers, A

can be considered as a constant. It is easy to see that $\mu = \mu(\mathbf{v})$ is actually a function of $s := \sqrt{\frac{1}{2}(\mathbf{D}(\mathbf{v}) : \mathbf{D}(\mathbf{v}))}$ only. Therefore we will write in the sequel $\mu = \mu(s)$. Moreover, when $m > 1$, it can be shown that μ is a positive, strictly decreasing function with respect to s [Colinge and Rappaz, 1999], [Rappaz and Glowinski, 2003]; the viscosity $\mu = \mu(s)$ is bounded from above by its value at $s = 0$ and has the following asymptotic behavior when s goes to the infinity:

$$\mu(s) = \mathcal{O}(s^{1/m-1}). \tag{9.18}$$

When $m = 1$, then μ is constant and the above problem corresponds to a Newtonian laminar fluid. In glaciology models, m is often taken equal to 3, see e.g. [Gudmundsson, 1999]. An example of this situation is visualized in Figure 9.10.

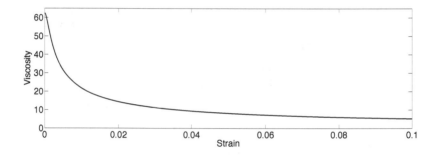

FIGURE 9.10: Viscosity (unit: [bar year]) with respect to the strain's norm $s = \sqrt{\frac{1}{2}(\mathbf{D}(\mathbf{v}) : \mathbf{D}(\mathbf{v}))}$ (unit: [year^{-1}]). The parameters are $A = 0.08$ [bar^{-3} year^{-1}], $m = 3$ and $\sigma_0 = \sqrt{0.1}$ [bar].

The boundary conditions corresponding to (9.15) are the following. Since the ice is surrounded by vacuum, no force is applied on the ice-air interface Γ_t:

$$-2\mu(s)\mathbf{D}(\mathbf{v})\mathbf{n} + p\mathbf{n} = \mathbf{0}, \tag{9.19}$$

where $\mathbf{n}$ is the unit outer normal vector on the boundary of the ice domain Ω_t. On the bedrock-ice interface $\Gamma_{B,t}$, ice may slip or not, according to the bedrock characteristics, and therefore a mix of slip and no-slip boundary conditions are applied on $\Gamma_{B,t}$. The no-slip conditions corresponds to imposing $\mathbf{v} = \mathbf{0}$ on the boundary. Many sliding laws have been proposed, see for instance [Gagliardini et al., 2007], [Schoof, 2005], [Vieli et al., 2000]. A nonlinear law that links the shear stress to the tangent velocity is considered here:

$\mathbf{v} \cdot \mathbf{n} = 0$ and $(2\mu(s)\mathbf{D}(\mathbf{v})\mathbf{n}) \cdot \mathbf{t}_i = -\alpha \mathbf{v} \cdot \mathbf{t}_i$, $i = 1, 2$, on a part of $\Gamma_{B,t}$,

$$(9.20)$$

where $\mathbf{t}_1$, $\mathbf{t}_2$ are two orthogonal vectors tangent to the boundary $\Gamma_{B,t}$, and α is the sliding coefficient. Following [Hutter, 1983, page 454] $\alpha = \alpha(\mathbf{v})$ is given by

$$\alpha = \frac{C}{(\|\mathbf{v}\| + s_0)^{1 - \frac{1}{m}}}, \tag{9.21}$$

where m is the Glen exponent, C is a positive value tuned from experiments and s_0 is a small numerical parameter which prevents $\alpha \to \infty$ when the velocity goes to zero.

The well-posedness of the nonlinear Stokes problem (9.15)-(9.16)-(9.17), supplemented by the boundary conditions (9.19) in a prescribed smooth domain Ω can be proved using the property (9.18), proceeding as in [Colinge and Rappaz, 1999], [Schoof, 2010].

The model for the volume fraction of ice is very similar to the one described in section 9.4.1 for the incompressible liquid flow, and is based on the characteristic function $\varphi : \Lambda \times (0, T) \to \mathbb{R}$ of the ice domain. The use of the characteristic function φ allows the description of topological changes that have been observed in glaciers during the last century.

In the absence of snow fall or melting, the volume fraction of ice satisfies (9.12), in a weak sense in the space-time domain. In other words, φ is constant along the trajectories of the fluid particles $\mathbf{X}(t)$ which are given by

$$\mathbf{X}'(t) = \mathbf{v}(\mathbf{X}(t), t).$$

However, in glaciers, the total mass constantly changes, due to snow fall and melting phenomena. *Accumulation* is the sum of all processes in which a glacier gains in mass, such as snow precipitation or snow redistribution due to wind. Conversely, *ablation* is the sum of all processes in which a glacier loses in mass, such as erosion or melting due to high temperature or solar radiation. The *mass balance* function $b(\mathbf{x}, t)$ is the water equivalent of ice height added or removed along the ice-air interface Γ_t within one year. It is a function of $(x, y, S(x, y, t), t)$ where $S(x, y, t)$ is the elevation of the ice-air free surface at point (x, y) at time t with respect to the horizontal plane Oxy. In any alpine glacier, two areas can be distinguished: the *accumulation zone* is the region where $b > 0$, and is usually localized in the high-elevation regions. The *ablation zone* is the region where $b < 0$, and is localized in the lower reaches. The elevation at which the mass balance satisfies $b = 0$ is called the *equilibrium line altitude* (ELA), see Figure 9.11.

The *mass balance* is mainly determined by the climate. It can be measured by glaciologists for instance using stakes drilled into the ice. In the applications of section 9.9, several models of various complexity are

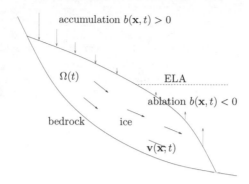

FIGURE 9.11: 2D section of a glacier and illustration of the mass-balance function b. When $b > 0$ (resp. $b < 0$), there is accumulation (resp. ablation) of ice at the ice-air interface. The line $b = 0$ defines the equilibrium line altitude (ELA).

proposed in order to obtain b. In section 9.9.1, the function b is given by a simple formula in which the equilibrium line altitude is the main parameter. In section 9.9.2, the function b accounts a more important set of parameters and data like precipitation and melting patterns based on daily weather data, wind patterns, and the topology of the bedrock. The coefficients involved are the result of a parameter identification problem [Huss et al., 2008a].

To account for the mass changes (accumulation and ablation processes), a source term must be added to the right-hand side of (9.12). Given the mass-balance function b and following [Jouvet et al., 2009], [Jouvet et al., 2008], (9.12) is modified as follows:

$$\frac{\partial \varphi}{\partial t} + \mathbf{v} \cdot \nabla \varphi = b \delta_{\Gamma_t}, \tag{9.22}$$

where δ_{Γ_t} is the Dirac function on the ice-air interface Γ_t which satisfies, by definition,

$$\int_V f \, \delta_{\Gamma_t} \, d\mathbf{x} = \int_{V \cap \Gamma_t} f \, dS,$$

for all volume V and all smooth function $f : V \rightarrow \mathbb{R}$. A physical interpretation can be obtained by writing the conservation of mass in an arbitrary volume V, as illustrated in Figure 9.12. Indeed, consider an arbitrary volume V contained in the cavity Λ and containing the ice-air interface Γ_t. Integrating (9.22) over V yields, using the divergence theorem:

$$\frac{d}{dt} \int_V \varphi d\mathbf{x} + \int_{\partial V} \mathbf{v} \cdot \mathbf{n} \varphi dS = \int_{\Gamma_t \cap V} b \, dS, \tag{9.23}$$

thus the time derivative of the volume of ice contained within V plus the flux of ice entering or leaving V equals the amount of ice added or removed by accumulation or ablation.

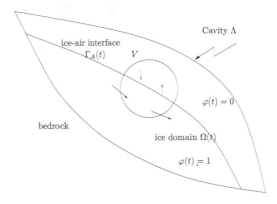

FIGURE 9.12: The mass conservation principle corresponding to (9.22) in an arbitrary volume V (2D figure) is given by (9.23).

Our goal is therefore to find the volume fraction of ice φ in the whole cavity Λ, together with the velocity $\mathbf{v}$ and pressure p in the ice domain only, which satisfy (9.15)-(9.16)-(9.22). Boundary conditions (9.19)-(9.20) are added, as well as the initial volume fraction of ice $\varphi(\mathbf{x}, 0)$, or equivalently the initial ice domain Ω_0.

REMARK 9.4 The ice thickness function $H : \mathbb{R}^{d-1} \times [0, T] \to \mathbb{R}_+$ can be associated to the volume fraction of ice φ (and vice versa) through the following relations:

$$H(x, y, t) = \int_{B(x,y)}^{S(x,y,t)} \varphi(\mathbf{x}, t) dz, \qquad (9.24)$$

where $B(x, y)$ is the bedrock elevation, and $S(x, y, t)$ is the elevation of the ice-air free surface with respect to the horizontal plane Oxy. Conversely,

$$\varphi(\mathbf{x}, t) = \begin{cases} 1, & \text{if } 0 \leq z - B(x, y) \leq H(x, y, t), \\ 0, & \text{otherwise.} \end{cases} \qquad (9.25)$$

This equivalence will be used for the design of a numerical technique to solve (9.22) (see section 9.6). ∎

9.5   Time splitting scheme

An implicit splitting algorithm is proposed to solve the liquid flow (9.10)-(9.12) and the ice flow (9.15)-(9.16)-(9.22) respectively, by decoupling the advection operators from the diffusion ones. Let $0 = t^0 < t^1 < t^2 < \ldots < t^N = T$ be a subdivision of the time interval $[0, T]$, define $\delta t^n = t^{n+1} - t^n$ the n-th time step, $n = 0, 1, 2, \ldots, N$, δt being the largest time step. Let φ^n, $\mathbf{v}^n$, p^n, Ω^n be approximations of φ, $\mathbf{v}$, p, Ω_t at time t^n, respectively. Then the approximations φ^{n+1}, $\mathbf{v}^{n+1}$, p^{n+1}, Ω^{n+1} at time t^{n+1} are computed by means of an implicit splitting algorithm, as illustrated in Figure 9.13 and in Figure 9.14. This splitting procedure is now detailed for both liquid and ice flow separately.

9.5.1   Liquid flow

Two advection problems are solved first, leading to a prediction of the new velocity $\mathbf{v}^{n+1/2}$ together with the new approximation of the characteristic function φ^{n+1} at time t^{n+1}, which allows to determine the new liquid domain Ω^{n+1} (and the new liquid interface Γ^{n+1}).

Then a generalized Stokes problem is solved on Ω^{n+1} with initial condition $\mathbf{v}^{n+1/2}$ and boundary condition (9.14) on the liquid interface Γ^{n+1}, Dirichlet, slip or Signorini-type conditions on the boundary of the cavity Λ, to obtain the velocity $\mathbf{v}^{n+1}$ and pressure p^{n+1} in the liquid.

This splitting algorithm is of order $\mathcal{O}(\delta t)$, see e.g. [Marchuk, 1990], and allows the motion of the free surface to be decoupled from the diffusion step, which consists in solving a Stokes problem in a fixed domain [Glowinski, 2003]. These two problems are detailed in the following.

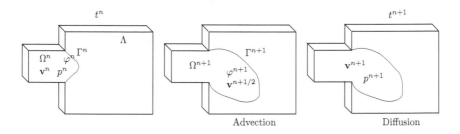

FIGURE 9.13: The splitting algorithm (from left to right). Two advection problems are solved to determine the new approximation of the characteristic function φ^{n+1}, the new liquid domain Ω^{n+1} and the predicted velocity $\mathbf{v}^{n+1/2}$. Then, a generalized Stokes problem is solved in the new liquid domain Ω^{n+1} in order to obtain the velocity $\mathbf{v}^{n+1}$ and pressure p^{n+1}.

Advection step for liquid flow We solve between times t^n and t^{n+1} the two advection problems:

$$\frac{\partial \mathbf{v}}{\partial t} + (\mathbf{v} \cdot \nabla)\mathbf{v} = 0, \qquad \frac{\partial \varphi}{\partial t} + \mathbf{v} \cdot \nabla \varphi = 0, \tag{9.26}$$

with initial conditions $\mathbf{v}^n$ and φ^n and boundary conditions on the inflow boundary. This step is solved exactly by the method of characteristics [Maury, 1996], [Pironneau, 1989] which yields a prediction of the velocity $\mathbf{v}^{n+1/2}$ and the characteristic function of the new liquid domain φ^{n+1}:

$$\mathbf{v}^{n+1/2}(\mathbf{x} + \delta t^n \mathbf{v}^n(\mathbf{x})) = \mathbf{v}^n(\mathbf{x}), \quad \varphi^{n+1}(\mathbf{x} + \delta t^n \mathbf{v}^n(\mathbf{x})) = \varphi^n(\mathbf{x}), \tag{9.27}$$

for all $\mathbf{x} \in \Omega^n$. Then, the new liquid domain Ω^{n+1} is defined as

$$\Omega^{n+1} = \{ \mathbf{x} \in \Lambda : \varphi^{n+1}(\mathbf{x}) = 1 \}.$$

Diffusion step for liquid flow The diffusion step consists in solving a generalized Stokes problem on the domain Ω^{n+1} using the predicted velocity $\mathbf{v}^{n+1/2}$ and the boundary condition (9.14). The following backward Euler scheme is used:

$$\rho \frac{\mathbf{v}^{n+1} - \mathbf{v}^{n+1/2}}{\delta t^n} - 2\nabla \cdot \left(\mu \mathbf{D}(\mathbf{v}^{n+1}) \right) + \nabla p^{n+1} = \mathbf{f}^{n+1} \quad \text{in } \Omega^{n+1}, \tag{9.28}$$

$$\nabla \cdot \mathbf{v}^{n+1} = 0 \quad \text{in } \Omega^{n+1}, \tag{9.29}$$

where $\mathbf{v}^{n+1/2}$ is the prediction of the velocity obtained with (9.27) after the advection step. The boundary conditions on the free surface are given by (9.14). The weak formulation corresponding to (9.28)-(9.29)-(9.14) therefore consists in finding $\mathbf{v}^{n+1}$ and p^{n+1} such that i) $\mathbf{v}^{n+1}$ satisfies the essential boundary conditions on $\partial \Lambda$ and ii)

$$\rho \int_{\Omega^{n+1}} \frac{\mathbf{v}^{n+1} - \mathbf{v}^{n+1/2}}{\delta t^n} \cdot \mathbf{w} dx + 2 \int_{\Omega^{n+1}} \mu \mathbf{D}(\mathbf{v}^{n+1}) : \mathbf{D}(\mathbf{w}) dx \tag{9.30}$$

$$- \int_{\Omega^{n+1}} p^{n+1} \nabla \cdot \mathbf{w} dx - \int_{\Omega^{n+1}} \mathbf{f}^{n+1} \cdot \mathbf{w} dx - \int_{\Omega^{n+1}} q \nabla \cdot \mathbf{v}^{n+1} dx = 0,$$

for all test functions $(\mathbf{w}, q)$ such that $\mathbf{w}$ is compatible with the essential boundary conditions on $\partial \Lambda$.

REMARK 9.5 The numerical treatment of the slip boundary condition $\mathbf{v} \cdot \mathbf{n} = 0$ is done in a weak sense by introducing a penalized term in the left-hand side of the weak formulation (9.30). Typically, this term reads

$$C_{\text{pen}} \int_{\Gamma_S^{n+1}} \left((\mathbf{v}^{n+1} \cdot \mathbf{n})(\mathbf{w} \cdot \mathbf{n}) \right) dS,$$

where C_{pen} is a very large constant, and $\Gamma_S^{n+1} \subset \partial \Lambda$ is the part of the boundary where the liquid is sliding.   □

9.5.2   Ice flow

The time discretization for the decoupling of the computations of φ, $\mathbf{v}$ and p is similar to the one detailed in section 9.5.1. The main differences reside in the iterative solution of the nonlinear Stokes problem and in the computation of the new domain at each time step; these differences are detailed hereafter. Figure 9.14 illustrates the splitting scheme and notations for the glacier geometry.

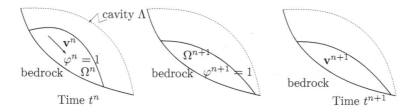

FIGURE 9.14: Time discretization. At time t^n, the previous volume fraction of ice φ^n is known in the cavity Λ which yields the ice domain Ω^n. Then, the transport problem is solved between t^n and t^{n+1} which yields φ^{n+1} and the new ice domain Ω^{n+1}. Finally, the nonlinear Stokes problem is solved in Ω^{n+1} to obtain the new velocity $\mathbf{v}^{n+1}$.

Advection step for ice flow   In order to incorporate (9.22) instead of (9.12), one has to modify the computation of the new domain at each time step. Let us assume that an approximation φ^n of $\varphi(t^n)$ is known in the cavity Λ. The ice region is defined by

$$\Omega^n = \{\mathbf{x} \in \Lambda; \varphi^n(\mathbf{x}) = 1\},$$

and the approximation of the bedrock-ice interface Γ_B^n and the ice-air interface Γ^n can also be identified. Then, the transport problem

$$\frac{\partial \varphi}{\partial t} + \mathbf{v}^n \cdot \nabla \varphi = b \delta_{\Gamma^n},$$

is solved between t^n and t^{n+1} to obtain the new volume fraction of ice φ^{n+1}. This transport problem is solved using an *order one splitting scheme in time*. The first step of this splitting scheme consists, starting from φ^n, in solving the homogeneous transport problem

$$\frac{\partial \varphi}{\partial t} + \mathbf{v}^n \cdot \nabla \varphi = 0,$$

between t^n and t^{n+1} to obtain a prediction $\varphi^{n+1/2}$ of the volume fraction of ice. A forward characteristics method is advocated as in (9.27), see section 9.5.1. The second step consists, starting from $\varphi^{n+1/2}$, in solving

$$\frac{\partial \varphi}{\partial t} = b \delta_{\Gamma^n}, \tag{9.31}$$

between t^n and t^{n+1} to obtain the new volume fraction of ice φ^{n+1}. For simplicity, the Euler scheme is used and yields

$$\varphi^{n+1} = \varphi^{n+1/2} + \delta t^n b^{n+1} \delta_{\Gamma^n}, \tag{9.32}$$

with $b^{n+1} = b(t^{n+1})$. The new ice domain Ω^{n+1} is obtained from the values of φ^{n+1}, and is defined as $\Omega^{n+1} = \{\mathbf{x} \in \Lambda : \varphi^{n+1}(\mathbf{x}) = 1\}$.

REMARK 9.6   Unlike in the model for liquid flows presented in section 9.5.1, there is no advection operator in the Stokes equations. Therefore, the advection equation is only used to compute the approximation φ^{n+1} of the volume fraction of ice; a prediction of the velocity $\mathbf{v}^{n+1/2}$ is not needed.
□

Diffusion step for ice flow   The nonlinear Stokes problem is then solved on the new ice domain, that is find $\mathbf{v}^{n+1} : \Omega^{n+1} \to \mathbb{R}^d$ and $p^{n+1} : \Omega^{n+1} \to \mathbb{R}$ such that

$$- 2\nabla \cdot (\mu^{n+1} \mathbf{D}(\mathbf{v}^{n+1})) + \nabla p^{n+1} = \rho \mathbf{g}, \tag{9.33}$$

$$\nabla \cdot \mathbf{v}^{n+1} = 0, \tag{9.34}$$

together with a zero force boundary condition (9.19) along Γ^{n+1} and slip or no-slip boundary conditions along Γ_B^{n+1} as in (9.20). Here, $\mu^{n+1} = \mu(s^{n+1})$ is the viscosity computed using (9.17) with velocity $\mathbf{v}^{n+1}$ instead of $\mathbf{v}$, and tensor $s^{n+1} = \sqrt{\frac{1}{2}(\mathbf{D}(\mathbf{v}^{n+1}) : \mathbf{D}(\mathbf{v}^{n+1}))}$ instead of s. The weak formulation corresponding to (9.33)-(9.34) with boundary conditions (9.19)-(9.20) therefore consists in finding $\mathbf{v}^{n+1}$ and p^{n+1} such that i) $\mathbf{v}^{n+1}$ satisfies the essential boundary conditions on the part of Γ_B^{n+1} where no-slip boundary conditions are applied, and ii) $\mathbf{v}^{n+1}$ and p^{n+1} satisfy the *nonlinear problem*:

$$2 \int_{\Omega^{n+1}} \mu^{n+1} \mathbf{D}(\mathbf{v}^{n+1}) : \mathbf{D}(\mathbf{w}) d\mathbf{x} - \int_{\Omega^{n+1}} p^{n+1} \nabla \cdot \mathbf{w} d\mathbf{x} - \rho \mathbf{g} \int_{\Omega^{n+1}} \mathbf{w} d\mathbf{x}$$

$$+ \int_{\Gamma_{B,S}^{n+1}} \alpha^{n+1} \Big((\mathbf{v}^{n+1} \cdot \mathbf{t}_1)(\mathbf{w} \cdot \mathbf{t}_1) + (\mathbf{v}^{n+1} \cdot \mathbf{t}_2)(\mathbf{w} \cdot \mathbf{t}_2) \Big) dS$$

$$- \int_{\Omega^{n+1}} \nabla \cdot \mathbf{v}^{n+1} q d\mathbf{x} = 0 \ , \tag{0.35}$$

for all test functions $(\mathbf{w}, q)$ such that $\mathbf{w}$ is compatible with the essential boundary conditions on Γ_B^{n+1}. Here $\Gamma_{B,S}^{n+1} \subset \Gamma_B^{n+1}$ is the part of the bedrock-ice interface where slip boundary conditions are imposed. Finally, note that $\alpha^{n+1} = \alpha(\mathbf{v}^{n+1})$ is defined by (9.21).

9.6   A two-grids method for space discretization

Advection and diffusion phenomena being now decoupled, two different space discretizations are used to solve each problem. A structured grid of small cells is used to solve (9.26) and (9.31), while an unstructured finite element mesh of tetrahedrons is used for the solution of (9.30) and (9.35). The introduction of the structured grid of small cells allows to treat more accurately the advection operators, including the computation of the interfaces.

The two meshes/discretizations are now detailed. On a first hand, the cavity Λ is embedded into a box that is meshed into a structured grid denoted by $\mathcal{C}_h$, made out of small cubic cells of size h, each cell being labeled by indices (ijk). Let us denote by N_c the number of cells in $\mathcal{C}_h$, and by C a generic element (cell) of $\mathcal{C}_h$. On the other hand, a family $\{\Lambda_H\}_H$ of polyhedral approximations of the domain Λ is introduced such that $\lim_{H \to 0} \Lambda_H = \Lambda$. Let us consider a discretization $\mathcal{T}_H$ of the cavity Λ_H satisfying the usual compatibility conditions between tetrahedra (see for instance [Glowinski, 2003] for a precise definition), where H is the typical size of the elements. Let us denote by N_e the number of elements of $\mathcal{T}_H$, and N_v the number of vertices of $\mathcal{T}_H$. Let K denote a generic element of $\mathcal{T}_H$, and P_j, $j = 1, \ldots, N_v$ be the vertices of $\mathcal{T}_H$.

Figure 9.15 visualizes the two meshes $\mathcal{C}_h$ and $\mathcal{T}_H$ for the two applications of interest. The grid of small cells is finer than the finite element mesh (actually $H \simeq 5h$). Again, liquid and ice flow space discretizations are detailed separately.

9.6.1   Liquid flow

Since the function φ is discontinuous across the interface, numerical diffusion is introduced. In order to reduce the numerical diffusion and to have an accurate approximation of the liquid region, (9.26) are first solved using the method of characteristics on a structured mesh of small cells. Then, (9.30) is solved on the unstructured finite element mesh with stabilized continuous piecewise linear finite element techniques.

Advection step for liquid flow   Let φ_{ijk}^n and $\mathbf{v}_{ijk}^n$ be the approximate values of φ^n and $\mathbf{v}^n$ at the center of cell number (ijk) at time t^n. The

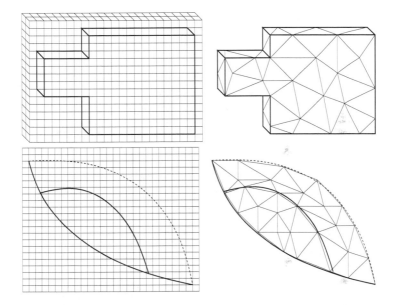

FIGURE 9.15: Two-grids method. The advection step is solved on a structured mesh $\mathcal{C}_h$ of small cubic cells composed of blocks whose union covers the cavity Λ (left), while the diffusion step is solved on a finite element unstructured mesh $\mathcal{T}_H$ of tetrahedra (right). Top: three-dimensional cavity to be filled with a liquid. Bottom: two-dimensional section of a glacier.

unknown φ^n_{ijk} is the volume fraction of liquid in the cell number (ijk); this leads to an approximation of the characteristic function φ at time t^n that is piecewise constant on each cell of the structured grid. Therefore, let $\mathbb{P}_k$ be the space of polynomials of degree k, and define

$$V^0_h = \left\{ v \in L^2(\Lambda) \; : \; v|_C \in \mathbb{P}_0, \, \forall C \in \mathcal{C}_h \right\}.$$

The *forward characteristics method* consists in advecting the values φ^n_{ijk} along the flow lines, with the flow velocities, to compute $\varphi^{n+1}_h \in V^0_h$, and $\mathbf{v}^{n+1/2}_h \in (V^0_h)^d$ such that $\varphi^{n+1}_h\big|_{C_{ijk}} = \varphi^{n+1}_{ijk}$. The advantage of using a characteristics method is to avoid the restriction by the CFL condition, as seen for instance in [Pironneau, 1989].

More precisely, the advection step for the cell number (ijk) consists in advecting φ^n_{ijk} and $\mathbf{v}^n_{ijk}$ by $\delta t^n \mathbf{v}^n_{ijk}$ and then projecting the values on the structured grid, to obtain φ^{n+1}_{ijk} and a prediction of the velocity $\mathbf{v}^{n+\frac{1}{2}}_{ijk}$. This step is illustrated in Figure 9.16 (left).

REMARK 9.7   Under the CFL condition, this characteristics method with projection corresponds exactly to the upwind finite differences scheme for the advection equation.   ⬚

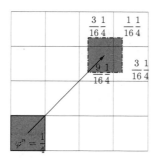
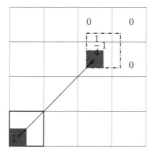

FIGURE 9.16: Advection of volume fractions with and without SLIC algorithm. An example of two dimensional advection and projection when the volume fraction of liquid in the cell is $\varphi^n_{ij} = \frac{1}{4}$. Left: without SLIC, the volume fraction of fluid is advected and projected on four cells, with contributions (from the top left cell to the bottom right cell) $\frac{3}{16}\frac{1}{4}$, $\frac{1}{16}\frac{1}{4}$, $\frac{9}{16}\frac{1}{4}$, $\frac{3}{16}\frac{1}{4}$. Right: with SLIC, the volume fraction of fluid is first pushed at one corner of the cell number (ij), then it is advected. Since the advected liquid is contained in one cell only, no projection occurs.

Advection of non-smooth functions, such as the characteristic function in the VOF method, is known to introduce *numerical diffusion*. This effect smoothes/blurs sharp fronts and makes the accurate approximation of the interfaces difficult.

Following section 9.2.8, a simple implementation of the SLIC (Simple Line Interface Calculation) algorithm, described in [Maronnier et al., 1999], [Maronnier et al., 2003] and inspired by [Noh and Woodward, 1976] allows to reduce the numerical diffusion of the domain occupied by the liquid, by pushing the fluid along the faces of the cell before advecting it. The choice of how to push the fluid depends on the volume fraction of fluid of the neighbor cells. An example in two dimensions of space is presented in Figure 9.16 (right). The critical point is then to decide how to push the volume fraction of fluid in a given cell along the sides of this cell. For a given cell, the choice depends on the volume fraction of fluid of the neighbor cells, and several cases are taken into account (when the liquid is pushed on one side, around one corner, or in the middle of the cell). Details are given in [Maronnier et al., 1999], [Maronnier et al., 2003] for the two-dimensional and three-dimensional cases respectively.

While the SLIC procedure allows to avoid numerical diffusion, the approximation of the volume fraction of fluid can be strictly larger than one after projection on the grid: it may happen that several cells of the small grid are transported at the same location. When neglected, this artificial effect leads to a loss of mass, for instance if the VOF function is truncated to 1.

We propose a post-processing technique, based on global repair algorithms (formalized in [Shaskov and Wendroff, 2004]), to guarantee the conservation of the mass of liquid. This algorithm is called a *decompression algorithm*. Its aim is to produce new values φ_{ijk}^{n+1} which are between zero and one. At each time step, the amount of liquid in the cells having values φ_{ijk}^{n+1} greater than one (strictly) is redistributed in the domain. When the redistribution is executed throughout the whole domain, the algorithm is called *global repair algorithm*; when the redistribution take into account local arguments, the algorithm is called *local repair algorithm*. For liquids, let us describe now a global decompression algorithm. A more local approach is detailed in the framework of glacier modeling.

The algorithm reads as follows: all the cells having values φ_{ijk}^{n+1} greater than one (strictly), or between zero and one (strictly) are sorted in descending order, according to their values φ_{ijk}^{n+1}. Cells advected outside of Λ are incorporated in the sorting. This can be done in an efficient way using for instance *quick-sort* algorithms. The cells having values φ_{ijk}^{n+1} greater than one are called the *dealer cells*, while the cells having values φ_{ijk}^{n+1} between zero and one are called the *receiver cells*. The algorithm then consists in moving the fraction of liquid in excess in the dealer cells to the receiver cells, starting with transferring liquid from the dealer cells with the most excess to the receiver cells with values closest to one.

Since the sorting of the cells is done independently of their geographic position in the domain, this algorithm globally guarantees that no fluid is lost, but does not contain any local information for the transfers. However, the method is still consistent since the number of cells involved decreases with the mesh size and the time step.

Projection operator: from the cells to the finite elements The advection step is solved on the grid of small cells, while the diffusion step is solved on a finite element grid of triangles (in two dimensions of space), or tetrahedra (in three dimensions of space). Before addressing the diffusion step, let us detail shortly the projection operator.

We consider in the sequel piecewise linear finite element approximations, and therefore we define

$$V_H^1 = \left\{ v \in C^0(\overline{\Lambda_H}) \ : \ v|_K \in \mathbb{P}_1, \ \forall K \in \mathcal{T}_H \right\}.$$

For any vertex P of $\mathcal{T}_H$, let ψ_P be the corresponding piecewise linear finite element basis function (i.e. the continuous, piecewise linear function having value one at P, zero at the other vertices).

Once values φ_{ijk}^{n+1} and $\mathbf{v}_{ijk}^{n+1/2}$ have been computed on the cells, a piecewise constant approximation $\varphi_h^{n+1} \in V_h^0$ is available. Then values of the fraction of liquid φ_P^{n+1} and of the velocity field $\mathbf{v}_P^{n+\frac{1}{2}}$ are computed at the vertices P of the finite element mesh, to obtain an approximation $\varphi_H^{n+1} \in V_H^1$ that is piecewise linear on each tetrahedron K of the unstructured finite element mesh $\mathcal{T}_H$. This multi-grid step is achieved with the *projection operator*

$$\mathcal{P}_{hH} : V_h^0 \to V_H^1,$$

that is defined as follows. We take advantage of the difference of refinement between a coarser finite element mesh and a finer structured grid of cells. Thus, $\varphi_P^{n+1} := \mathcal{P}_{hH}(\varphi_h^{n+1})(P)$, the volume fraction of fluid at vertex P and time t^{n+1} is computed as follows:

$$\varphi_P^{n+1} = \frac{\displaystyle\sum_{\substack{K \in \mathcal{T}_H \\ P \in K}} \sum_{\substack{ijk \\ C_{ijk} \in K}} \psi_P(C_{ijk}) \varphi_{ijk}^{n+1}}{\displaystyle\sum_{\substack{K \in \mathcal{T}_H \\ P \in K}} \sum_{\substack{ijk \\ C_{ijk} \in K}} \psi_P(C_{ijk})}, \tag{9.36}$$

where C_{ijk} is the center of the cell number (ijk), and where ψ_P is the usual piecewise linear finite element basis function associated with the vertex P. The same kind of formula is used to obtain a predicted velocity $\mathbf{v}^{n+\frac{1}{2}}$ at the vertices of the finite element mesh. Figure 9.17 (left) visualizes the projection operator.

REMARK 9.8 Relation (9.36) is a weighted average of the values on the cells; thus $\mathcal{P}_{hH}$ is an approximate L^2-projection of φ_h^{n+1} onto V_H^1 obtained with the relation

$$\int_\Lambda \varphi_h^{n+1}\psi d\mathbf{x} = \int_\Lambda \varphi_H^{n+1}\psi d\mathbf{x}, \quad \forall \psi \in V_H^1,$$

where the integrals are approximated by trapezoidal formulas. ▯

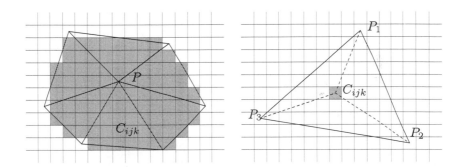

FIGURE 9.17: Multi-grid method: Left: Projection operator $\mathcal{P}_{hH}$ from the cells to the finite elements; right: interpolation operator $\mathcal{I}_{Hh}$ from the finite elements to the cells.

When the values φ_P^{n+1} are available at the vertices of the finite element mesh, the approximation Ω_H^{n+1} of the liquid region Ω^{n+1} used for solving (9.30) is defined as the union of all tetrahedra of the mesh $K \in \mathcal{T}_H$ with (at least) one of its four vertices P are such that $\varphi_P^{n+1} > d_{liq}$, where $d_{liq} = 0.5$ is a given threshold. The approximation of the free surface is denoted by Γ_H^{n+1}. The threshold d_{liq} is arbitrary, but numerical results show that the method is robust with respect to this parameter.

REMARK 9.9 Numerical experiments reported in [Maronnier et al., 1999], [Maronnier et al., 2003] have shown that choosing the size of the cells h of the structured mesh approximately 5 to 10 times smaller than the size of the finite elements H is a good choice to reduce numerical diffusion of the interface. Furthermore, since the characteristics method is used, the time step is not restricted by the CFL condition. Numerical results in [Maronnier et al., 1999], [Maronnier et al., 2003] have shown that a good choice generally consists in choosing CFL numbers (ratio between the time step δt times the maximal velocity divided by the mesh size h) ranging from 1 to 5. ☐

Finite element techniques for liquid flow The diffusion step consists in solving the Stokes problem (9.30) with finite element techniques on the unstructured grid of tetrahedra.

Since Ω_H^{n+1} is defined as the union of all elements $K \in T_H$ with (at least) one vertex P such that $\varphi_P^{n+1} > d_{liq}$, one can denote by Γ_H^{n+1} the approximation (with a polynomial line) of Γ^{n+1}. Assuming $\Gamma_H^{n+1} \neq \emptyset$, let us define the following functional spaces:

$$V_H^{1,n+1} = \left\{ \mathbf{v} \in C^0(\overline{\Omega_H^{n+1}})^d \; : \; \mathbf{v}|_K \in (\mathbb{P}_1)^d, \forall K \in T_H \right\},$$

$$Q_H^{n+1} = \left\{ p \in L^2(\Omega_H^{n+1}) \; : \; p|_K \in \mathbb{P}_1, \forall K \in T_H \right\}.$$

The generalized Stokes problem is solved with stabilized $\mathbb{P}_1 - \mathbb{P}_1$ finite elements (Galerkin Least Squares, see [Franca and Frey, 1992]). It consists in finding the velocity $\mathbf{v}_H^{n+1} \in V_H^{1,n+1}$ and the pressure $p_H^{n+1} \in Q_H^{n+1}$ such that i) $\mathbf{v}_H^{n+1}$ satisfies the essential boundary conditions on $\partial \Lambda_H$ and ii)

$$\rho \int_{\Omega_H^{n+1}} \frac{\mathbf{v}_H^{n+1} - \mathbf{v}_H^{n+1/2}}{\delta t^n} \cdot \mathbf{w} dx + 2 \int_{\Omega_H^{n+1}} \mu \mathbf{D}(\mathbf{v}_H^{n+1}) : \mathbf{D}(\mathbf{w}) dx$$

$$- \int_{\Omega_H^{n+1}} \mathbf{f}^{n+1} \cdot \mathbf{w} dx - \int_{\Omega_H^{n+1}} p_H^{n+1} \nabla \cdot \mathbf{w} dx - \int_{\Omega_H^{n+1}} q \nabla \cdot \mathbf{v}_H^{n+1} dx \quad (9.37)$$

$$+ \sum_{K \in T_H} \gamma_K \int_K \left(\frac{\mathbf{v}_H^{n+1} - \mathbf{v}_H^{n+1/2}}{\delta t^n} + \nabla p_H^{n+1} - \mathbf{f}^{n+1} \right) \cdot \nabla q dx = 0,$$

for all $\mathbf{w} \in V_H^{1,n+1}$ and $q \in Q_H^{n+1}$, such that $\mathbf{w}$ are compatible with the essential boundary conditions. Following [Maronnier et al., 1999], [Maronnier et al., 2003], the value of the parameter γ_K is given by

$$\gamma_K = \begin{cases} \dfrac{1}{12} \dfrac{H_K^2}{\mu}, & \text{if } Re_K \leq 3, \\[2mm] \dfrac{1}{4 Re_K} \dfrac{H_K^2}{\mu}, & \text{otherwise,} \end{cases} \quad (9.38)$$

where H_K is the diameter of the element K and Re_K is the local Reynolds number, defined as $Re_K = \dfrac{\rho h_K \max_{\mathbf{x} \in K} \left| \mathbf{v}_H^{n+1/2}(\mathbf{x}) \right|}{2\mu}$.

Interpolation operator: from the finite elements to the cells The diffusion step is solved on a finite element grid of triangles (in two dimensions of space), or tetrahedra (in three dimensions of space), while the advection step is solved on the grid of small cells. In order to resume with the advection step at the next time step, the solution obtained by the Stokes solver on the finite element mesh is interpolated back to the grid of small cells.

Therefore, when the value of the velocity $\mathbf{v}_H^{n+1}$ is known at the vertices of the finite element mesh, it has to be projected back onto the grid of cells. Figure 9.17 (right) visualizes the corresponding interpolation operator

$$\mathcal{I}_{Hh} : V_H^1 \to V_h^0;$$

the interpolation of the continuous piecewise linear approximation $\mathbf{v}_H^{n+1}$ back on the cell number (ijk) is obtained by interpolation of the piecewise linear finite element approximation at the center C_{ijk} of the cell. When the cell number (ijk) is located inside the tetrahedron K, the value is given by:

$$\mathbf{v}_{ijk}^{n+1} = \sum_{P \in K} \mathbf{v}_P^{n+1} \psi_P(C_{ijk}).$$

It allows to obtain an approximation $\mathbf{v}_h^{n+1} \in (V_h^0)^d$, i.e. a value of the velocity $\mathbf{v}_{ijk}^{n+1}$ on each cell number (ijk) of the structured grid for the next time step.

REMARK 9.10 In number of industrial mold filling applications, the shape of the cavity containing the liquid (the mold) is complex. Therefore, a special, hierarchical, data structure has been implemented in order to reduce the memory requirements needed to store the cells, see [Maronnier et al., 2003], [Rappaz et al., 2000]. The cavity is meshed into tetrahedra for the resolution of the diffusion problem. For the advection part, a hierarchical structure of blocks, which cover the cavity and are glued together, is defined. A computation is performed inside a block if and only if it contains cells with liquid. Otherwise the whole block is deactivated and the memory corresponding to the cells is not used. ☐

9.6.2 Ice flow

As in section 9.6.1, two different space discretizations are used for the advection and diffusion operators. Figure 9.15 (bottom) visualizes these two grids for the framework of a glacier domain. The two post-processing procedures described earlier (a SLIC method to avoid numerical diffusion, and a decompression algorithm) are incorporated to the algorithm. Interpolation and projection operators are the same as the ones described in section 9.6.1.

Advection step for ice flow Again, assume that the old values φ_{ijk}^n of the volume fraction of ice are available at each cubic cell with coordinates of center (x_i, y_j, z_k) contained in the cavity Λ.

We now present more details of how to compute the new values φ_{ijk}^{n+1}, by assuming that the values of the velocity $\mathbf{v}_{ijk}^n$ are available at the center of each cell ijk in the structured grid.

The two formulas (9.27) and (9.32) are discretized on the structured grid to obtain new values $\varphi_{ijk}^{n+1/2}$, φ_{ijk}^{n+1}, respectively. The discretization of (9.27) has been addressed in section 9.6.1. Let us turn to the solution of (9.32).

For each vertical column (ij) of the structured grid, the first task consists in computing the height of ice $H_{ij}^{n+\frac{1}{2}}$, which is an approximation of $H(x_i, y_j, t^{n+1/2})$, H being defined in (9.24). Using the rectangle formula to evaluate (9.24), we obtain

$$H_{ij}^{n+\frac{1}{2}} = h \sum_k \varphi_{ijk}^{n+\frac{1}{2}}.$$

According to (9.32), the amount of ice that has to be added or removed in the column (i, j) is

$$\left| b(x_i, y_j, B(x_i, y_j) + H_{ij}^{n+\frac{1}{2}}, t^n) \right| \Delta t,$$

where (x_i, y_j) are the horizontal coordinates of the center of the column (ij). Thus

$$I_{ij} := \frac{b(x_i, y_j, B(x_i, y_j) + H_{ij}^{n+\frac{1}{2}}, t^n) \Delta t}{h},$$

denotes the number (not necessarily an integer) of cells to be filled (if $I_{ij} > 0$) or to be emptied (if $I_{ij} < 0$), and we determine the largest vertical index k such that $\varphi_{ijk-1}^{n+\frac{1}{2}} = 1$ and $\varphi_{ijk}^{n+\frac{1}{2}} < 1$. With this information, the filling is carried out from bottom to top while the emptying is carried out from top to bottom, as visualized in Figure 9.18.

More precisely, the filling algorithm reads:

- If $(I_{ij} > 0)$, then : while $(I_{ij} > 0)$ do

$$\varphi_{ijk}^{n+1} = \min(\varphi_{ijk}^{n+\frac{1}{2}} + I_{ij}, 1),$$

$$I_{ij} \leftarrow I_{ij} - (\varphi_{ijk}^{n+1} - \varphi_{ijk}^{n+\frac{1}{2}}),$$

$$k \leftarrow k + 1,$$

- If $(I_{ij} < 0)$, then : while $(I_{ij} < 0)$ do

$$\varphi_{ijk}^{n+1} = \max(\varphi_{ijk}^{n+\frac{1}{2}} + I_{ij}, 0),$$

$$I_{ij} \leftarrow I_{ij} - (\varphi_{ijk}^{n+1} - \varphi_{ijk}^{n+\frac{1}{2}}),$$

$$k \leftarrow k - 1.$$

The solution of (9.12) with the proposed algorithm avoid artificial compression in a natural way, by redistributing the mass of ice column by column. This *local repair algorithm* exploits the structure of the ice domain to provide a better approximation of the ice domain than for the general case of liquid flow.

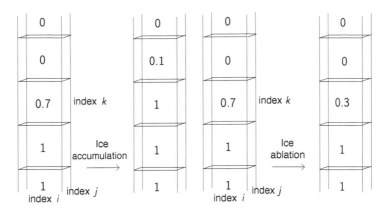

FIGURE 9.18: VOF filling (on the left) and VOF emptying (on the right) of cells having index (ij). On the left, $I_{ij} = 0.4$ means that 0.4 cells has to be added starting from cell number (ijk). In the same way, $I_{ij} = -0.4$ on the right means that 0.4 cells has to be removed.

Finite element techniques for ice flow   Once the approximation φ_{ijk}^{n+1} is known on the structured grid, the ice domain is defined on the finite element mesh with the same techniques as in section 9.6.1. The bedrock-ice interface $\Gamma_{B,H}^{n+1}$ and the ice-air interface Γ_H^{n+1} can be identified easily.

We look for a solution to the nonlinear problem (9.35) in the space of continuous functions, piecewise linear on the tetrahedrons of the finite element mesh $\mathcal{T}_H$, for both the velocity and the pressure. Since the combination of piecewise linear approximations of the velocity and pressure is not stable, we introduce additional stabilized terms as in (9.37). Consequently, we find the velocity $\mathbf{v}_H^{n+1} \in V_H^{1,n+1}$ and the pressure $p_H^{n+1} \in Q_H^{n+1}$ such that i) $\mathbf{v}_H^{n+1}$ satisfies the essential boundary conditions on (some part of) $\Gamma_{B,H}^{n+1}$ and ii)

$$
2 \int_{\Omega_H^{n+1}} \mu_H^{n+1} \mathbf{D}(\mathbf{v}_H^{n+1}) : \mathbf{D}(\mathbf{w}) dx - \int_{\Omega_H^{n+1}} p_H^{n+1} \nabla \cdot \mathbf{w} dx - \rho \mathbf{g} \int_{\Omega_H^{n+1}} \mathbf{w} dx
$$
$$
+ \int_{\Gamma_{B,S,H}^{n+1,S}} \alpha_H^{n+1} \Big((\mathbf{v}_H^{n+1} \cdot \mathbf{t}_1)(\mathbf{w} \cdot \mathbf{t}_1) + (\mathbf{v}_H^{n+1} \cdot \mathbf{t}_2)(\mathbf{w} \cdot \mathbf{t}_2) \Big) dS
$$
$$
- \int_{\Omega_H^{n+1}} q \nabla \cdot \mathbf{v}_H^{n+1} dx
$$
$$
+ \sum_{K \subset \mathcal{T}_H} \gamma_K \int_{\Omega_H^{n+1}} \nabla p_H^{n+1} \cdot \nabla q dx = \sum_{K \subset \mathcal{T}_H} \gamma_K \int_{\Omega_H^{n+1}} \rho \mathbf{g} \cdot \nabla q dx,
$$

$$(9.39)$$

for all $\mathbf{w} \in V_H^{1,n+1}$ and $q \in Q_H^{n+1}$, such that $\mathbf{w}$ is compatible with the essential boundary conditions. Here $\Gamma_{B,S,H}^{n+1} \subset \Gamma_{B,H}^{n+1}$ is the part of the bedrock-ice interface where slip boundary conditions are imposed. The coefficients are given by $\mu_H^{n+1} = \mu(s_H^{n+1})$ (with

$s_H^{n+1} = \sqrt{\frac{1}{2}(\mathbf{D}(\mathbf{v}_H^{n+1}):\mathbf{D}(\mathbf{v}_H^{n+1}))}$ and $\alpha_H^{n+1} = \alpha(\mathbf{v}_H^{n+1})$ where μ and α are defined respectively by (9.17) and (9.21). The nonlinearities appearing in the coefficients μ_H^{n+1} and α_H^{n+1} are taken into account via an iterative algorithm. A *fixed point algorithm*, and a *Newton method* are detailed in the following.

A fixed point iteration method consists in freezing the nonlinearities in (9.39), and solving a sequence of linear Stokes problems. At each time step, let $\mathbf{v}_H^{n+1,0} = \mathbf{v}_H^n$ be a given velocity, and $s_H^{n+1,0} = \sqrt{\frac{1}{2}(\mathbf{D}(\mathbf{v}_H^{n+1,0}):\mathbf{D}(\mathbf{v}_H^{n+1,0}))}$. Then, for each $k \geq 0$, define $\mu_H^{n+1,k} = \mu(s_H^{n+1,k})$ and $\alpha_H^{n+1,k} = \alpha(\mathbf{v}_H^{n+1,k})$, and solve

$$2\int_{\Omega_H^{n+1}} \mu_H^{n+1,k}\mathbf{D}(\mathbf{v}_H^{n+1,k+1}):\mathbf{D}(\mathbf{w})dx - \int_{\Omega_H^{n+1}} p_H^{n+1,k+1}\nabla\cdot\mathbf{w}dx - \rho\mathbf{g}\int_{\Omega_H^{n+1}}\mathbf{w}dx$$

$$+\int_{\Gamma_{B,S,H}^{n+1}} \alpha_H^{n+1,k}\left((\mathbf{v}_H^{n+1,k+1}\cdot\mathbf{t}_1)(\mathbf{w}\cdot\mathbf{t}_1) + (\mathbf{v}_H^{n+1,k+1}\cdot\mathbf{t}_2)(\mathbf{w}\cdot\mathbf{t}_2)\right)dS$$

$$-\int_{\Omega_H^{n+1}}\nabla\cdot\mathbf{v}_H^{n+1,k+1}qdx$$

$$+\sum_{K\subset\mathcal{T}_H}\gamma_K\int_{\Omega_H^{n+1}}\nabla p_H^{n+1,k+1}\cdot\nabla qdx = \sum_{K\subset\mathcal{T}_H}\gamma_K\int_{\Omega_H^{n+1}}\rho\mathbf{g}\cdot\nabla qdx.$$

$$(9.40)$$

to obtain $(\mathbf{v}_H^{n+1,k+1}, p_H^{n+1,k+1})$, and repeat until convergence.

A Newton iteration method consists of the replacement of (9.39) by a linearized version. For the case without sliding, the following term is added to the left-hand side of (9.40)

$$r\int_{\Omega_H^{n+1}} \frac{\mu'(s_H^{n+1,k})}{s_H^{n+1,k}}(\mathbf{D}(\mathbf{v}_H^{n+1,k}):\mathbf{D}(\mathbf{v}_H^{n+1,k+1}))(\mathbf{D}(\mathbf{v}_H^{n+1,k}):\mathbf{D}(\mathbf{w}))dx, \quad (9.41)$$

and the following term to the right-hand side of (9.40)

$$r\int_{\Omega_H^{n+1}} \frac{\mu'(s_H^{n+1,k})}{s_H^{n+1,k}}(\mathbf{D}(\mathbf{v}_H^{n+1,k}):\mathbf{D}(\mathbf{v}_H^{n+1,k}))(\mathbf{D}(\mathbf{v}_H^{n+1,k}):\mathbf{D}(\mathbf{w}))dx. \quad (9.42)$$

In this general method, $r = 0$ corresponds to the fixed point iteration method, while $r = 1$ corresponds to the Newton method. Any choice of r in $(0, 1)$ allows to obtain another, hybrid, approximation method.

In order to evaluate the difference between the two methods, a two-dimensional test problem is addressed, in which only the viscosity μ is a nonlinear function of the velocity. We consider a known vertical cut of the Gries glacier (Switzerland) in year 1961 [Kirner, 2007]. We assume that there is no sliding on the bedrock-ice interface, and standard parameters: $m = 3$ [Gudmundsson, 1999], $\sigma_0 = \sqrt{0.1}$ [bar] and $A = 0.08$ [bar^{-3} year^{-1}].

Starting with $(\mathbf{v}_H^0, p_0) := (\mathbf{0}, 0)$, (9.39) together with (9.41)-(9.42) is considered for $r = 0$ (fixed point iteration), $r = 1$ (Newton method), and $r = 1/2$ (hybrid method). The stopping criterion for all algorithms applies with the norm $||\mathbf{v}||_{\ell^2} = \left(\sum_{j=1}^{N_v} \mathbf{v}_j^T \mathbf{v}_j\right)^{1/2}$, where $\mathbf{v}_j$ is the value of the field $\mathbf{v}$ at the node number j. After 50 iterations, both solutions differ from less than 10^{-8}, which defines an exact reference solution, denoted by $\bar{\mathbf{u}}$.

Thus we can compute the (normalized) error

$$E_k := \frac{||\mathbf{u}_k - \bar{\mathbf{u}}||_{\ell^2}}{||\bar{\mathbf{u}}||_{\ell^2}}.$$

Figure 9.19 visualizes the results at each iteration for the three values of r. As expected, the order 2 Newton method ($r = 1$) is very accurate. Typically, around 5 iterations are enough to converge. The convergence is not influenced by a refinement of the mesh, as noticed in [Colinge and Rappaz, 1999]. The fixed point iteration is also robust but the convergence is slower.

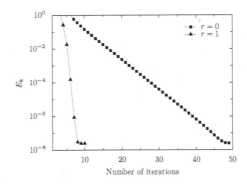

FIGURE 9.19: Convergence error E_k as a function of the number of iterations (semi-log graph in the y-coordinate).

9.7 Modeling of interfacial effects

The assumption that the liquid/ice is surrounded by vacuum leads to a vanishing force on the free boundary, see equation (9.14). Extensions due to the external has pressure and surface tension effects are briefly presented for Newtonian flows (see [Caboussat et al., 2005], [Caboussat, 2006] for more details).

9.7.1 Modeling of gas pressure

When neglecting surface tension effects, (9.14) is replaced by

$$[-p\mathbf{n} + 2\mu D(\mathbf{u}) \cdot \mathbf{n}] = 0, \qquad (9.43)$$

where $[\cdot]$ denotes the jump across the interface, and the velocity and pressure in the gas domain have to be computed outside the liquid domain, i.e. in Λ. This can be computationally expensive, since it requires to solve the incompressible Navier-Stokes or the compressible Euler equations in the gas.

In order to avoid such an important task, we have proposed a *simplified model* [Caboussat et al., 2005] that assumes that the velocity in the gas domain is zero, and takes into account only the pressure in the gas. We assume that the gas is a compressible, ideal gas. Adding the pressure in the gas allows to take into account the resistance of the gas onto the fluid. For instance, when a bubble of gas is trapped by the liquid, the gas pressure prevents the bubble to collapse rapidly, as it is the case for vacuum.

The simplified model is as follows: the pressure in the gas is denoted by P, and assumed to be constant in space in each bubble of gas, i.e. in each connected component of the gas domain. Let $k(t)$ be the number of bubbles of gas at time t and $B_i(t) \subset \Lambda$ the domain occupied by bubble number i (the ith connected component). Let $P_i(t)$ denote the constant pressure in $B_i(t)$, with initial pressure $P_i(0)$ constant in each bubble. If $V_i(t)$ is defined as the volume of $B_i(t)$, the pressure in each bubble at time t is thus computed by using the law of ideal gases at constant temperature:

$$P_i(t)V_i(t) = \text{ constant } \quad i = 1, \ldots, k(t). \qquad (9.44)$$

This simplified model requires the *tracking of the positions* of the bubbles of gas between two time steps. The additional unknowns in our model are the locations and pressure of the bubbles of gas $B_i(t)$. The equations (9.10)-(9.12) are to be solved together with $-p\mathbf{n} + 2\mu D(\mathbf{u}) \cdot \mathbf{n} = P_i$ on the interface between the liquid and bubble number i.

The tracking of the bubbles of gas and the computation of their internal pressure introduce an additional step in our time splitting scheme. This procedure is inserted between the advection step and the diffusion step, in order to compute an approximation of the force on the interface. The underlying idea is as follows [Caboussat et al., 2005]: given the liquid domain Ω^{n+1}, the key point is to find the number of bubbles (that is to say the number of connected components) and the bubbles B_j^{n+1} iteratively. Given a point P in the gas domain $\Lambda \backslash \overline{\Omega^{n+1}}$, we search for a function u such that $-\Delta u = \delta_P$ in $\Lambda \backslash \Omega^{n+1}$, with $u = 0$ on Ω^{n+1}, and u continuous (where δ_P is the Dirac measure at point P). By using the maximum principle, the solution u to this problem is strictly positive in the connected component containing point P and vanishes outside (in particular in the other connected components of $\Lambda \backslash \Omega^{n+1}$). Therefore it allows to track the first bubble. The procedure is repeated iteratively until all the bubbles are recognized.

Let us denote by k^n the number of bubbles at $t = t^n$, and P_i^n, B_i^n, $i = 1, 2, \ldots, k^n$, the approximations of P_i, B_i, $i = 1, 2, \ldots, k$, respectively at time t^n. Let $\xi^{n+1} : \Lambda \to \mathbb{R}$ be a *bubble numbering function*, negative in the liquid region Ω^{n+1} and equal to i in bubble B_i^{n+1}. The approximations k^{n+1}, P_i^{n+1}, B_i^{n+1}, $i = 1, 2, \ldots, k^{n+1}$ and ξ^{n+1} are computed as follows. Given the new liquid domain Ω^{n+1}, the number of bubbles k^{n+1} (i.e. the number of connected components) and the bubbles B_i^{n+1}, $i = 1, \ldots, k^{n+1}$ are obtained with the following algorithm: Set $k^{n+1} = 0$, $\xi^{n+1} = 0$ in $\Lambda \backslash \Omega^{n+1}$ and $\xi^{n+1} = -1$ in Ω^{n+1}, and $\Theta^{n+1} = \{ \mathbf{x} \in \Lambda : \xi^{n+1}(\mathbf{x}) = 0 \}$. While $\Theta^{n+1} \neq \emptyset$, do:

1. Choose a point P in Θ^{n+1};

2. Solve the following problem: Find $u : \Lambda \to \mathbb{R}$ which satisfies:

$$\begin{cases} -\Delta u = \delta_P, & \text{in } \Theta^{n+1}, \\ u = 0, & \text{in } \Lambda \backslash \Theta^{n+1}, \\ [u] = 0, & \text{on } \partial \Theta^{n+1}, \end{cases} \qquad (9.45)$$

where δ_P is Dirac delta function at point P, $[u]$ is the jump of u through $\partial \Theta^{n+1}$. The discretization of (9.45) is addressed with continuous piecewise linear finite elements, based on the triangulation $\mathcal{T}_H$;

3. Increase the number of bubbles k^{n+1} at time t^{n+1}: $k^{n+1} = k^{n+1} + 1$;

4. Define the bubble of gas number k^{n+1}: $B_{k^{n+1}}^{n+1} = \{ \mathbf{x} \in \Theta^{n+1} : u(\mathbf{x}) \neq 0 \}$;

5. Update the bubble numbering function $\xi^{n+1}(\mathbf{x}) = k^{n+1}$, $\forall \mathbf{x} \in B_{k^{n+1}}^{n+1}$;

6. Update Θ^{n+1} for the next iteration: $\Theta^{n+1} = \{ \mathbf{x} \in \Lambda : \xi^{n+1}(\mathbf{x}) = 0 \}$.

REMARK 9.11   The cost of this original numbering algorithm is bounded by the cost of solving k^{n+1} Poisson problems in the gas domain. The corresponding CPU time used to solve the Poisson problems is usually much less expensive than solving a full model in the gas domain, and much less expensive than solving the Stokes problem in the liquid domain. ☐

Once the connected components of gas are numbered, an approximation P_i^{n+1} of the constant pressure in bubble i at time t^{n+1} has to be computed with (9.44). In the case of a single bubble in the liquid, (9.44) yields $P_1^{n+1} V_1^{n+1} = P_1^n V_1^n$. In the case when two bubbles merge between times t^n and t^{n+1}, this relation becomes $P_1^{n+1} V_1^{n+1} = P_1^n V_1^n + P_2^n V_2^n$. When a bubble B_1^n splits onto two, each of its parts at time t^n contributes to bubbles B_1^{n+1} and B_2^{n+1}. Details of the implementation require to distinguish several situations, and are given in [Caboussat et al., 2005].

The value of the pressure can be inserted as a boundary term in (9.30) for the resolution of the generalized Stokes problem (9.28)-(9.29). By using the divergence theorem in the variational formulation (9.30) and the fact that P^{n+1} is piecewise constant, the integral on the free surface Γ_H^{n+1} is transformed into an integral on Ω_H^{n+1} (see [Caboussat et al., 2005]).

9.7.2   Modeling of surface tension

Independently of the presence or not of gas around the liquid, surface tension effects can be incorporated into (9.13), by adding a normal force $\sigma \kappa \mathbf{n}$ on the interface. This requires the approximation of the curvature κ and the normal vector $\mathbf{n}$. An additional step is added in the time splitting scheme to compute these two unknowns before the diffusion part. When incorporating the surface tension, the left-hand-side of (9.37) must be supplemented by

$$\int_{\Gamma_H^{n+1}} \sigma \kappa_H^{n+1} \mathbf{w} \cdot \mathbf{n}_H^{n+1} dS. \tag{9.46}$$

Several strategies can be considered for the modeling and computation of surface tension effects.

Continuum surface force model   The *Continuum Surface Force* (CSF) model has been detailed e.g. [Brackbill et al., 1992], [Caboussat, 2006], [Rider and Kothe, 1998], [Williams et al., 1999]. Let us denote by κ^{n+1} and $\mathbf{n}^{n+1}$ the approximations of κ and $\mathbf{n}$ respectively, at time t^{n+1}. Since the characteristic function φ^{n+1} is not smooth, it is first regularized by convolution, see e.g. [Caboussat, 2006], [Williams et al., 1999], in order to obtain a smooth approximation $\tilde{\varphi}^{n+1}$:

$$\tilde{\varphi}^{n+1}(\mathbf{x}) = \int_\Lambda \varphi^{n+1}(\mathbf{y}) K_\varepsilon(\mathbf{x} - \mathbf{y}) d\mathbf{y}, \quad \forall \mathbf{x} \in \Lambda. \tag{9.47}$$

The parameter ε is the smoothing parameter that describes the size of the support of K_ε, i.e. the size of the smoothing layer around the interface, and K_ε is a *smoothing kernel*.

The liquid-gas interface Γ^{n+1} is given by the level line $\{\mathbf{x} \in \Lambda : \tilde{\varphi}^{n+1}(\mathbf{x}) = d_{liq}\}$, with $\tilde{\varphi}^{n+1}(\mathbf{x}) < d_{liq}$ in the gas domain and $\tilde{\varphi}^{n+1}(\mathbf{x}) > d_{liq}$ in the liquid domain ($d_{liq} \in (0,1)$ is the threshold between liquid and gas phases). At each time step, the normal vector $\mathbf{n}^{n+1}$ and the curvature κ^{n+1} on the liquid-gas interface are given respectively by

$$\mathbf{n}^{n+1} = -\frac{\nabla \tilde{\varphi}^{n+1}}{||\nabla \tilde{\varphi}^{n+1}||} \text{ and } \kappa^{n+1} = -\nabla \cdot \left(\frac{\nabla \tilde{\varphi}^{n+1}}{||\nabla \tilde{\varphi}^{n+1}||} \right),$$

see e.g. [Osher and Fedkiw, 2001]. The discrete approximation of κ^{n+1} is achieved on the finite element mesh, in order to use the variational framework of the finite element approximation. The normal vector $\mathbf{n}_H^{n+1}$ is given by the normalized gradient of $\tilde{\varphi}_H^{n+1}$ at each grid point P_j, $j = 1, \ldots, N_v$ where N_v denotes the number of nodes in the finite element discretization. Details can be found in [Caboussat, 2006]. The curvature κ^{n+1} is approximated by the L^2-projection of the divergence of the normal vector on the piecewise linear finite elements space with *mass lumping* and is denoted by κ_H^{n+1}, given by the relation

$$\int_{\Lambda_H} \kappa_H^{n+1} \psi_P d\mathbf{x} = \int_{\Lambda_H} -\nabla \cdot \frac{\nabla \tilde{\varphi}_H^{n+1}}{||\nabla \tilde{\varphi}_H^{n+1}||} \psi_P d\mathbf{x}, \qquad (9.48)$$

for all vertices P of $\mathcal{T}_H$, where ψ_P are the piecewise linear finite element basis functions. The left-hand side of this relation is computed with *mass lumping*, while the right-hand side is integrated by parts. Explicit values of the curvature of the level lines of $\tilde{\varphi}_H^{n+1}$ (the continuous piecewise linear approximation of $\tilde{\varphi}^{n+1}$) are obtained at the vertices of the finite element mesh being in a layer around the free surface. The restriction of κ_H^{n+1} to the nodes lying on Γ_H^{n+1} is used to compute (9.13).

Geometric surface tension model   Following, [Bonito et al., 2010], [Gerbeau and Lelièvre, 2009], the surface tension effects can also be written in terms of a *surface operator*. By using the divergence theorem, see equation (20) in [Gerbeau and Lelièvre, 2009], the boundary term (9.46) can be expressed as

$$\int_{\Gamma_H^{n+1}} \sigma \kappa_H^{n+1} \mathbf{w} \cdot \mathbf{n}_H^{n+1} dS = -\int_{\Gamma_H^{n+1}} \sigma \text{Tr} \left(\nabla_{\Gamma_H^{n+1}} \mathbf{w} \right) dS, \qquad (9.49)$$

where $\nabla_{\Gamma_H^{n+1}}$ is the *surface gradient*, defined for each vector field $\mathbf{X}$ on the surface Γ_H^{n+1} as

$$\nabla_{\Gamma_H^{n+1}} \mathbf{X} = P_{\Gamma_H^{n+1}} \nabla \mathbf{X} = \nabla \mathbf{X} - \mathbf{n}_H^{n+1} \otimes \mathbf{n}_H^{n+1} \nabla \mathbf{X}, \qquad (9.50)$$

and where $P_{\Gamma^{n+1}}(\mathbf{x})$ is the orthogonal projector on the tangential direction of Γ_H^{n+1} at point $\mathbf{x}$. This approach allows to reduce the computational cost for the determination of the surface tension effects (compared to the CSF model) without sacrificing the accuracy. If the liquid-gas interface is in contact with the boundary of the cavity, this formulation allows to take into account the *contact angle* between the Γ_t and $\partial \Lambda$. An additional boundary term is therefore added to (9.49).

9.8   Numerical results for liquid flow

Originally designed for mold filling applications, the proposed algorithm actually allows to cover a wider range of applications, as illustrated by the following numerical experiments having Reynolds numbers ranging from 0 to 10^5! Benchmark problems have been presented in [Caboussat et al., 2005], [Maronnier et al., 1999], [Maronnier et al., 2003] to validate i) the accuracy of the advection operator in stretching flow, and ii) the accuracy of the position of the interface and the computation of the velocity and pressure. In the following, examples from mold casting, sloshing problems, and bubbly flow, are presented.

9.8.1   Casting problems

An S-shaped channel lying between two horizontal plates is filled. The channel is contained in a 0.17 m×0.24 m rectangle. The distance between the two horizontal plates is 0.008 m. Water is injected at one end with constant velocity 9.2 m/s. Density and viscosity are taken to be respectively $\rho = 1000$ kg/m^3 and $\mu = 0.01$ kg/(ms). Slip boundary conditions are enforced to avoid boundary layers. Since the ratio between Capillary number and Reynolds number is very small, surface tension effects are neglected. The final time is $T = 0.0054$ [s] and the time step is $\tau = 0.0001$ [s]. The finite element mesh is made out of 96030 elements.

In Figure 9.20 (top), 3D computations are presented when the liquid is surrounded with vacuum (no external pressure due to the gas bubbles). The CPU time for the simulations in three space dimensions is approximately 120 minutes for 540 time steps (Intel Core 2 Ghz CPU, with less than 2Gb memory). Most of the CPU time is spent to solve the Stokes problem.

FIGURE 9.20: S-shaped channel: 3D results when the cavity is initially filled with vacuum (top); when the cavity is initially filled with compressible gas at atmospheric pressure (bottom). Time equals 8.0 ms, 22.0 ms, 35.0 ms and 50 ms.

A comparison with experimental results shows that the bubbles of gas trapped by the liquid vanish too rapidly. In order to obtain more realistic results, the effect of the gas compressibility onto the liquid must be considered. The same S-shaped channel is therefore initially filled with compressible gas at atmospheric pressure $P = 101300$ [Pa]. A valve is located at the upper extremity of the channel allowing gas to escape, so that the gas domain connected to this valve does not apply any pressure on the liquid. Numerical results in Figure 9.20 (bottom) show the persistence of the bubbles. The CPU time for the simulations is approximately 130 [minutes] with the bubbles computations; thus the overhead for computing the compressibility effects is less than 10% of the total CPU time.

9.8.2 Sloshing simulations

The simulation of horizontal sloshing is usually difficult, due to the numerical diffusion introduced at the interface. Here we show two examples that involve nearly horizontal interfaces, namely the breaking dam and the impact of a droplet on an horizontal surface.

We consider a cavity with size $0.09 \times 0.045 \times 0.045$ [m^3] and place a block of water with size $0.03 \times 0.02 \times 0.04$ [m^3] along one of the vertical edges. The finite element mesh contains 168750 elements. The water is released at initial time and free to move under action of gravity. It is surrounded by vacuum and surface tension effects are neglected, implying a zero boundary force condition on the free surface. It splashes against the opposite walls of the cavity, before *sloshing* from one end to the other. The final time is 0.5 [s] and 500 time steps were used. Figure 9.21 visualizes snapshots of the liquid domain at several times.

We consider the cubic cavity $(0, 0.1)^3$. An horizontal liquid layer is located at the bottom of the cavity below height $z = 0.04$, while a spherical droplet of initial radius $r = 0.015$ center $(0.05, 0.05, 0.06)$, and velocity $\mathbf{v}_0 = (0, 0, -5)$ is falling under gravity forces. The finite element mesh contains 367200 elements. The droplet falls on the liquid layer under gravity forces, and splashes on the horizontal surface. Interfacial effects are neglected.

Figure 9.22 visualizes snapshots of the liquid domain at several times of the impact, before reflections are introduced by the boundaries of Λ. One can observe that symmetries are perfectly respected, that the droplet is swallowed by the water at impact, and forms a circular wave on the free surface.

9.8.3 Bubbles simulations with surface tension

Finally, we address the simulation of bubbly flow. In these cases, the Reynolds numbers are very small, and the interfacial effects play a more important role.

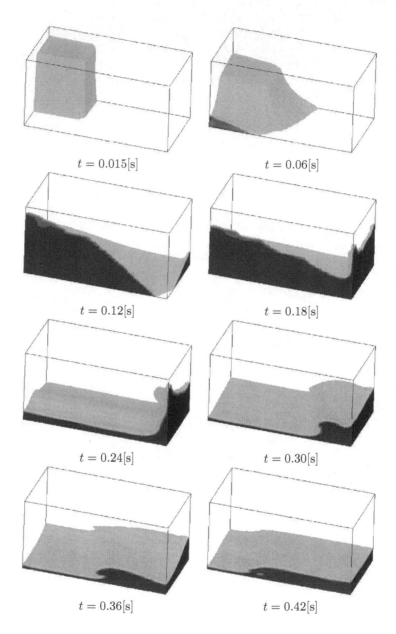

$t = 0.015[\mathrm{s}]$ $t = 0.06[\mathrm{s}]$

$t = 0.12[\mathrm{s}]$ $t = 0.18[\mathrm{s}]$

$t = 0.24[\mathrm{s}]$ $t = 0.30[\mathrm{s}]$

$t = 0.36[\mathrm{s}]$ $t = 0.42[\mathrm{s}]$

FIGURE 9.21: Breakage of a dam in a confined cavity.

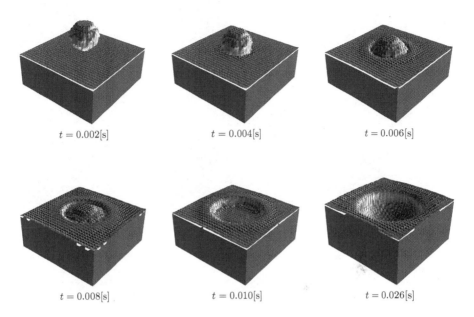

FIGURE 9.22: Impact of a liquid droplet onto an horizontal liquid surface.

We consider a bubble of gas of radius $r = \sqrt{0.0004}$ at the bottom of the cylinder $(0, 0.05) \times (0, 0.05) \times (0, 0.1)$ filled with liquid up to the surface $z = 0.07$, under gravity forces. The bubble rises and reaches the free surface between water and air, see Figure 9.23. The compressible effects of the air, and the surface tension effects are taken into account. The physical constants are $\mu = 0.01$ kg/(ms), $\rho = 1000$ kg/m^3 and $\sigma = 0.0738$ N/m. The mesh made out of 115200 tetrahedra. The smoothing parameter for the smoothing of the volume fraction of liquid in the CSF model is $\varepsilon = 0.005$.

Numerical simulations for one bubble of gas have been benchmarked in [Caboussat, 2006]. Let us consider then two bubbles of gas under gravity forces starting at the bottom the cylinder $(0, 0.05) \times (0, 0.05) \times (0, 0.1)$, that is completely filled with liquid. The two bubbles have the same size ($r = \sqrt{0.0002}$). Due to the pressure difference, the bottom bubble rises faster, and coalesces with the top one. Figure 9.24 visualizes the rising and interactions between bubbles at several time steps when the two bubbles are *on-axis* [Chen and Li, 1998], [Hua et al., 2008], i.e. the center of both bubbles is located on the symmetry axis of the cylinder. Numerical results capture accurately the coalescence of the two bubbles (see [Chen and Li, 1998] and experimental results therein).

Figure 9.25 visualizes the rising and interactions between bubbles at several time steps when the two bubbles are *off-axis*; the center of the top bubble

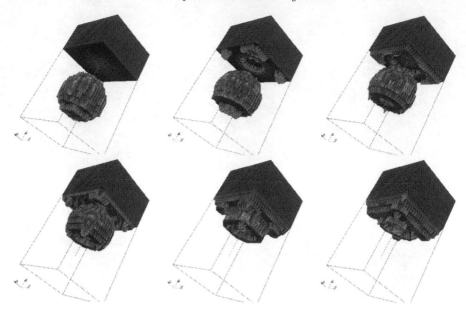

FIGURE 9.23: Three-dimensional rising bubble under a free surface: Representation of the gas domain at times $t = 50.0$, 75.0, 100.0, 150.0, 200.0 and 250.0 [ms] (left to right, top to bottom).

is initially located at $(0.0175, 0.0175, 0.06)$, while the center of the bottom one is at $(0.0275, 0.0275, 0.01)$. One observes that the bottom bubble is first attracted in the wake of the top one, before the rising and the coalescence.

9.9   Numerical results for ice flow

We present numerical results for two Alpine glaciers located in Switzerland, the Muragl glacier, and the Rhone glacier (Rhonegletscher). Several mass balance models of various complexity are used. In the first application, a simple parametrization of b is considered because of the lack of data. In the second application, a more complex mass balance model which involves daily data of temperature and precipitation is used.

9.9.1   Muragl glacier

Presently, Vadret Muragl is a small glacier $200 - 300$ [m] long in the highest reaches of the Val de Muragl, in the Eastern Swiss Alps. Based on geomorphological evidence, several positions of the glacier terminus have been recon-

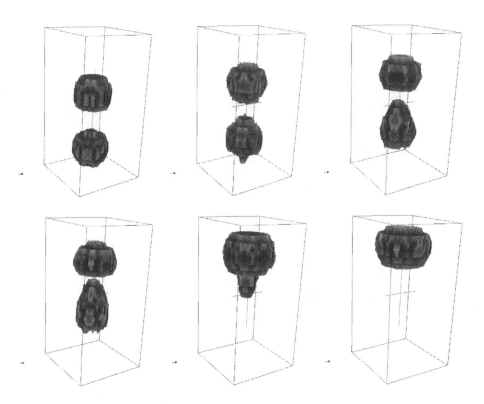

FIGURE 9.24: Rising and coalescence of two on-axis bubbles of same size. Representation of the gas domain at times $t = 10.0, 50.0, 80.0, 100.0., 150.0$ and 180.0 [ms] (left to right, top to bottom).

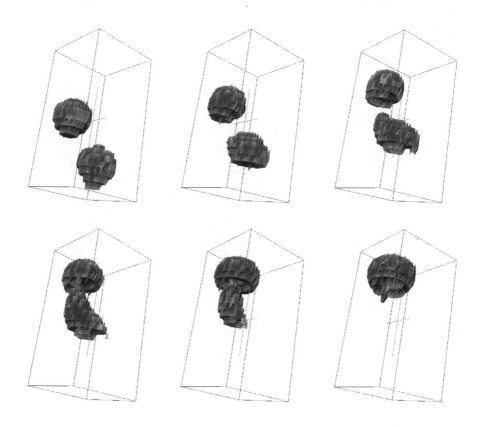

FIGURE 9.25: Rising and coalescence of two off-axis bubbles of same size. Representation of the gas domain at times $t = 5.0, 75.0, 150.0, 200.0., 225.0$ and 250.0 [ms] (left to right, top to bottom).

structed throughout the ages. Among them, the "Margun" position with a length of about 3.65 km, was close to the position of Vadret Muragl about 10,000 years ago [Imbaumgarten, 2005], [Rothenbühler, 2000]. Figure 9.26 visualizes the reconstructed "Margun" position. It can be used to illustrate the numerical simulation of ice flows and to test climatological hypotheses.

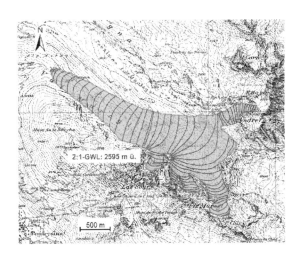

FIGURE 9.26: Reconstructed position Punt Muragl for Vadret Muragl, from T. Imbaumgarten master's thesis (2005).

A (simple) mass balance mode is considered [Jouvet et al., 2008]. It is considered to be only elevation dependent, i.e. $b(x, y, z, t) = b(z)$. This function is defined by the melt gradient, a_m, the equilibrium line altitude, z_{ELA}, and the maximum accumulation rate, a_c :

$$b(z) = \min [a_m (z - z_{ELA}), a_c]. \tag{9.51}$$

Figure 9.27 illustrates the function $b(z)$. This simplified parametrization follows the experience that the mass balance increases approximately linearly with altitude in the ablation zone and often levels out in the accumulation zone [Kuhn, 1981], [Stroeven, 1996]. In this model the mass balance is independent of time.

For a given set of parameters (a_m, z_{ELA}, a_c), the equilibrium shape of the glacier is reconstructed, starting from two different initial shapes. Both computations lead to the same equilibrium shape, illustrated in Figure 9.28.

The numerical values of the physical parameters in both runs were $m = 3$ for the flow law exponent, rate factor $A = 0.08$ [bar^{-3}year^{-1}], and regularization parameter $\sigma_0^2 = 0.1$ [bar^2]. The time step is $\Delta t = 1$[year] and the grid spacing for the advection cells is $h - 5$ [m].

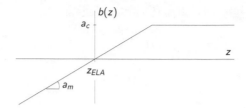

FIGURE 9.27: Mass balance function $b(z)$.

Figure 9.28 shows that, after a few hundred years, the two simulations coincide within 50 [m], i.e. within the finite element grid resolution ($H = 50$ [m]).

It is then possible to use the model in an inverse problem framework in order to find the appropriate mass balance distribution (a_m, z_{ELA}, a_c) such that the modeled glacier fits a given tongue position.

We therefore can use the reconstructed Margun position to determine the set of parameters a_c, a_m, z_{ELA} appearing in (9.51), for which the computed steady state terminus position fits the given reconstructed glacier terminus. The well-posedness of this inverse problem is not guaranteed, but, to favor physical solutions, we restrict the values of the parameters to realistic intervals.

For fixed values of the two parameters a_c and a_m, the equilibrium line altitude z_{ELA} can be found by an iterative process. Since the glacier length is a monotonous function of z_{ELA}, a given tongue position can be obtained by applying a *secant method*, as illustrated in Figure 9.29.

By repeating this procedure with a set of realistic parameters a_c and a_m, it is shown in [Jouvet et al., 2008] that the equilibrium line altitude of Muragl Glacier that corresponds to the Margun position is about 2700.

REMARK 9.12   The numerical simulation of such glaciers obviously relies on several sets of parameters, each estimated or calibrated from experiments or measurements. Sensitivity of the simulations with respect to the parameters must be discussed. Several examples of sensitivity studies can be found in [Jouvet et al., 2008], for instance w.r.t the rate factor A, or the climatic conditions.   ⬚

9.9.2   Rhone glacier

A second example is presented here to illustrate the prediction capabilities of numerical simulations of glaciers. The Rhone glacier has been thoroughly investigated by glaciologists throughout the last century. As many Alpine glaciers, it has significantly retreated since the end of the Little Ice Age around

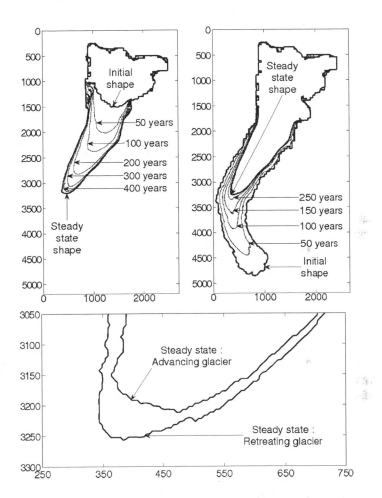

FIGURE 9.28: Evolution of the glacier starting from a short position (top left) and from a long position (top right). The parameter values for the steady state corresponding approximately to the Margun position are $z_{\mathrm{ELA}} = 2700[\mathrm{m}$ a.s.l.$]$, $a_c = 0.5[\mathrm{m\ year}^{-1}]$ and $a_m = 0.004$. Below: difference between the two tongues' location, the discrepancy is of the order of the mesh size $H = 50$ $[\mathrm{m}]$.

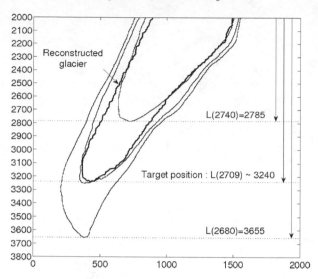

FIGURE 9.29: Computations of steady shapes for $z_{ELA} = 2740$ and $z_{ELA} = 2680$ provides two abscissa of the tongue's end: $L(2740) = 2785$ and $L(2680) = 3655$. Using a secant method, $z_{ELA} = 2709$ is a judicious choice for the next computation; the corresponding steady shape almost fits the target's position.

1850, as shown in Figure 9.30. Measurements are actually available starting already in 1874.

In this section, several numerical simulations of Rhone glacier, from 1874 to 2100, are presented. Firstly, the measurements available between 1874 and 2007 allow to calibrate the physical parameters. Secondly, simulations between 2008 and 2100 are run, by setting the mass balance parameters to probable climatic trends [Frei, 2007]. In all the simulations, the Glen's exponent is set to $m = 3$ (e.g. [Gudmundsson, 1999]), and the regularization parameters are set to $\sigma_0 = \sqrt{0.1}$ [bar] and $s_0 = 0.01$ [m year^{-1}]. Let us now describe the complex mass balance model used for the simulation of Rhone glacier. The parameters involved in this model are calibrated to fit observations optimally [Huss et al., 2008a].

The mass balance b is a function of (x, y, t) defined by

$$b(x, y, t) = P(x, y, t) - M(x, y, t), \tag{9.52}$$

where P corresponds to solid precipitation (snow) and M to melt processes. The function P is given by

$$P = P(x, y, t) = P_{ws}(t) \left(1 + P_z(S(x, y, t) - z_{ws})\right) c_{prec} D(x, y, t), \tag{9.53}$$

where P_{ws} is the precipitation measured at a nearby weather station at elevation z_{ws}, P_z and c_{prec} are constant coefficients and D is the distribution

FIGURE 9.30: Left: Pictures of the Rhone glacier tongue in 1856 (from http://www.unifr.ch/geosciences/geographie/glaciers/); right: in 2008.

pattern of solid precipitations, calculated from the curvature and slope of the terrain. The function $D(x, y, t)$ varies between 0 (complete snow erosion due to the wind) and 2 (snow deposition). To distinguish between solid and liquid precipitation, a threshold temperature is given. The function $M(x, y, t)$ is given [Hock, 1999] by

$$M(x, y, t) = \begin{cases} (f_{\mathrm{M}} + r_{\mathrm{ice/snow}} I(x, y, t)) T(x, y, t) & \text{if } T(x, y, t) > 0, \\ 0 & \text{otherwise,} \end{cases} \quad (9.54)$$

where f_{M} and $r_{\mathrm{ice/snow}}$ are constant coefficients, I is the potential direct solar radiation [Hock, 1999] and T is the temperature field computed from measured air temperature T_{ws} at the nearby weather station:

$$T(x, y, t) = T_{\mathrm{ws}}(t) - G\left(S(x, y, t) - z_{\mathrm{ws}}\right),$$

where G is a constant temperature gradient. Note that mass balance is first computed in a daily resolution by using (9.52) and next integrated over the year to provide the function b. The five unknown coefficients f_{M}, r_{ice}, r_{snow}, c_{prec}, P_z involved in (9.53) and (9.54) are calibrated according to i) ice volume changes ii) point-based mass balance measurements and iii) water discharge [Huss et al., 2008a]. The mass balance function is depicted in Figure 9.31 for two extreme years in terms of accumulation and ablation.

Let us shortly describe the numerical implementation of this test case. At each vertex (x_i, y_j) of a structured grid in the horizontal rectangle $(0, 4000) \times (0, 10000)$, the bedrock elevation $B(x_i, y_j)$ and the initial ice surface $S(x_i, y_j, 0)$ are provided [Farinotti et al., 2009]. The cell size in the x, y directions is 50 [m]. A triangular finite element mesh of the bedrock is generated. A triangular finite element mesh of the top surface of the cavity Λ is also generated by adding 150 [m] to the initial ice thickness. Then, a Delaunay unstructured mesh of tetrahedrons is generated between the two surfaces using

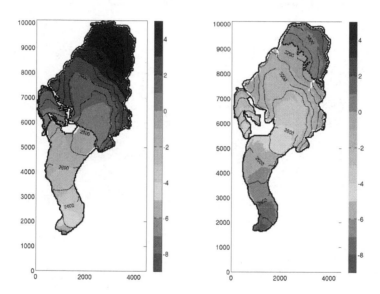

FIGURE 9.31: Mass balance function b for the Rhone glacier in (left) 1977 (mass gain of the glacier - cold year) and (right) 2003 (mass loss - warm year). The unit of b is meter water equivalent. The results are shown using a local system of reference. The abscissa of the lower left corner is 671250 while the ordinate is 157400 (in the Swiss referential).

TetMesh-GHS3D [Frey and George, 2008], filling the cavity Λ with tetrahedrons of typical size 50 [m]. The MeshAdapt remesher [MeshAdapt, 2003] is used in order to refine the mesh in the Oz direction only (anisotropic mesh of typical size 10 [m]). The final mesh of the cavity has 240147 vertices. The number of vertices of the cavity contained in the initial ice region Ω^0 is 84161. The block $(0, 4000) \times (0, 10000) \times (1700, 3600)$ containing the cavity Λ is cut into $400 \times 1000 \times 200$ structured cells. The time step is half a year. About 10 days are required for performing the simulation from 1874 to 2007 (AMD Opteron 242 CPU with less than 8Gb memory).

Numerical simulation from 1874 to 2007 The sliding law (9.20) is subject to higher uncertainties than the flow model (9.15)-(9.17), which has been used in several studies in the past. The influence of the slip boundary conditions on the bedrock is an important point to evaluate. We first simulate the retreat of the glacier from 1874 to 2007 with no-slip conditions along the bedrock (no sliding, $A = 0.08$ [bar^{-3} year^{-1}] [Gudmundsson, 1999]). Figures 9.32 and 9.33 show that the retreat of the glacier from 1874 to 1900 is significantly too fast in comparison to observations. This can be explained by the following uncertainties in the model:

1. the slip boundary conditions are not taken into account on the bedrock;

2. the value of the coefficient A is inaccurate;

3. the bedrock location is inaccurate in some regions of the glacier;

4. the ice flow model does not correctly describe the glacier dynamics.

Experimental data show that the sliding effect is a very important process ahead the glacier, see e.g. [Jouvet et al., 2009], [Mercanton, 1916], [Nishimura, 2008]. The sole introduction of slip boundary conditions on the bedrock allows to obtain a much better fit with the measurements. We consider the physical parameters $A = 0.1$ and $C = 0.3$ (adjusted to the best fit), and a sliding zone defined as in [Jouvet et al., 2009] (the ice located near the tongue of the glacier - the lower part of the glacier - is sliding).

We compare the position of the tip of the glacier's tongue, with no-slip boundary conditions along the bedrock Γ_B, or slip boundary conditions on some portion of the bedrock-ice interface Γ_B. Figure 9.32 visualizes the comparison with experimental data, showing that slip boundary conditions provide a convincing fit with measurements. Simulation results are given in Figure 9.33.

Future predictions for the Rhone glacier Finally we emphasize the role of numerical simulations for the prediction of the future behavior of the Rhone glacier. This task requires the *extrapolation* of current and past data in the future, according to various *scenarios*. The calibration of future data

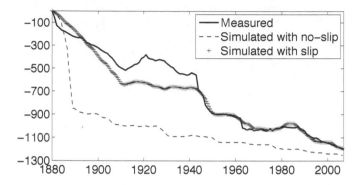

FIGURE 9.32: Comparison of measured and simulated retreat of the glacier tongue between 1880 and 2007. Model runs using no-slip boundary conditions on Γ_B with $A = 0.08$ and runs using slip boundary conditions of some part of Γ_B with $A = 0.10$ are shown.

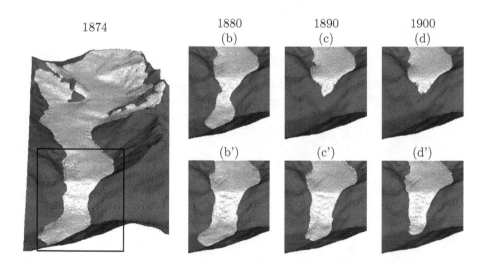

FIGURE 9.33: Simulation of the Rhone glacier tongue over the period 1874-1900. Left (a): initial glacier extent in 1874. Right, evolution from 1880 to 1900. Top (b, c, d): no-slip on Γ_B and $A = 0.08$; bottom (b', c', d'): slip on some part of Γ_B, $A = 0.10$.

is achieved as follows. Ice velocity measures are obtained from aerial photographs [Nishimura, 2008]; the factor A is given by $A = 0.8$ [bar^{-3} year^{-1}]; no-slip boundary conditions are enforced on $\Gamma_{B,t}$.

Three types of simulations are achieved. In a first step, mass balance terms are calculated, according to three different climatic scenarios, in terms of temperature and precipitations (denoted hereafter as median scenario S2 and extreme scenarios S1 and S3). These scenarios are given explicitly in [Frei, 2007].

In order to emphasize the influence of the temperature and precipitations, we perform two additional types of simulations. In the first one, several temperature scenarios are considered for fixed precipitation levels (T0 through T5); conversely, we finally change the precipitations level with a fixed temperature (P1 through P3).

Simulations with three fixed scenarios Neutral data in terms of temperature and precipitations are considered, i.e. random data in past periods that do not contain major climatic changes. Seasonal effects are incorporated into these time series [Frei, 2007]. The first one (S1) is 'cold and humid'. This is the optimistic case, in terms of glacier life expectation. The second one (S2) is 'median'. The last one (S3) is 'hot and dry', and is the pessimistic one. These three scenarios corresponds to the average and almost extreme predictions among a pool of model simulations for future climatic conditions [Frei, 2007]. Figure 9.34 visualizes S1, S2 and S3, while the simulation results for the three major scenarios in terms of temperature and precipitation trends are illustrated in Figure 9.35.

These results confirm the trend of the retreat of Alpine glaciers in the 21st century. The simulation based on the most realistic assumptions for future climate change predicts a dramatic retreat of Rhonegletscher during the 21st century. However, the uncertainty of climatic projections is still high and allows for a wide range of possible glacier changes in the near future.

Simulations according to several temperature tendencies Let us consider a fixed time series of precipitations, and six temperature projections, called T0 through T5. They correspond to an increase of temperature of $0, 1, 2, 3, 4$ and $5°$ between 2090 and 1990. Numerical simulations are presented in Figure 9.36, while the volumes of ice are represented in Figure 9.37b. One degree increase of temperature implies a significant loss of ice in 2100.

Simulations according to several precipitations tendencies Conversely let us consider a fixed time series of (average) temperatures, and three precipitation projections, called P1 through P3. They correspond to the variation depicted on Figure 9.34, bottom, already used for defining scenario S1, S2 and S3. Numerical simulations are presented in Figure 9.36, while the

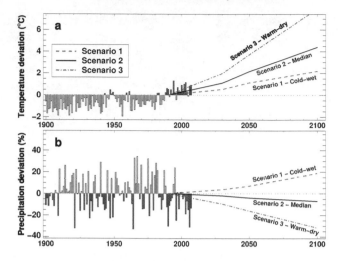

FIGURE 9.34: Deviations of annual (top) mean temperature and (bottom) precipitation from the climate in 1990. Measured data for the 20th century are displayed by bars, annual changes as assumed by the three climate scenarios, that are shown until 2100.

volumes of ice are represented in Figure 9.37 (c). One observes that the variations of the volume of ice in 2100 between P1, P2 and P3 are less important than the variations due to a temperature increase.

9.10   Concluding remarks

In this chapter, an efficient computational model for the simulation of two-phases flows has been presented. It allows to consider liquid and ice flows with complex free surfaces and within a large range of Reynolds numbers. It relies on an Eulerian framework and couples finite element techniques with a forward characteristics method. It allows to incorporate many different features, such as interfacial effects (surface tension), bubbles of gas, addition of mass in the system, or various definitions of viscosity.

Numerical results illustrate the large range of applications covered by the model, going from liquid modeling to ice modeling. Simulation of liquid flow show the flexibility of the approach for large and small Reynolds numbers. This allows to tackle casting problems, and bubbles simulations, although the features of the volume-of-fluid method are more appropriate for large Reynolds numbers.

Simulations of ice flow have been used to predict future behavior of glaciers.

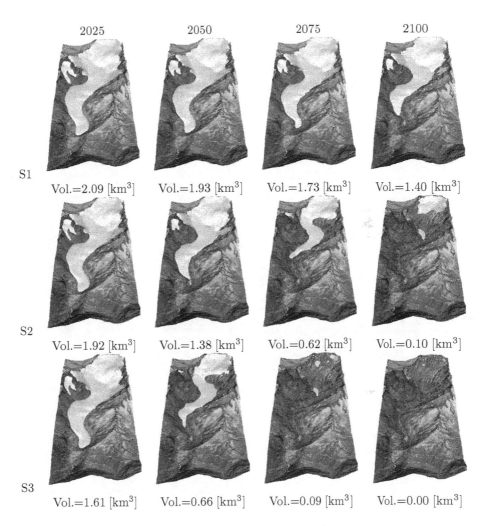

FIGURE 9.35: Simulation on the time interval 2007-2100 (from left to right) for the scenarios S1, S2, and S3 (top to bottom). The volume of ice is indicated on each figure.

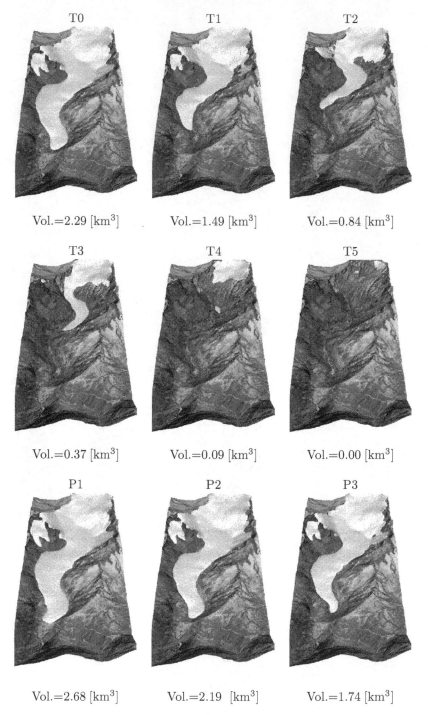

FIGURE 9.36: Numerical simulation of the Rhone glacier in 2100, according to the various temperature and precipitations scenarios T0, T1, T2, T3, T4, T5, P1, P2 and P3 (left to right, top to bottom. The volume of ice is indicated on each figure.

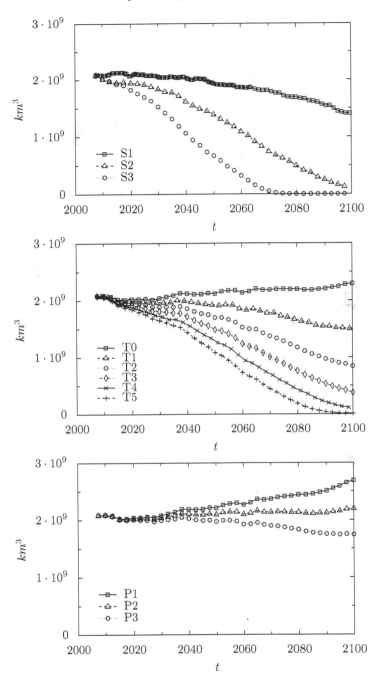

FIGURE 9.37: Volume of ice in the Rhone glacier during the time interval 2007-2100 according to scenarios S1, S2 and S3 (top), scenarios T0, T1, T2, T3, T4 and T5 (middle), and scenarios P1, P2 and P3 (bottom).

For such a slow flow, a nonlinear Stokes model has proved to be appropriate, and benchmarked against real data. The same real data have been extrapolated in the future to design climate scenarios, on the behavior and extinction of glaciers. The model presented in this chapter has a high potential to predict the impact of climate change on alpine glaciers during the coming decades. This is of immediate interest for scientists, but also for tourism, electricity production and economy in alpine environments (most of the drinkable water supply is stocked as ice); it emphasizes once again the crucial role of numerical simulations for policy-makers.

Acknowledgments

It would not have been possible to write this book without the help and support of the many people or institutions, to only some of whom it is possible to give particular mention here.

Chapter by Koobus et al.

This work was granted access to the HPC resources of CINES/IDRIS under the allocations 2009-c2009025067 and 2009-x2009025044 made by GENCI (Grand Equipement National de Calcul Intensif).

Chapter by Ouvrard et al.

This work was granted access to the HPC resources of CINES/IDRIS under the allocations 2009-c2009025067 and 2009-x2009025044 made by GENCI (Grand Equipement National de Calcul Intensif).

CINECA (Bologna, Italy) is also gratefully acknowledged for having provided computational resources.

Finally, the authors wish to thank E. Lamballais (Université de Poitiers, France) for kindly sending the experimental data and G. Pagano and M. Meldi (Université de Pise, Italy) for their contributions to the hybrid simulations.

Chapters by Caboussat et al.

The authors wish to thank V. Maronnier and A. Bonito (Texas A&M University) for their contributions to this project. The authors acknowledge the contribution of R. Lehaire (Polytech, Clermont-Ferrand) and its implementation of the surface tension geometric model during his master thesis.

The authors are also grateful to H. Blatter, M. Funk and M. Huss (ETH Zürich) for their fundamental input in the field of glaciology and experimental data that are crucial for the simulations, as well as for fruitful discussions and comments.

The authors thank D. Cohen (ETH Zürich), Th. Zwinger (CSC-Scientific Computing Ltd., Espoo, Finland) for relevant advice concerning slip boundary conditions, D. Farinotti and A. Bauder (ETH Zürich), for providing data on the bedrock topography and the glacier surface elevation. The authors are grateful to D. Nishimura (Hokkaido University, Sapporo) for providing data for the ice surface velocity of Rhone glacier, and M. Maisch, T. Imbaumgarten and C. Rothenbühler for the permission to use the map in Figure 9.26.

The authors thank M. Francois, J. Sicilian (Los Alamos National Laboratory), D. Kothe (Oak Ridge National Laboratory), R. Glowinski, and T.W. Pan (University of Houston) for fruitful discussions.

The first author, A. Caboussat, gratefully acknowledges the support of the Institute of Analysis and Scientific Computing at the Ecole Polytechnique Fédérale de Lausanne (EPFL) and of Ycoor systems SA, Sierre, Switzerland; part of this work has been achieved during his sabbatical leave in 2009-2010 at EPFL. The second author, G. Jouvet, is supported by the Swiss National Science Foundation.

The numerical results have been processed by using the Paraview (Kitware, Inc.), Ensight (CEI), and XD3D softwares.

Bibliography

[Abalakin et al., 2001] Abalakin, I., Bobkov, V., Dervieux, A., Kozubskaya, T., and Shiryaev, V. (2001). Study of high accuracy order schemes and non-reflecting boundary conditions for noise propagation problems. Research Report 99-02, Liapunov Institute. Available online at `http://www-sop.inria.fr/tropics/Alain.Dervieux/liapunov.html` (accessed July 1, 2010).

[Abalakin et al., 2002a] Abalakin, I., Dervieux, A., and Kozubskaya, T. (2002a). A vertex centered high order MUSCL scheme applying to linearised Euler acoustics. Research Report RR-4459, INRIA.

[Abalakin et al., 2002b] Abalakin, I., Dervieux, A., and Kozubskaya, T. (2002b). Computational study of mathematical models for noise DNS. *AIAA*, 2002-2585.

[Abalakin et al., 2002c] Abalakin, I., Dervieux, A., and Kozubskaya, T. (2002c). High accuracy study of mathematical models for DNS of noise around steady mean flow. In Zeitoun, D., Periaux, J., Desideri, J., and Marini, M., editors, *Proceedings of West East High Speed Flow Field Conference, CIMNE, Barcelona, 2002.*

[Abalakin et al., 2004] Abalakin, I., Dervieux, A., and Kozubskaya, T. (2004). On accuracy of noise direct calculation based on Euler model. *International Journal of Acoustics and Vibration*, 3(2):157–180.

[Abgrall, 1994] Abgrall, R. (1994). On essentially non-oscillatory schemes on unstructured meshes, analysis and implementation. *Journal of Computational Physics*, 114:45–58.

[Abgrall, 1996] Abgrall, R. (1996). How to prevent pressure oscillations in multicomponent flows: a quasi conservative approach. *Journal of Computational Physics*, 125:150–160.

[Abgrall et al., 2003] Abgrall, R., Nkonga, B., and Saurel, R. (2003). Efficient numerical approximation of compressible multi-material flow for unstructured meshes. *Computers and Fluids*, 32:571–605.

[Abgrall and Saurel, 2003] Abgrall, R. and Saurel, R. (2003). Discrete equations for physical and numerical compressible multiphase mixtures. *Journal of Computational Physics*, 186:361–396.

[Absil et al., 2008] Absil, P.-A., Mahony, R., and Sepulchre, R. (2008). *Optimization algorithms on matrix manifolds*. Princeton University Press, Princeton, NJ.

[Achenbach, 1968] Achenbach, E. (1968). Distribution of local pressure and skin friction around a circular cylinder in cross-flow up to $Re = 5 \times 10^6$. *Journal of Fluid Mechanics*, 34(4):625–639.

[Agouzal et al., 1999] Agouzal, A., Lipnikov, K., and Vassilevski, Y. (1999). Adaptive generation of quasi-optimal tetrahedral meshes. *East-West Journal*, 7(4):879–882.

[Ahmadi, 1984] Ahmadi, G. (1984). On thermodynamics of turbulence. *Bulletin of the American Physical Society*, 29:1529.

[Alauzet, 2009] Alauzet, F. (2009). Size gradation control of anisotropic meshes. *Finite Elements in Analysis and Design*, 46:181–202.

[Alauzet et al., 2007] Alauzet, F., Frey, P., George, P.-L., and Mohammadi, B. (2007). 3D transient fixed point mesh adaptation for time-dependent problems. Application to CFD simulations. *Journal of Computational Physics*, 222:592–623.

[Alauzet and Loseille, 2009] Alauzet, F. and Loseille, A. (2009). High order sonic boom modeling by adaptive methods. Research Report RR-6845, INRIA.

[Alauzet et al., 2006] Alauzet, F., Loseille, A., Dervieux, A., and Frey, P. (2006). Multi-dimensional continuous metric for mesh adaptation. In *Proceedings of the 15th International Meshing Roundtable, Birmingham, AL, USA*, pages 191–214. Springer.

[Anderson and Bonhaus, 1994] Anderson, W. and Bonhaus, D. (1994). An implicit upwind algorithm for computing a turbulent flow on unstructured grids. *Computer and Fluids*, 23(1):1–21.

[Apel, 1999] Apel, T. (1999). *Anisotropic finite elements: local estimates and applications*. Advances in Numerical Mathematics. Teubner, Stuttgart.

[Arminjon and Dervieux, 1993] Arminjon, P. and Dervieux, A. (1993). Construction of TVD-like artificial viscosities on two-dimensional arbitrary FEM grids. *Journal of Computational Physics*, 106(1):176–198.

[Arminjon and Viallon, 1999] Arminjon, P. and Viallon, M. (1999). Convergence of a finite volume extension of the Nessyahu-Tadmor scheme on unstructured grids for a two-dimensional linear hyperbolic equation. *SIAM Journal on Numerical Analysis*, 36(3):738–771.

[Arnold et al., 2002] Arnold, D., Brezzi, F., Cockburn, B., and Marini, L. (2002). Unified nalysis of discontinuous galerkin methods for elliptic problems. *SIAM Journal on Numerical Analysis*, 39:1749–1779.

[Arsigny et al., 2006] Arsigny, V., Fillard, P., Pennec, X., and Ayache, N. (2006). Log-Euclidean metrics for fast and simple calculus on diffusion tensors. *Magnetic Resonance in Medicine*, 56(2):411–421.

[Aulisa et al., 2003] Aulisa, E., Manservisi, S., and Scardovelli, R. (2003). A mixed markers and volume-of-fluid method for the reconstruction and advection of interfaces in two-phase and free-boundary flows. *Journal of Computational Physics*, 188:611–639.

[Aulisa et al., 2004] Aulisa, E., Manservisi, S., and Scardovelli, R. (2004). A surface marker algorithm coupled to an area-preserving marker redistribution method for three-dimensional interface tracking. *Journal of Computational Physics*, 197(2):555–584.

[Autrique and Magoulès, 2006a] Autrique, J.-C. and Magoulès, F. (2006a). Numerical analysis of a coupled finite-infinite element method for exterior Helmholtz problems. *Journal of Computational Acoustics*, 14(1):21–43.

[Autrique and Magoulès, 2006b] Autrique, J.-C. and Magoulès, F. (2006b). Studies of an infinite element method for acoustical radiation. *Applied Mathematical Modelling*, 30(7):641–655.

[Autrique and Magoulès, 2007] Autrique, J.-C. and Magoulès, F. (2007). Analysis of a conjugated infinite element method for acoustic scattering. *Computers and Structures*, 85(9):518–525.

[Baba and Tabata, 1981] Baba, K. and Tabata, M. (1981). On a conservative upwind finite element scheme for convective diffusion equations. *Revue Française d'Automatique Informatique Recherche Opérationnelle*, 15:3–25.

[Babuška and Strouboulis, 2001] Babuška, I. and Strouboulis, T. (2001). *The finite element method and its reliability*. Oxford Scientific Publications, New York.

[Baldwin and Lomax, 1978] Baldwin, B. and Lomax, H. (1978). Thin layer approximation and algebraic model for separated turbulent flows. *AIAA*, 78-257.

[Bänsch, 2001] Bänsch, E. (2001). Finite element discretization of the Navier-Stokes equations with a free capillary surface. *Numerische Mathematik*, 88(2):203–235.

[Barth, 1994] Barth, T. (1994). Aspects of unstructured grids and finite volume solvers for the Euler and Navier-Stokes equations. Research report, Von Karman Institute Lecture Notes.

[Batten et al., 2004] Batten, P., Goldberg, U., and Chakravarthy, S. (2004). Interfacing statistical turbulence closures with large-eddy simulation. *AIAA Journal*, 42(3):485–492.

[Bearman and Obasaju, 1982] Bearman, P. and Obasaju, E. (1982). An experimental study of pressure fluctuations on fixed and oscillating square-section cylinders. *Journal of Fluid Mechanics*, 119:297–321.

[Becker and Rannacher, 1996] Becker, R. and Rannacher, R. (1996). A feedback approach to error control in finite element methods: basic analysis and examples. *East-West Journal of Numerical Mathematics*, 4:237–264.

[Belytschko et al., 1998] Belytschko, T., Krongauz, Y., Dolbow, J., and Gerlach, C. (1998). On the completeness of meshfree particle methods. *International Journal for Numerical Methods in Engineering*, 43:785–819.

[Ben Moussa and Vila, 2000] Ben Moussa, B. and Vila, J. (2000). Convergence of SPH method for scalar nonlinear conservation laws. *SIAM Journal on Numerical Analysis*, 37(3):863–887.

[Berger, 2003] Berger, M. (2003). *A panoramic view of Riemannian geometry.* Springer Verlag, Berlin.

[Blatter, 1995] Blatter, H. (1995). Velocity and stress fields in grounded glaciers: a simple algorithm for including deviatoric stress gradients. *Journal of Glaciology*, 41(138):333–344.

[Boffi and Gastaldi, 2003] Boffi, D. and Gastaldi, L. (2003). A finite element approach for the immersed boundary method. *Computer and Structures*, 81:491–501.

[Boffi and Gastaldi, 2004] Boffi, D. and Gastaldi, L. (2004). Stability and geometric conservation laws for ALE formulations. *Computer Methods in Applied Mechanics and Engineering*, 193:4717–4739.

[Bonet and Lock, 1999] Bonet, J. and Lock, T.-S. (1999). Variational and momentum preservation aspects of smooth particle hydrodynamic formulations. *Computer Methods in Applied Mechanics and Engineering*, 180:97–115.

[Bonito et al., 2010] Bonito, A., Nochetto, R. H., and Pauletti, M. S. (2010). Parametric FEM for geometric biomembranes. *Journal of Computational Physics*, 229:3171–3188.

[Bonito et al., 2006] Bonito, A., Picasso, M., and Laso, M. (2006). Numerical simulation of 3D viscoelastic flows with free surfaces. *Journal of Computational Physics*, 215(2):691–716.

[Bottasso, 2004] Bottasso, C. (2004). Anisotropic mesh adaption by metric-driven optimization. *International Journal for Numerical Methods in Engineering*, 60:597–639.

[Boudet et al., 2010] Boudet, J., Caro, J., and Jacob, M. (2010). Large-eddy simulation of a single airfoil tip clearance flow. In *Proceedings of 16th AIAA/CEAS Aeroacoustics Conference, Stockholm, Nederlands*.

[Boyd, 2001] Boyd, J. (2001). *Chebyshev and Fourier spectral methods*. Dover Publications.

[Brackbill et al., 1992] Brackbill, J., Kothe, D., and Zemach, C. (1992). A continuum method for modeling surface tension. *Journal of Computational Physics*, 100:335–354.

[Breuer, 1998] Breuer, M. (1998). Numerical and modeling on large-eddy simulation for the flow past a circular cylinder. *International Journal of Heat and Fluid Flow*, 19:512–521.

[Bueler et al., 2007] Bueler, E., Brown, J., and Lingle, C. (2007). Exact solutions to the thermomechanically coupled Shallow-ice approximation: effective tools for verification. *Journal of Glaciology*, 53(182):499–516.

[Bunner and Tryggvason, 2002a] Bunner, B. and Tryggvason, G. (2002a). Dynamics of homogeneous bubbly flows. I. Rise velocity and microstructure of the bubbles. *Journal of Fluid Mechanics*, 466:17–52.

[Bunner and Tryggvason, 2002b] Bunner, B. and Tryggvason, G. (2002b). Dynamics of homogeneous bubbly flows. II. Velocity fluctuations. *Journal of Fluid Mechanics*, 466:53–84.

[Burger and Osher, 2005] Burger, M. and Osher, S. (2005). A survey on level set methods for inverse problems and optimal design. *European Journal of Applied Mathematics*, 16(2):263–301.

[Burman et al., 2004] Burman, E., Jacot, A., and Picasso, M. (2004). Adaptive finite elements with high aspect ratio for the computation of coalescence using a phase-field model. *Journal of Computational Physics*, 195(1):153–174.

[Caboussat, 2005] Caboussat, A. (2005). Numerical simulation of two-phase free surface flows. *Archives of Computational Methods in Engineering*, 12(2):165–210.

[Caboussat, 2006] Caboussat, A. (2006). A numerical method for the simulation of free surface flows with surface tension. *Computers and Fluids*, 35(10):1205–1216.

[Caboussat et al., 2008] Caboussat, A., François, M., Glowinski, R., Kothe, D., and Sicilian, J. (2008). A numerical method for interface reconstruction of triple points within a volume tracking algorithm. *Mathematical and Computer Modelling*, 48:1957–1971.

[Caboussat et al., 2005] Caboussat, A., Picasso, M., and Rappaz, J. (2005). Numerical simulation of free surface incompressible liquid flows surrounded by compressible gas. *Journal of Computational Physics*, 203(2):626–649.

[Cai and Sarkis, 1999] Cai, X.-C. and Sarkis, M. (1999). A restricted additive Schwarz preconditioner for general sparse linear systems. *SIAM Journal on Scientific Computing*, 21(2):792–797.

[Caiden et al., 2001] Caiden, R., Fedkiw, R., and Anderson, C. (2001). A numerical method for two-phase flow consisting of separate compressible and incompressible regions. *Journal of Computational Physics*, 166:1–27.

[Camarri et al., 2004] Camarri, S., Koobus, B., Salvetti, M., and Dervieux, A. (2004). A low diffusion MUSCL scheme for LES on unstructured grids. *Computers and Fluids*, 33:1101–1129.

[Camarri and Salvetti, 1999] Camarri, S. and Salvetti, M. (1999). Towards the large-eddy simulation of complex engineering flows with unstructured grids. Research Report RR-3844, INRIA.

[Camarri et al., 2001a] Camarri, S., Salvetti, M., Koobus, B., and Dervieux, A. (2001a). Numerical diffusion based on high-order derivatives in MUSCL schemes for LES on unstructured grids. In *Proceedings of the Direct and Large-eddy Simulation IV Conference, Twente, Holland, July 18-20*.

[Camarri et al., 2002a] Camarri, S., Salvetti, M., Koobus, B., and Dervieux, A. (2002a). Large-eddy simulation of a bluff-body flow on unstructured grids. *International Journal for Numerical Methods in Fluids*, 40:1431–1460.

[Camarri et al., 2005] Camarri, S., Salvetti, M., Koobus, B., and Dervieux, A. (2005). Hybrid RANS/LES simulations of a bluff-body flow. *Wind and Structures*, 8:407–426.

[Camarri et al., 2001b] Camarri, S., Salvetti, M., and Lombardi, G. (2001b). Large-eddy simulation of the flow over a forward-swept wing at high angle of attack. In *Proceedings of Atti del XVI Congresso dell'AIDAA*.

[Camarri et al., 2002b] Camarri, S., Salvetti, M.-V., Dervieux, A., and Koobus, B. (2002b). A low diffusion MUSCL scheme for LES on unstructured grids. Research Report RR-4412, INRIA.

[Canuto et al., 1987] Canuto, C., Hussaini, M., Quarteroni, A., and Zang, T. A. J. (1987). *Spectral methods in fluid dynamics*. Springer, 1st edition.

[Carpentier, 1995] Carpentier, R. (1995). *Approximation d'écoulements instationnaires. Application à des instabilités tourbillonnaires.* PhD thesis, University of Nice-Sophia Antipolis, France.

[Castro-Diaz et al., 1997] Castro-Diaz, M., Hecht, F., and Mohammadi, B. (1997). Anisotropic unstructured mesh adaption for flow simulations. *International Journal for Numerical Methods in Fluids*, 25:475–491.

[Catalano, 2002] Catalano, L. A. (2002). A new reconstruction scheme for the computation of inviscid compressible flows on 3D unstructured grids. *International Journal for Numerical Methods in Fluids*, 40(1-2):273–279.

[Cattabriga, 1961] Cattabriga, L. (1961). Su un problema al contorno relativo al sistema di equazioni di Stokes. *Rend. Sem. Mat. Univ. Padova*, 31:308–340.

[Ceniceros and Roma, 2004] Ceniceros, H. and Roma, A. (2004). A multiphase flow method with a fast, geometry-based fluid indicator. *Journal of Computational Physics*, 205:391–400.

[Chen and Li, 1998] Chen, L. and Li, Y. (1998). A numerical method for two-phase flows with an interface. *Environmental Modelling and Software*, 13:247–255.

[Choi and Bussman, 2006] Choi, B. and Bussman, M. (2006). A piecewise linear approach to volume tracking a triple point. *International Journal for Numerical Methods in Fluids*, 53(6):1005–1018.

[Chung, 2002] Chung, T. (2002). *Computational fluid dynamics.* Cambridge University Press, 1st edition.

[Cockburn, 2003] Cockburn, B. (2003). Discontinuous Galerkin methods. *ZAMM - Journal of Applied Mathematics and Mechanics / Zeitschrift für Angewandte Mathematik und Mechanik*, 83(11):731–754.

[Cockburn and Shu, 1989] Cockburn, B. and Shu, C.-W. (1989). TVB Runge-Kutta local projection discontinuous Galerkin finite element method for conservation laws. II. General framework. *Mathematics of Computation*, 52:411–435.

[Codina and Soto, 2002] Codina, R. and Soto, O. (2002). A numerical model to track two-fluid interfaces based on a stabilized finite element method and a level set technique. *International Journal for Numerical Methods in Fluids*, 40:293–301.

[Colagrossi and Landrini, 2003] Colagrossi, A. and Landrini, M. (2003). Numerical simulation of interfacial flows by smoothed particle hydrodynamics. *Journal of Computational Physics*, 191:448.

[Colinge and Rappaz, 1999] Colinge, J. and Rappaz, J. (1999). A strongly nonlinear problem arising in glaciology. *Mathematical Modelling and Numerical Analysis*, 33(2):395–406.

[Collis, 2001] Collis, S. (2001). Monitoring unresolved scales in multiscale turbulence modeling. *Physics of Fluids*, 13:1800–1806.

[Collis, 2002] Collis, S. (2002). The DG/VMS method for unified turbulence simulation. *AIAA*.

[Coupez, 2000] Coupez, T. (2000). Génération de maillages et adaptation de maillage par optimisation locale. *Revue Européenne des Éléments Finis*, 9:403–423.

[Cournède et al., 2006] Cournède, P.-H., Koobus, B., and Dervieux, A. (2006). Positivity statements for a mixed-element-volume scheme on fixed and moving grids. *European Journal of Computational Mechanics*, 15(7-8):767–798.

[Courty et al., 2006] Courty, F., Leservoisier, D., George, P.-L., and Dervieux, A. (2006). Continuous metrics and mesh adaptation. *Applied Numerical Mathematics*, 56(2):117–145.

[Cummins et al., 2005] Cummins, S., François, M., and Kothe, D. (2005). Reconstructing distance functions from volume fractions: a better way to estimate interface topology? *Computers and Structures*, 83:425–434.

[Dacorogna et al., 2003] Dacorogna, B., Glowinski, R., and Pan, T.-W. (2003). Numerical methods for the solution of a system of Eikonal equations with Dirichlet boundary conditions. *Comptes Rendus à l'Académie des Sciences, Paris, Série I*, 336:511–518.

[Davidson, 2004] Davidson, P. (2004). *Turbulence, an introduction for scientists and engineers*. Oxford University Press.

[de Hart, 2002] de Hart, J. (2002). *Fluid-structure interaction in the aortic heart valve: a three-dimensional computational analysis*. PhD thesis, Technische Universiteit Eindhoven, Nederlands.

[de la Bourdonnaye et al., 1998] de la Bourdonnaye, A., Farhat, C., Macedo, A., Magoulès, F., and Roux, F.-X. (1998). A non-overlapping domain decomposition method for the exterior Helmholtz problem. *Contemporary Mathematics*, 218:42–66.

[Debiez, 1996] Debiez, C. (1996). *Approximation et linéarisation d'écoulements aérodynamiques instationnaires*. PhD thesis, University of Nice, France.

[Debiez and Dervieux, 1999] Debiez, C. and Dervieux, A. (1999). Mixed element volume MUSCL methods with weak viscosity for steady and unsteady flow calculation. *Computer and Fluids*, 29:89–118.

[Debiez et al., 1998] Debiez, C., Dervieux, A., Mer, K., and NKonga, B. (1998). Computation of unsteady flows with mixed finite/finite element upwind methods. *International Journal for Numerical Methods in Fluids*, 27:193–206.

[Deconinck et al., 1993] Deconinck, H., Roe, P., and Struijs, R. (1993). A multidimensional generalization of Roe's flux difference splitter for the Euler equations. *Computers and Fluids*, 22(23):215–222.

[Deponti et al., 2006] Deponti, A., Pennati, V., de Biase, L., Maggi, V., and Berta, F. (2006). A new fully three-dimensional numerical model for ice dynamics. *Journal of Glaciology*, 52(178):365–376.

[Dervieux, 1987] Dervieux, A. (1987). *Partial differential equations of hyperbolique type and applications*, chapter Steady Euler simulations using unstructured meshes, pages 33–111. World Scientific.

[Dervieux et al., 2003] Dervieux, A., Leservoisier, D., George, P.-L., and Coudiere, Y. (2003). About theoretical and practical impact of mesh adaptations on approximation of functions and of solution of PDE. *International Journal for Numerical Methods in Fluids*, 43:507–516.

[Dervieux et al., 2007] Dervieux, A., Loseille, A., and Alauzet, F. (2007). High-order adaptive method applied to high speed flows. In *Proceedings of WEHSFF'2007, Moscow, Russia*.

[Desideri et al., 1987] Desideri, J.-A., Goudjo, A., and Selmin, V. (1987). Third-order numerical schemes for hyperbolic problems. Research Report RR-607, INRIA.

[Dilts, 1999] Dilts, G. (1999). Moving least-squares particle hydrodynamics. I. Consistency and stability. *International Journal for Numerical Methods in Engineering*, 44(8):1115–1155.

[Dilts, 2000] Dilts, G. (2000). Moving least-squares particle hydrodynamics. II. Conservation and boundaries. *International Journal for Numerical Methods in Engineering*, 48(10):1503–1524.

[Dobrzynski and Frey, 2008] Dobrzynski, C. and Frey, P. (2008). Anisotropic Delaunay mesh adaptation for unsteady simulations. In *Proceedings of the 17th International Meshing Roundtable, Pittsburgh, PA, USA*, pages 177–194. Springer.

[Dompierre et al., 1997] Dompierre, J., Vallet, M., Fortin, M., Bourgault, Y., and Habashi, W. (1997). Anisotropic mesh adaptation: towards a solver and user independent CFD. *AIAA*, 97-0861.

[Donea et al., 1987] Donea, J., Quartapelle, L., and Selmin, V. (1987). An analysis of time discretization in the finite element solution of hyperbolic problems. *Journal of Computational Physics*, 70:463–499.

[Dong et al., 2006] Dong, S., Karniadakis, G., Ekmekci, A., and Rockwell, D. (2006). A combined direct numerical simulation-particle image velocimetry study of the turbulent near wake. *Journal of Fluid Mechanics*, 569:185–207.

[Doring, 2006] Doring, M. (2006). *Développement d'une méthode SPH pour les applications à surface libre en hydrodynamique*. PhD thesis, Ecole Centrale de Nantes, France.

[Dubois, 2001] Dubois, F. (2001). Partial Riemann problem, boundary conditions, and gas dynamics. In *Absorbing boundaries and layers, domain decomposition methods: applications to large scale computations*, pages 16–77. Nova Science Publishers, Inc., NY.

[Dubuc et al., 2000] Dubuc, L., Cantariti, F., Woodgate, M., Gribben, B., Badcock, K., and Richards, B. (2000). A grid deformation technique for unsteady flow computations. *International Journal for Numerical Methods in Fluids*, 32:285–311.

[Einfeldt et al., 1991] Einfeldt, B., Munz, C., Roe, P., and Sjogrcen, B. (1991). On Godunov-type methods near low densities. *Journal of Computational Physics*, 92:273–295.

[Erlebacher et al., 1992] Erlebacher, G., Hussaini, M., Speziale, C., and Zang, T. (1992). Toward the large-eddy simulation of compressible flows. *Journal of Fluid Mechanics*, 238:155–185.

[Farhat, 1995] Farhat, C. (1995). High performance simulation of coupled nonlinear transient aeroelastic problems. Special course on parallel computing in CFD R-807, NATO AGARD Report.

[Farhat et al., 2001] Farhat, C., Geuzaine, P., and Grandmont, C. (2001). The discrete geometric conservation law and the nonlinear stability of ALE schemes for the solution of flow problems on moving grids. *Journal of Computational Physics*, 174:669–694.

[Farhat and Lesoinne, 2000] Farhat, C. and Lesoinne, M. (2000). Two efficient staggered procedures for the serial and parallel solution of three-dimensional nonlinear transient aeroelastic problems. *Computer Methods in Applied Mechanics and Engineering*, 182:499–516.

[Farhat et al., 2000] Farhat, C., Macedo, A., Lesoinne, M., Roux, F.-X., Magoulès, F., and de la Bourdonnaye, A. (2000). Two-level domain decomposition methods with Lagrange multipliers for the fast iterative solution of acoustic scattering problems. *Computer Methods in Applied Mechanics and Engineering*, 184(2–4):213–240.

[Farhat et al., 2006] Farhat, C., Rajasekharan, A., and Koobus, B. (2006). A dynamic variational multiscale method for large-eddy simulations on unstructured meshes. *Computer Methods in Applied Mechanics and Engineering*, 195:1667–1691.

[Farinotti et al., 2009] Farinotti, D., Huss, M., Bauder, A., Funk, M., and Truffer, M. (2009). A method to estimate ice volume and ice thickness distribution of Alpine glaciers. *Journal of Glaciology*, 55(191):422–430.

[Fedkiw et al., 1998] Fedkiw, R., Merriman, B., and Osher, S. (1998). Numerical methods for a one-dimensional interface separating compressible and incompressible flows. In Venkatakrishnan, V., Salas, M., and Chakravarthy, S., editors, *Barriers and challenges in computational fluid dynamics*, pages 155–194. Kluwer Academic Publishers.

[Fezoui, 1985] Fezoui, L. (1985). Résolution des équations d'Euler par un schéma de Van Leer en éléments finis. Research Report RR-0358, INRIA.

[Fezoui and Dervieux, 1989] Fezoui, L. and Dervieux, A. (1989). Finite-element non oscillatory schemes for compressible flows. In *Symposium on Computational Mathematics and Applications, Italy*, number 730. Publications of university of Pavie.

[Fezoui and Stoufflet, 1989] Fezoui, L. and Stoufflet, B. (1989). A class of implicit schemes for Euler simulations with unstructured meshes. *Journal of Computational Physics*, 84(1):174–206.

[Fletcher, 1991] Fletcher, C. A. (1991). *Computational techniques for fluid dynamics. I. Fundamental and general techniques.* Springer, 2nd edition.

[Flück et al., 2009] Flück, M., Hofer, T., Picasso, M., Rappaz, J., and Steiner, G. (2009). Scientific computing for aluminum production. *International Journal of Numerical Analysis and Modeling*, 6:489–504.

[Formaggia et al., 2004] Formaggia, L., Micheletti, S., and Perotto, S. (2004). Anisotropic mesh adaptation in computational fluid dynamics: application to the advection-diffusion-reaction and the Stokes problems. *Applied Numerical Mathematics*, 51(4):511–533.

[Formaggia and Perotto, 2001] Formaggia, L. and Perotto, S. (2001). New anisotropic a priori error estimates. *Numerische Mathematik*, 89:641–667.

[Fortin et al., 1996] Fortin, M., Vallet, M.-G., Dompierre, J., Bourgault, Y., and Habashi, W. (1996). Anisotropic mesh adaptation: theory, validation and applications. In *Proceedings of ECCOMAS CFD*.

[Fournier et al., 1998] Fournier, L., Carré, G., and Lanteri, S. (1998). Parallel multigrid algorithms for the acceleration of compressible flow calculations. *Calculateurs Parallèles, Réseaux et Systèmes Répartis*, 10(4):381–389.

[Franca and Frey, 1992] Franca, L. and Frey, S. (1992). Stabilized finite element method. II. The incompressible Navier-Stokes equations. *Computer Methods in Applied Mechanics and Engineering*, 99:209–233.

[François et al., 2006] François, M., Cummins, S., Dendy, E., Kothe, D., Sicilian, J., and Williams, M. (2006). A balanced-force algorithm for continuous and sharp interfacial surface tension models within a volume tracking framework. *Journal of Computational Physics*, 213(1):141–173.

[Francescatto and Dervieux, 1998] Francescatto, J. and Dervieux, A. (1998). A semi-coarsening strategy for unstructured multigrid based on agglomeration. *International Journal for Numerical Methods in Fluids*, 26:927–957.

[Frei, 2007] Frei, C. (2007). Die Klimazukunft der Schweiz. In *Klimaänderung und die Schweiz 2050. Erwartete Auswirkungen auf Umwelt, Gesellschaft und Wirtschaft*, pages 12–16. Beratendes Organ für Fragen der Klimaänderung (OcCC). Available online at: http://www.occc.ch (accessed July 1, 2010).

[Frey, 2000] Frey, P. (2000). About surface remeshing. In *Proceedings of the 9th International Meshing Roundtable, New Orleans, LA, USA*, pages 123–136.

[Frey, 2001] Frey, P. (2001). Yams: a fully automatic adaptive isotropic surface remeshing procedure. Research Report RR-0252, INRIA.

[Frey and Alauzet, 2005] Frey, P. and Alauzet, F. (2005). Anisotropic mesh adaptation for CFD computations. *Computer Methods in Applied Mechanics and Engineering*, 194(48-49):5068–5082.

[Frey and George, 2008] Frey, P. and George, P.-L. (2008). *Mesh generation. Application to finite elements*. ISTE Ltd and John Wiley & Sons, 2nd edition.

[Frohlich and von Terzi, 2008] Frohlich, J. and von Terzi, D. (2008). Hybrid LES/RANS methods for the simulation of turbulent flows. *Progress in Aerospace Sciences*, 44:349–377.

[Fureby, 2008] Fureby, C. (2008). Towards the use of large-eddy simulation in engineering. *Progress in Aerospace Sciences*, 44:381–396.

[Fureby et al., 2000] Fureby, C., Tabor, G., Weller, H. G., and Gosman, A. D. (2000). Large-eddy simulation of the flow around a square prism. *AIAA Journal*, 38(3):442–452.

[Gagliardini et al., 2007] Gagliardini, O., Cohen, D., Råback, P., and Zwinger, T. (2007). Finite-element modeling of subglacial cavities and related friction law. *Journal of Geophysical Research*, 112.

[Gander et al., 2007a] Gander, M., Halpern, L., and Magoulès, F. (2007a). An optimized Schwarz method with two-sided Robin transmission conditions for the Helmholtz equation. *International Journal for Numerical Methods in Fluids*, 55(2):163–175.

[Gander et al., 2007b] Gander, M., Halpern, L., Magoulès, F., and Roux, F.-X. (2007b). Analysis of patch substructuring methods. *International Journal of Applied Mathematics and Computer Science*, 17(3):395–402.

[Gander et al., 2002] Gander, M., Magoulès, F., and Nataf, F. (2002). Optimized Schwarz methods without overlap for the Helmholtz equation. *SIAM Journal on Scientific Computing*, 24(1):38–60.

[Garnier et al., 1999] Garnier, E., Sagaut, P., Comte, P., and Deville, M. (1999). On the use of shock-capturing schemes for large-eddy simulation. *Journal of Computational Physics*, 153:273–311.

[George, 1999] George, P.-L. (1999). Tet meshing: construction, optimization and adaptation. In *Proceedings of the 8th International Meshing Roundtable, South Lake Tahoe, CA, USA*.

[George and Borouchaki, 1998] George, P.-L. and Borouchaki, H. (1998). *Delaunay triangulation and meshing. Application to finite elements*. Hermès, Paris.

[Gerbeau and Lelièvre, 2009] Gerbeau, J.-F. and Lelièvre, T. (2009). Generalized Navier boundary condition and geometric conservation law for surface tension. *Computer Methods in Applied Mechanics and Engineering*, 198(5-8):644–656.

[Gerbeau et al., 2003] Gerbeau, J.-F., Lelievre, T., and Le Bris, C. (2003). Simulations of MHD flows with moving interfaces. *Journal of Computational Physics*, 184(1):163–191.

[Germano et al., 1991] Germano, M., Piomelli, U., Moin, P., and Cabot, W. (1991). A dynamic subgrid-scale eddy viscosity model. *Physics of Fluids*, 3(7):1760–1765.

[Geurts, 2006] Geurts, B. (2006). Interacting errors in large-eddy simulation: a review of recent developments. *Journal of Turbulence*, 7(1).

[Giles, 1997] Giles, M. (1997). On adjoint equations for error analysis and optimal grid adaptation in CFD. Research Report NA-97/11, Oxford.

[Giles and Pierce, 1999] Giles, M. and Pierce, N. (1999). Improved lift and drag estimates using adjoint Euler equations. *AIAA*, 99-3293.

[Giles and Suli, 2002] Giles, M. and Suli, E. (2002). *Adjoint methods for PDEs: a posteriori error analysis and postprocessing by duality*, pages 145–236. Cambridge University Press.

[Gingold and Monaghan, 1977] Gingold, R. and Monaghan, J. (1977). Smoothed particle hydrodynamics: theory and application to non spherical stars. *Monthly Notices of the Royal Astronomical Society*, pages 181–375.

[Gingold and Monaghan, 1983] Gingold, R. and Monaghan, J. (1983). Shock simulation by the particle method SPH. *Journal of Computational Physics*, 52:374–389.

[Glen, 1958] Glen, J. (1958). The flow law of ice. In *Proceedings of the IUGG/IAHS Symposium of Chamonix*, pages 171–183. IAHS Publication.

[Glowinski, 2003] Glowinski, R. (2003). *Finite Element Method for Incompressible Viscous Flow*, volume 9 of *Handbook of Numerical Analysis*, pages 3–1176. Elsevier, Amsterdam.

[Glowinski et al., 2001] Glowinski, R., Pan, T.-W., Hesla, T., Joseph, D., and Periaux, J. (2001). A fictitious domain approach to the direct numerical simulation of incompressible viscous flow past moving rigid bodies: application to particulate flow. *Journal of Computational Physics*, 169(2):363–427.

[Godlewski and Raviart, 1996] Godlewski, E. and Raviart, P.-A. (1996). *Numerical approximation of hyperbolic systems of conservation law.* Springer.

[Godunov, 1959] Godunov, S. (1959). A finite difference method for the computation of discontinuous solutions of the equations of fluid dynamics. *Matematicheskii Sbornik*, pages 271–306.

[Goldberg and Ota, 1990] Goldberg, U. and Ota, D. (1990). A $k-\varepsilon$ near-wall formulation for separated flows. *AIAA*, 91-1482.

[Goldberg et al., 1998] Goldberg, U., Peroomian, O., and Chakravarthy, S. (1998). A wall-distance-free $k-\varepsilon$ model with enhanced near-wall treatment. *Journal of Fluids Engineering*, 120:457–462.

[Gomes and Faugeras, 2000] Gomes, J. and Faugeras, O. (2000). Reconciling distance functions and level sets. *Journal of Vizualisation Communication and Image Representation*, 11:209–223.

[Gottlieb and Orszag, 1977] Gottlieb, D. and Orszag, S. (1977). *Numerical analysis of spectral methods.* SIAM.

[Gourvitch et al., 2004] Gourvitch, N., Roge, G., Abalakin, I., Dervieux, A., and Kozubskaya, T. (2004). A tetrahedral-based superconvergent scheme for aeroacoustics. Research Report RR-5212, INRIA.

[Grandmont and Maday, 2000] Grandmont, C. and Maday, Y. (2000). Existence for an unsteady fluid-structure interaction problem. *Mathematical Modelling and Numerical Analysis*, 34(3):609–636.

[Gravemeier et al., 2006] Gravemeier, V., Lenz, S., and Wall, W. (2006). Variational multiscale methods for incompressible flows. In *Proceedings of the International Conference on Boundary and Interior Layers, Bail, 2006*.

[Gravemeier et al., 2004] Gravemeier, V., Wall, W., and Ramm, E. (2004). A three-level finite element method for the instationary incompressible Navier-Stokes equations. *Computer Methods in Applied Mechanics and Engineering*, 193:1323–1366.

[Gravemeier et al., 2005] Gravemeier, V., Wall, W., and Ramm, E. (2005). Large-eddy simulation of turbulent incompressible flows by a three-level finite element method. *International Journal for Numerical Methods in Fluids*, 48:1067–1099.

[Grinstein and Fureby, 2006] Grinstein, F. and Fureby, C. (2006). From canonical to complex flows: recent progress on monotonically integrated LES. *Computing in Science and Engineering*, 6(2):36–49.

[Gudmundsson, 1999] Gudmundsson, G. (1999). A three-dimensional numerical model of the confluence area of Unteraargletscher, Bernese Alps, Switzerland. *Journal of Glaciology*, 45(150):219–230.

[Guégan, 2007] Guégan, D. (2007). *Modélisation numérique d'écoulements bifluides 3D instationnaires avec adaptation de maillage*. PhD thesis, Université de Nice Sophia Antipolis, France.

[Guidoboni et al., 2009] Guidoboni, G., Glowinski, R., Cavallini, N., and Canic, S. (2009). Stable loosely-coupled-type algorithm for fluid-structure interaction in blood flow. *Journal of Computational Physics*, 228(18):6916–6937.

[Guillard and Viozat, 1999] Guillard, H. and Viozat, C. (1999). On the behaviour of upwind schemes in the low Mach number limit. *Computers and Fluids*, 28:63–86.

[Hackbusch, 1985] Hackbusch, W. (1985). *Multi-grid methods and applications*. Springer-Verlag, Berlin.

[Hanjalic and Launder, 1976] Hanjalic, K. and Launder, B. (1976). Contribution toward a Reynolds-stress closure for low Reynolds-number turbulence. *Journal of Fluid Mechanics*, 74:593–610.

[Harari and Magoulès, 2004] Harari, I. and Magoulès, F. (2004). Numerical investigations of stabilized finite element computations for acoustics. *Wave Motion*, 39(4):339–349.

[Harlow, 1964] Harlow, F. (1964). The particle-in-cell computing method for fluid dynamics. *Methods of Computational Physics*, 3:313–343.

[Harlow and Welch, 1965] Harlow, F. and Welch, E. (1965). Numerical calculation of time-dependent viscous incompressible flow of fluids with free surface. *Physics of Fluids*, 8:2182.

[Harten, 1983] Harten, A. (1983). High resolution schemes for hyperbolic conservation laws. *Journal of Computational Physics*, 49:357–393.

[Harten et al., 1987] Harten, A., Engquist, B., Osher, S., and Chakravarthy, S. (1987). Uniformly high-order accurate essentially non oscillatory schemes. III. *Journal of Computational Physics*, 71:231–303.

[Hinze, 1959] Hinze, J. (1959). *Turbulence.* McGraw-Hill.

[Hirsch, 1991] Hirsch, C. (1991). *Numerical computation of internal and external flows*, volume 1. J. Wiley & Sons.

[Hirsch, 2007] Hirsch, C. (2007). *Numerical computation of internal and external flows.* Elsevier, 2nd edition.

[Hirt, 1968] Hirt, C. (1968). Heuristic stability theory for finite difference equations. *Journal of Computational Physics*, 2:339–355.

[Hirt et al., 1974] Hirt, C., Amsden, A., and Cook, J. (1974). An arbitrary Lagrangian-Eulerian computing method for all speeds. *Journal of Computational Physics*, 14:227–253.

[Hirt and Nichols, 1981] Hirt, C. and Nichols, B. (1981). Volume of fluid (VOF) method for the dynamics of free boundaries. *Journal of Computational Physics*, 39:201–225.

[Hock, 1999] Hock, R. (1999). A distributed temperature-index ice- and snowmelt-model including potential direct solar radiation. *Journal of Glaciology*, 45(149):101–111.

[Holmen et al., 2004] Holmen, J., Hughes, T., Oberai, A., and Well, G. (2004). Sensitivity of the scale partition for variational multiscale large-eddy simulation of channel flow. *Physics of Fluids*, 16:824–827.

[Hou, 1995] Hou, T. (1995). Numerical solutions to free boundary problems. *Acta Numerica*, 4:335–415.

[Hu and Adams, 2006] Hu, X. and Adams, N. (2006). A multi-phase SPH method for macroscopic and mesoscopic flows. *Journal of Computational Physics*, 213:844.

[Hua et al., 2008] Hua, J., Stene, J., and Lin, P. (2008). Numerical simulation of 3D bubbles rising in viscous liquids using a front tracking method. *Journal of Computational Physics*, 227:3358–3382.

[Huang, 2005] Huang, W. (2005). Metric tensors for anisotropic mesh generation. *Journal of Computational Physics*, 204(2):633–665.

[Hughes and Mallet, 1986] Hughes, T. and Mallet, M. (1986). A new finite element formulation for computational fluid dynamics. III. The generalized streamline operator for multidimensional advective-diffusive systems. *Computer Methods in Applied Mechanics and Engineering*, 58(3):305–328.

[Hughes et al., 2000] Hughes, T., Mazzei, L., and Jansen, K. (2000). Large-eddy simulation and the variational multiscale method. *Comput. Vis. Sci.*, 3:47–59.

[Hughes et al., 2001a] Hughes, T., Mazzei, L., Oberai, A., and Wray, A. (2001a). The multiscale formulation of large-eddy simulation: decay of homogeneous isotropic turbulence. *Physics of Fluids*, 13:505–512.

[Hughes et al., 2001b] Hughes, T., Oberai, A., and Mazzei, L. (2001b). Large-eddy simulation of turbulent channel flows by the variational multiscale method. *Physics of Fluids*, 13:1784–1799.

[Huss et al., 2008a] Huss, M., Bauder, A., Funk, M., and Hock, R. (2008a). Determination of the seasonal mass balance of four Alpine glaciers since 1865. *Journal of Geophysical Research*, 113:F01015.

[Huss et al., 2008b] Huss, M., Farinotti, D., Bauder, A., and Funk, M. (2008b). Modelling runoff from highly glacierized Alpine drainage basins in a changing climate. *Hydrological Processes*, 22(19):3888–3902.

[Hutter, 1983] Hutter, K. (1983). *Theoretical glaciology*. Reidel.

[Imbaumgarten, 2005] Imbaumgarten, T. (2005). Kartierung und GIS-basierte darstellung der geomorphologie im gebiet Val Bever/Val Saluver (GR) sowie modellierung spaet und postglazialer gletscherstaende in der Val Muragl (GR). Master's thesis, University of Zurich, Switzerland.

[IPCC, 2007] IPCC (2007). Climate Change 2007. The physical scientific basis. Contributions of working group I to the fourth assessment report of the intergovernmental panel on climate change. Technical report, WMO/UNEP, Cambridge University Press.

[Issa, 2005] Issa, R. (2005). *Numerical assessment of the smoothed particle hydrodynamics gridless method for incompressible flows and its extension to turbulent fows*. PhD thesis, University of Manchester, Institute of Science and Technology.

[Ivings et al., 1998] Ivings, M., Causon, D., and Toro, E. (1998). On Riemann solvers for compressible liquids. *International Journal for Numerical Methods in Fluids*, 28:395–418.

[James et al., 1980] James, W., Paris, S., and Malcolm, G. (1980). Study of viscous cross flow effects on circular cylinders at high Reynolds numbers. *AIAA Journal*, 18:1066–1072.

[Jameson, 1987] Jameson, A. (1987). Successes and challenges in computational aerodynamics. volume 1184.

[Jameson, 1993] Jameson, A. (1993). Artificial diffusion, upwind biasing, limiters and their effect on accuracy and multigrid convergence in transonic and hypersonic flows. *AIAA*, 93-3359.

[Jameson et al., 1981] Jameson, A., Schmidt, W., and Turkel, E. (1981). Numerical solution of the euler equations by finite volume methods using Runge-Kutta time-stepping schemes. In *Proceedings of the AIAA 14th Fluid and Plasma Dynamic Conference, Palo Alto, California*.

[Jiang and Peng, 2000] Jiang, G.-S. and Peng, D. (2000). Weighted ENO schemes for Hamilton-Jacobi equations. *SIAM Journal on Scientific Computing*, 21:2126.

[John and Kaya, 2005] John, V. and Kaya, S. (2005). A finite element variational multiscale method for the Navier-Stokes equations. *SIAM Journal of Scientific Computing*, 26:57–80.

[Johnson and Beissel, 1996] Johnson, G. and Beissel, S. (1996). Normalized smoothing functions for impact computations. *International Journal for Numerical Methods in Engineering*.

[Jones and Launder, 1972] Jones, W. and Launder, B. (1972). The prediction of laminarization with a two-equation model of turbulence. *International Journal of Heat and Mass Transfer*, 15:301–314.

[Jones et al., 2006] Jones, W., Nielsen, E., and Park, M. (2006). Validation of 3D adjoint based error estimation and mesh adaptation for sonic boom reduction. *AIAA*, 2006-1150.

[Jongen and Marx, 1997] Jongen, T. and Marx, Y. (1997). Design of an unconditionally stable, positive scheme for the $k - \epsilon$ and two-layer turbulence model. *Computers and Fluids*, 26(5):469–487.

[Josserand and Zaleski, 2003] Josserand, C. and Zaleski, S. (2003). Droplet splashing on a thin liquid film. *Physics of Fluids*, 15(6):1650–1657.

[Jouvet et al., 2009] Jouvet, G., Huss, M., Picasso, M., Rappaz, J., and Blatter, H. (2009). Numerical simulation of Rhonegletscher from 1874 to 2100. *Journal of Computational Physics*, 228(17):6426–6439.

[Jouvet et al., 2008] Jouvet, G., Picasso, M., Rappaz, J., and Blatter, H. (2008). A new algorithm to simulate the dynamics of a glacier: theory and applications. *Journal of Glaciology*, 54(188):801–811.

[Junk and Struckmeier, 2002] Junk, M. and Struckmeier, J. (2002). Consistency analysis for meshfree methods for conservation laws. *Mitteilungen der Gesellschaft für Angewandte Mathematik und Mechanik*, 24:99–126.

[Keck and Hietel, 2005] Keck, R. and Hietel, D. (2005). A projection technique for incompressible flow in the meshless finite volume particle method. *Advances in Computational Mathematics*, 23:143–169.

[Kim et al., 1987] Kim, J., Moin, P., and Moser, R. (1987). Turbulence statistics in fully developed channel flow at low Reynolds number. *Journal of Fluid Mechanics*, 177:133–166.

[Kim and Lee, 2003] Kim, M. and Lee, W. (2003). A new VOF-based numerical scheme for the simulation of fluid flow with free surface. I. New free surface-tracking algorithm and its verification. *International Journal for Numerical Methods in Fluids*, 42:765–790.

[Kim et al., 2003] Kim, M., Park, J., and Lee, W. (2003). A new VOF-based numerical scheme for the simulation of fluid flow with free surface. II. Application to the cavity filling and sloshing problems. *International Journal for Numerical Methods in Fluids*, 42:791–812.

[Kirner, 2007] Kirner, P. (2007). *Modélisation mathématique et simulation numérique des phenomènes dynamiques et thermiques apparaissant dans un glacier*. PhD thesis, EPF Lausanne, Switzerland.

[Knoll and Keyes, 2004] Knoll, D. A. and Keyes, D. E. (2004). Jacobian-free Newton-Krylov methods: a survey of approaches and applications. *Journal of Computational Physics*, 193(2):357–397.

[Koobus et al., 2007] Koobus, B., Camarri, S., Salvetti, M., Wornom, S., and Dervieux, A. (2007). Parallel simulation of three-dimensional flows: application to turbulent wakes and two-phase compressible flows. *Advances in Engineering Software*, 38:328–337.

[Koobus and Farhat, 2004] Koobus, B. and Farhat, C. (2004). A variational multiscale method for the large-eddy simulation of compressible turbulent flows on unstructured meshes-application to vortex shedding. *Computer Methods in Applied Mechanics and Engineering*, 193:1367–1383.

[Kothe, 1998] Kothe, D. (1998). *Perspective on Eulerian finite volume methods for incompressible interfacial flows*, pages 267–331. Springer.

[Kothe et al., 1998] Kothe, D., Juric, D., Lam, K., and Lally, B. (1998). Numerical recipes for mold filling simulation. In *Proceedings of the Eighth International Conference on Modeling of Casting, Welding, and Advanced Solidification Processes, San Diego, CA.*

[Kothe et al., 1999] Kothe, D., Williams, M., Lam, K., Korzewka, D., Tubesing, P., and Puckett, E. (1999). A second-order accurate, linearity-preserving volume tracking algorithm for free surface flows on 3D unstructured meshes. In *Proceedings of the 3rd ASME/JSME Joint Fluids Engineering Conference, San Francisco, CA.*

[Kravchenko and Moin, 1999] Kravchenko, A. and Moin, P. (1999). Numerical studies of flow over a circular cylinder at $Re_d = 3900$. *Physics of Fluids*, 12:403–417.

[Kuhn, 1981] Kuhn, M. (1981). Climate and glaciers. *Journal of the International Association of Hydrological Sciences (IAHS)*, 131:3–20.

[Kuzmin and Turek, 2004] Kuzmin, D. and Turek, S. (2004). *Bubbly flows: analysis, modelling and calculation*, chapter Finite element discretization tools for gas-liquid flow, pages 191–201. Springer.

[Labourasse and Sagaut, 2002] Labourasse, E. and Sagaut, P. (2002). Reconstruction of turbulent fluctuations using a hybrid RANS/LES approach. *Journal of Computational Physics*, 182:301–336.

[Labourasse and Sagaut, 2004] Labourasse, E. and Sagaut, P. (2004). Advances in RANS-LES coupling, a review and an insight on the NLDE approach. *Archives of Computational Methods in Engineering*, 11:199–256.

[Lagüe and Hecht, 2006] Lagüe, J.-F. and Hecht, F. (2006). Optimal mesh for P_1 interpolation in H^1 semi-norm. In *Proceedings of the 15th International Meshing Roundtable, Birmingham, AL, USA*, pages 259–270. Springer.

[Lallemand et al., 1992] Lallemand, M., Steve, H., and Dervieux, A. (1992). Unstructured multigridding by volume agglomeration: current status. *Computers and Fluids*, 21:397–433.

[Larrouturou, 1991] Larrouturou, B. (1991). How to preserve the mass fraction positivity when computing compressible multi-component flows. *Journal of Computational Physics*, 95:59–84.

[Launder and Spalding, 1972] Launder, B. and Spalding, D. (1972). *Mathematical models of turbulence*. Academic Press, New York.

[Launder and Spalding, 1979] Launder, B. and Spalding, D. (1979). The numerical computation of turbulent flows. *Computer Methods in Applied Mechanics and Engineering*, 3:269–289.

[Lax and Wendroff, 1972] Lax, P. and Wendroff, B. (1972). Systems of conservation laws. *Communications on Pure and Applied Mathematics*, 13:217–237.

[le Meur et al., 2004] le Meur, E., Gagliardini, O., Zwinger, T., and Ruokolainen, J. (2004). Glacier flow modelling: a comparison of the Shallow ice approximation and the full-Stokes solution. *Comptes Rendus Physique*, 5(7):709–722.

[Le Ribault et al., 2006] Le Ribault, C., Le Penven, L., and Buffat, M. (2006). LES of a compressed Taylor vortex flow using a finite volume/finite element method on unstructured grids. *International Journal for Numerical Methods in Fluids*, 52:355–379.

[Lee et al., 2006] Lee, J., Park, N., Lee, S., and Choi, H. (2006). A dynamical subgrid-scale eddy viscosity model with a global model coefficient. *Physics of Fluids*.

[Lerat and Peyret, 1974] Lerat, A. and Peyret, R. (1974). Noncentered schemes and shock propagation problems. *Comput. Fluids*, 2:35–52.

[Lesieur and Comte, 1997] Lesieur, M. and Comte, P. (1997). Large-eddy simulations of compressible turbulent flows. Research Report R-819, AGARD.

[Lesieur et al., 2005] Lesieur, M., Métais, O., and Comte, P. (2005). *Large-eddy simulation of turbulence*. Cambridge University Press, Cambridge.

[Lesoinne and Farhat, 1996] Lesoinne, M. and Farhat, C. (1996). Geometric conservation laws for flow problems with moving boundaries and deformable meshes and their impact on aeroelastic computations. *Computer Methods in Applied Mechanics and Engineering*, 134:71–90.

[Lewandowski and Mohammadi, 1993] Lewandowski, R. and Mohammadi, B. (1993). Existence and positivity results for a $\phi-\theta$ model and a modified $k-\varepsilon$ turbulence model. *Mathematical Models and Methods in Applied Sciences*, 3(2):195–215.

[Li and Renardy, 2000] Li, J. and Renardy, Y. (2000). Numerical study of flows of two immiscible liquids at low Reynolds number. *SIAM Review*, 42(3):417–439.

[Li et al., 2005] Li, X., Shephard, M., and Beal, M. (2005). 3D anisotropic mesh adaptation by mesh modification. *Computer Methods in Applied Mechanics and Engineering*, 194(48-49):4915–4950.

[Li and Ito, 2006] Li, Z. and Ito, K. (2006). *The immersed interface method: numerical solutions of PDEs involving interfaces and irregular domains.* Number 33 in Frontiers in Applied Mathematics. SIAM, Philadelphia.

[Linde and Roe, 1998] Linde, T. and Roe, P. (1998). On multidimensional positively conservative high resolution schemes. *AIAA*, 97-2098.

[Liu and Liu, 2006] Liu, M. and Liu, G. (2006). Restoring particle consistency in smoothed particle hydrodynamics. *Applied Numerical Mathematics*, 56:19–36.

[Liu et al., 2003] Liu, M., Liu, G., and Lam, K. (2003). Constructing smoothing functions in smoothed particle hydrodynamics with applications. *Journal of Computational and Applied Mathematics*, 155(2):263–284.

[Liu et al., 1995] Liu, W., Jun, S., and Zhang, Y. (1995). Reproducing kernel particle methods. *International Journal for Numerical Methods in Fluids*, 20:1081–1106.

[Lo et al., 2005] Lo, S.-C., Hofmann, K., and Dietiker, J.-F. (2005). Numerical investigation of high Reynolds number flows over square and circular cylinder. *Journal of Thermophysics and Heat Transfer*, 19:72–80.

[Löhner, 2001] Löhner, R. (2001). *Applied CFD techniques. An introduction based on finite element methods.* John Wiley & Sons, Ltd, New York.

[Lombardi, 1993] Lombardi, G. (1993). Experimental study on the aerodynamic effects of a forward sweep angle. *Journal of Aircraft*, 30(5):629–635.

[Lombardi et al., 1998] Lombardi, G., Salvetti, M., and Morelli, M. (1998). Appraisal of numerical methods in predicting the aerodynamics of forward-swept wings. *Journal of Aircraft*, 35(4):561–568.

[Loseille, 2008] Loseille, A. (2008). *Adaptation de maillage anisotrope 3D multi-échelles et ciblée à une fonctionnelle pour la mécanique des fluides. Application à la prédiction haute-fidélité du bang sonique.* PhD thesis, Université Pierre et Marie Curie, Paris VI, Paris, France.

[Loseille and Alauzet, 2009] Loseille, A. and Alauzet, F. (2009). Continuous mesh model and well-posed continuous interpolation error estimation. Research Report RR-6846, INRIA.

[Loseille et al., 2007] Loseille, A., Dervieux, A., Frey, P., and Alauzet, F. (2007). Achievement of global second-order mesh convergence for discontinuous flows with adapted unstructured meshes. *AIAA*, 2007-4186.

[Lourenco and Shih, 1993] Lourenco, L. and Shih, C. (1993). Characteristics of the plane turbulent near weak of a circular cylinder: a particle image velocimetry study.

[Lucy, 1977] Lucy, L. (1977). A numerical approach to the testing of the fission hypothesis. *Astronomical Journal*, 82:1013.

[Luo et al., 1994] Luo, S., Yazdani, M., Chew, Y., and Lee, T. (1994). Effects of incidence and afterbody shape on flow past bluff cylinders. *J. Ind. Aerodyn.*, 53:375–399.

[Lyn et al., 1995] Lyn, D., Einav, S., Rodi, W., and Park, J. (1995). A laser-doppler velocimeter study of ensemble-averaged characteristics of the turbulent near wake of a square cylinder. *Journal of Fluid Mechanics*, 304:285–319.

[Lyn and Rodi, 1994] Lyn, D. and Rodi, W. (1994). The flapping shear layer formed by flow separation from the forward corner of a square cylinder. *Journal of Fluid Mechanics*, 267:353–376.

[Maday and Magoulès, 2005] Maday, Y. and Magoulès, F. (2005). Non-overlapping additive Schwarz methods tuned to highly heterogeneous media. *Comptes Rendus à l'Académie des Sciences, Mathématiques*, 341(11):701–705.

[Maday and Magoulès, 2006a] Maday, Y. and Magoulès, F. (2006a). Absorbing interface conditions for domain decomposition methods: a general presentation. *Computer Methods in Applied Mechanics and Engineering*, 195(29–32):3880–3900.

[Maday and Magoulès, 2006b] Maday, Y. and Magoulès, F. (2006b). Improved ad hoc interface conditions for Schwarz solution procedure tuned to highly heterogeneous media. *Applied Mathematical Modelling*, 30(8):731–743.

[Maday and Magoulès, 2007a] Maday, Y. and Magoulès, F. (2007a). Optimal convergence properties of the FETI domain decomposition method. *International Journal for Numerical Methods in Fluids*, 55(1):1–14.

[Maday and Magoulès, 2007b] Maday, Y. and Magoulès, F. (2007b). Optimized Schwarz methods without overlap for highly heterogeneous media. *Computer Methods in Applied Mechanics and Engineering*, 196(8):1541–1553.

[Magoulès, 2007] Magoulès, F., editor (2007). *Mesh partitioning techniques and domain decomposition methods*. Saxe-Coburg Publications, Stirlingshire, UK.

[Magoulès, 2008] Magoulès, F., editor (2008). *Computational methods for acoustics problems*. Saxe-Coburg Publications, Stirlingshire, UK.

[Magoulès, 2010] Magoulès, F., editor (2010). *Substructuring techniques and domain decomposition methods.* Saxe-Coburg Publications, Stirlingshire, UK.

[Magoulès and Autrique, 2001] Magoulès, F. and Autrique, J.-C. (2001). Calcul parallèle et méthodes d'éléments finis et infinis pour des problèmes de radiation acoustique. *Calculateurs Parallèles, Réseaux et Systèmes Répartis*, 13(1):11–34.

[Magoulès and Benelmir, 2007] Magoulès, F. and Benelmir, R., editors (2007). *Techniques of scientific computing for energy and the environment.* Nova Science Publishers, Inc., New York, USA.

[Magoulès et al., 2004a] Magoulès, F., Iványi, P., and Topping, B. (2004a). Convergence analysis of Schwarz methods without overlap for the Helmholtz equation. *Computers and Structures*, 82(22):1835–1847.

[Magoulès et al., 2004b] Magoulès, F., Iványi, P., and Topping, B. (2004b). Non-overlapping Schwarz methods with optimized transmission conditions for the Helmholtz equation. *Computer Methods in Applied Mechanics and Engineering*, 193(45–47):4797–4818.

[Magoulès and Kako, 2006] Magoulès, F. and Kako, T., editors (2006). *Domain decomposition methods: theory and applications*, volume 25 of *Mathematical Sciences and Applications*. Gakkotosho, Tokyo, Japan.

[Magoulès et al., 2000] Magoulès, F., Meerbergen, K., and Coyette, J.-P. (2000). Application of a domain decomposition method with Lagrange multipliers to acoustic problems arising from the automotive industry. *Journal of Computational Acoustics*, 8(3):503–521.

[Magoulès and Putanowicz, 2005] Magoulès, F. and Putanowicz, R. (2005). Optimal convergence of non-overlapping Schwarz methods for the Helmholtz equation. *Journal of Computational Acoustics*, 13(3):525–545.

[Magoulès and Putanowicz, 2006] Magoulès, F. and Putanowicz, R. (2006). Large-scale data visualization using multi-language programming applied to environmental problems. *International Journal of Energy, Environment and Economics*, 13(2):45–75.

[Magoulès and Roux, 2006] Magoulès, F. and Roux, F.-X. (2006). Lagrangian formulation of domain decomposition methods: a unified theory. *Applied Mathematical Modelling*, 30(7):593–615.

[Magoulès et al., 1998] Magoulès, F., Roux, F.-X., and de la Bourdonnaye, A. (1998). Méthode de décomposition de domaine pour des problèmes hyperboliques. *Calculateurs Parallèles, Réseaux et Systèmes Répartis*, 10(4):353–361.

[Magoulès et al., 2004c] Magoulès, F., Roux, F.-X., and Salmon, S. (2004c). Optimal discrete transmission conditions for a non-overlapping domain decomposition method for the Helmholtz equation. *SIAM Journal on Scientific Computing*, 25(5):1497–1515.

[Magoulès et al., 2005] Magoulès, F., Roux, F.-X., and Series, L. (2005). Algebraic way to derive absorbing boundary conditions for the Helmholtz equation. *Journal of Computational Acoustics*, 13(3):433–454.

[Magoulès et al., 2006a] Magoulès, F., Roux, F.-X., and Series, L. (2006a). Algebraic approximation of Dirichlet-to-Neumann maps for the equations of linear elasticity. *Computer Methods in Applied Mechanics and Engineering*, 195(29–32):3742–3759.

[Magoulès et al., 2006b] Magoulès, F., Roux, F.-X., and Series, L. (2006b). Algebraic Dirichlet-to-Neumann mapping for linear elasticity problems with extreme contrasts in the coefficients. *Applied Mathematical Modelling*, 30(8):702–713.

[Magoulès et al., 2007] Magoulès, F., Roux, F.-X., and Series, L. (2007). Algebraic approach to absorbing boundary conditions for the Helmholtz equation. *International Journal of Computer Mathematics*, 84(2):231–240.

[Marchuk, 1990] Marchuk, G. (1990). *Splitting and alternating direction methods*, volume 1 of *Handbook of Numerical Analysis*, pages 197–462. Elsevier.

[Marongiu, 2007] Marongiu, J. (2007). *Méthode numérique Lagrangienne pour la simulation d'écoulements à surface libre. Application aux turbines Pelton*. PhD thesis, Ecole Centrale de Lyon, France.

[Marongiu et al., 2009a] Marongiu, J., Leboeuf, F., Caro, J., Parkinson, E., and Jang, Y. (2009a). Preconditioning of the SPH-ALE meshless method for the simulation of free surface flows in Pelton turbines. In *Proceedings of the 8th European Turbomachinery Conference, Graz, Austria*.

[Marongiu et al., 2009b] Marongiu, J., Leboeuf, F., Lance, M., and Parkinson, E. (2009b). Riemann solvers and efficient boundary treatments: an hybrid SPH-finite volume numerical method. In *Proceedings of the 4th International SPHERIC SPH Workshop, Nantes, France*. Ecole Centrale de Nantes.

[Marongiu et al., 2006] Marongiu, J., Leboeuf, F., and Parkinson, E. (2006). Simulation of Lagrangian particles adapted to hydraulic surfaces. In *Proceedings of the SPHERIC ERCOFTAC Workshop, Rome, May*. La Sapienza University.

[Marongiu et al., 2007a] Marongiu, J., Leboeuf, F., and Parkinson, E. (2007a). A new simple solid boundary treatment for the meshless method SPH. Application to Pelton turbine flows. In *Proceedings of the 8th International Symposium on Experimental and Computational Aerothermodynamics of Internal Flows, ISAIF8, Ecully, France, 2-5 July.*

[Marongiu et al., 2007b] Marongiu, J., Leboeuf, F., and Parkinson, E. (2007b). A new treatment of solid boundaries for the SPH method. Application to Pelton turbine flows. In *Proceedings of the SPHERIC ERCOFTAC Workshop, Madrid, May.* Universidad Politechnica de Madrid.

[Marongiu et al., 2007c] Marongiu, J., Leboeuf, F., and Parkinson, E. (2007c). Numerical simulation of the flow in a Pelton turbine using the meshless method smoothed particle hydrodynamics: a new simple solid boundary treatment. In *Proceedings of the Institution of Mechanical Engineers*, volume 221:6, pages 849–856. Professional Engineering Publishing.

[Maronnier et al., 1999] Maronnier, V., Picasso, M., and Rappaz, J. (1999). Numerical simulation of free surface flows. *Journal of Computational Physics*, 155:439–455.

[Maronnier et al., 2003] Maronnier, V., Picasso, M., and Rappaz, J. (2003). Numerical simulation of three dimensional free surface flows. *International Journal for Numerical Methods in Fluids*, 42(7):697–716.

[Martin and Guillard, 1996] Martin, R. and Guillard, H. (1996). A second-order defect correction scheme for unsteady problems. *Computers and Fluids*, 25(1):9–27.

[Martinat et al., 2008] Martinat, G., Hoarau, Y., Braza, M., Vos, J., and Harran, G. (2008). Numerical simulation of the dynamic stall of a NACA 0012 airfoil using DES and advanced OES/URANS modelling. *Notes on Numerical Fluid Mechanics and Multidisciplinary Design*, 97:271–278.

[Mason, 1994] Mason, P. (1994). Large-eddy simulation: a critical review of the technique. *Quarterly Journal of the Royal Meteorological Society*, 120:1–26.

[Mathieu and Scott, 2000] Mathieu, J. and Scott, J. (2000). *An introduction to turbulent flow.* Cambridge University Press.

[Maury, 1996] Maury, B. (1996). Characteristics ALE method for the 3D Navier-Stokes equations with a free surface. *International Journal of Computational Fluid Dynamics*, 6:175–188.

[Maury, 1999] Maury, B. (1999). Direct simulations of 2D fluid-particle flows in bi-periodic domains. *Journal of Computational Physics*, 156:325–351.

[Mavriplis, 1997] Mavriplis, D. (1997). Unstructured grid techniques. *Annual Review of Fluid Mechanics*, 29:473–514.

[McKee et al., 2004] McKee, S., Tome, M., Cuminato, J., Castelo, A., and Ferreira, V. (2004). Recent advances in the marker and cell method. *Archives of Computational Methods in Engineering*, 11(2):107–142.

[Meneveau and Katz, 2000] Meneveau, C. and Katz, J. (2000). Scale-invariance and turbulence models for large-eddy simulation. *Annual Review of Fluid Mechanics*, 32:1–32.

[Mer, 1998a] Mer, K. (1998a). Variational analysis of a mixed-element-volume scheme with fourth-order viscosity on general triangulations. *Computer Methods in Applied Mechanics and Engineering*, 153:45–62.

[Mer, 1998b] Mer, K. (1998b). Variational analysis of a mixed element/volume scheme with fourth-order viscosity on general triangulations. *Computer Methods in Applied Mechanics and Engineering*, 153:45–62.

[Mercanton, 1916] Mercanton, P. (1916). Vermessungen am Rhonegletscher, 1874–1915. *Neue Denkschriften der Schweizerischen Naturforschenden Gesellschaft*, 52.

[MeshAdapt, 2003] MeshAdapt (2003). *MeshAdapt: a mesh adaptation tool. User's manual version 3.0*. Distene S.A.S., Pôle Teratec - BARD-1, Domaine du Grand Rué, 91680 Bruyères-le-Chatel, France.

[Meyers et al., 2008] Meyers, J., Geurts, B., and Sagaut, P., editors (2008). *Quality and reliability of large-eddy simulations*, volume 12 of *ERCOFTAC Series*. Springer.

[Moin and Mahesh, 1998] Moin, P. and Mahesh, K. (1998). Direct numerical simulation. A tool in turbulence research. *Annual Review of Fluid Mechanics*, 30:539–578.

[Moin et al., 1991] Moin, P., Squires, K., Cabot, W., and Lee, S. (1991). A dynamic subgrid-scale model for compressible turbulence and scalar transport. *Physics of Fluids*, A3:2746–2757.

[Monaghan, 1985] Monaghan, J. (1985). A refined method for astrophysical problems. *Astronomy and Astrophysics*, 149:135–143.

[Monaghan, 1992] Monaghan, J. (1992). Smoothed particle hydrodynamics. *Annual review of Astronomy and Astrophysics*, 30:543–574.

[Monaghan, 1995a] Monaghan, J. (1995a). Simulating gravity currents with SPH. I. Lock gates. *Mathematics Reports and Preprints*, 95(5).

[Monaghan, 1995b] Monaghan, J. (1995b). Simulating gravity currents with SPH. III. Boundary forces. *Mathematics Reports and Preprints*.

[Monaghan, 1997] Monaghan, J. (1997). SPH and Riemann solvers. *Journal of Computational Physics*, pages 298–307.

[Monaghan and Kajtar, 2009] Monaghan, J. and Kajtar, J. (2009). SPH boundary forces. In *Proceedings of the 4th International SPHERIC Workshop, Nantes, France, May 27-29*.

[Monaghan and Lattanzio, 1985] Monaghan, J. and Lattanzio, J. (1985). A refined method for astrophysical problems. *Astronomy and Astrophysics*, 149:135–143.

[Monaghan et al., 1994] Monaghan, J., Thompson, M. C., and Hourigan, K. (1994). Simulation of free surface flows with SPH. In *Proceedings of the ASME Symposium on Computational Methods in Fluid, Dynamics, Lake Tahoe, USA*.

[Morris et al., 1997] Morris, J., Fox, P., and Yi, Z. (1997). Modeling low Reynolds number incompressible flows using SPH. *Journal of Computational Physics*, 136:214–226.

[Munts et al., 2007] Munts, E., Hulshoff, S., and Borst, R. (2007). A modal-based multiscale method for large-eddy simulation. *Journal of Computational Physics*, 224:389–402.

[Murrone and Guillard, 2005] Murrone, A. and Guillard, H. (2005). A five equation model for compressible two-phase flow computations. *Journal of Computational Physics*, 202(2):664–698.

[Nestor Ruairi et al., 2009] Nestor Ruairi, M., Basa, M., Lastiwka, M., and Quinlan Nathan, J. (2009). Extension of the finite volume particle method to viscous flow. *Journal of Computational Physics*.

[Newman and Launder, 1981] Newman, G. and Launder, B. (1981). Modelling the behavior of homogeneous scalar turbulence. *Journal of Fluid Mechanics*, 111:217–232.

[Nicoud and Ducros, 1999] Nicoud, F. and Ducros, F. (1999). Subgrid-scale stress modelling based on the square of the velocity gradient tensor. *Flow Turbulence and Combustion*, 62(3):183–200.

[Nielsen and Anderson, 2002] Nielsen, E. and Anderson, W. (2002). Recent improvement in aerodynamic design optimization on unstructured meshes. *AIAA Journal*, 40(6):1155–1163.

[Nishimura, 2008] Nishimura, D. (2008). Changes in surface flow speed over the past 100 years (Rhonegletscher, Swiss Alps). Master's thesis, Graduate School of Environmental Science, Hokkaido University, Japan.

[Nkonga and Guillard, 1994] Nkonga, B. and Guillard, H. (1994). Godunov type method on non-structured meshes for three-dimensional moving boundary problems. *Computer Methods in Applied Mechanics and Engineering*, 113(1-2):183–204.

[Noh and Woodward, 1976] Noh, W. and Woodward, P. (1976). *Simple line interface calculation (SLIC)*, volume 59 of *Lectures Notes in Physics*, pages 330–340. Springer-Verlag.

[Norberg, 1993] Norberg, C. (1993). Flow around rectangular cylinders: pressure forces and wake frequencies. *Journal of Wind Engineering and Industrial Aerodynamics*, 49:187–196.

[Oger, 2006] Oger, G. (2006). *Aspects théoriques de la méthode SPH et applications à l'hydrodynamique à surface libre*. PhD thesis, Ecole Centrale de Nantes, France.

[Ong and Wallace, 1996] Ong, L. and Wallace, J. (1996). The velocity field of the turbulent very near wake of a circular cylinder. *Experiments in Fluids*, 20:441–453.

[Osher and Fedkiw, 2001] Osher, S. and Fedkiw, R. (2001). Level set methods: an overview and some recent results. *Journal of Computational Physics*, 169:463–502.

[Osher and Fedkiw, 2003] Osher, S. and Fedkiw, R. (2003). *Level set methods and dynamic implicit surfaces*. Applied Mathematical Sciences. Springer-Verlag.

[Osher and Sethian, 1988] Osher, S. and Sethian, J. (1988). Fronts propagating with curvature dependent speed: algorithms based on Hamilton-Jacobi formulations. *Journal of Computational Physics*, 79:12–49.

[Pagano et al., 2006] Pagano, G., Camarri, S., Salvetti, M., Koobus, B., and Dervieux, A. (2006). Strategies for RANS/VMS-LES coupling. Research Report RR-5954, INRIA.

[Pain et al., 2001] Pain, C., Humpleby, A., de Oliveira, C., and Goddard, A. (2001). Tetrahedral mesh optimisation and adaptivity for steady-state and transient finite element calculations. *Computer Methods in Applied Mechanics and Engineering*, 190:3771–3796.

[Pan and Wang, 2009] Pan, T. and Wang, T. (2009). Dynamical simulation of red blood cell rheology in microvessels. *International Journal of Numerical Analysis and Modeling*, 6:455–473.

[Parneaudeau et al., 2008] Parneaudeau, P., Carlier, J., Heitz, D., and Lamballais, E. (2008). Experimental and numerical studies of the flow over a circular cylinder at Reynolds number 3900. *Physics of Fluids*, 20(085101).

[Parolini and Quarteroni, 2004] Parolini, N. and Quarteroni, A. (2004). Mathematical models and numerical simulations for the America's cup. *Computer Methods in Applied Mechanics and Engineering*, 194:1001–1026.

[Pasquetti, 2005] Pasquetti, R. (2005). Spectral vanishing viscosity method for LES: sensitivity to the SVV control parameters. *Journal of Turbulence*, 6.

[Pasquetti, 2006a] Pasquetti, R. (2006a). Spectral vanishing viscosity method for high-order LES: computation of the dissipation rates. In Wesseling, P., Onate, E., and Periaux, J., editors, *Proceedings of the ECCOMAS CFD Conference*.

[Pasquetti, 2006b] Pasquetti, R. (2006b). Spectral vanishing viscosity method for large-eddy simulation of turbulent flows. *Journal of Scientific Computing*, 27:365–375.

[Paterson, 1994] Paterson, W. (1994). *The physics of glaciers*. Pergamon, New York, 3rd edition.

[Pattyn, 2003] Pattyn, F. (2003). A new three-dimensional higher-order thermomechanical ice sheet model: basic sensitivity, ice stream development, and ice flow across subglacial lakes. *Journal of Geophysical Research*, 106(B8).

[Perthame and Khobalate, 1992] Perthame, B. and Khobalate, B. (1992). Maximum principle on the entropy and minimal limitations for kinetic scheme. Research Report RR-1628, INRIA.

[Perthame and Shu, 1996] Perthame, B. and Shu, C. (1996). On positivity preserving finite-volume schemes for Euler equations. *Numerical Mathematics*, 73:119–130.

[Peskin, 1977] Peskin, C. (1977). Numerical analysis of blood flow in the heart. *Journal of Computational Physics*, 25:220–252.

[Peskin, 1980] Peskin, C. (1980). The immersed boundary method. *Acta Numerica*, 11:479–517.

[Peyret, 1996] Peyret, R. (1996). *Handbook of computational fluid dynamics*. Academic Press.

[Peyret and Krause, 2000] Peyret, R. and Krause, E. (2000). *Advanced turbulent flow computations*. Number 395 in CISM courses and lectures. Springer Wein, New York.

[Picasso, 2003] Picasso, M. (2003). An anisotropic error indicator based on Zienkiewicz-Zhu error estimator: application to elliptic and parabolic problems. *SIAM Journal on Scientific Computing*, 24(4):1328–1355.

[Picasso et al., 2004] Picasso, M., Rappaz, J., Reist, A., Funk, M., and Blatter, H. (2004). Numerical simulation of the motion of a two-dimensional glacier. *International Journal for Numerical Methods in Engineering*, 60(5):995–1009.

[Pierce and Giles, 2000] Pierce, N. and Giles, M. (2000). Adjoint recovery of superconvergent functionals from PDE approximations. *SIAM Review*, 42(2):247–264.

[Piperno and Depeyre, 1998] Piperno, S. and Depeyre, S. (1998). Criteria for the design of limiters yielding efficient high resolution TVD schemes. *Computers and Fluids*, 27(2):183–197.

[Piperno and Farhat, 2000] Piperno, S. and Farhat, C. (2000). Design of efficient partitioned procedures for the transient solution of aeroelastic problems. *Revue Européenne des Eléments Finis*, 9(6-7):655–680.

[Pironneau, 1989] Pironneau, O. (1989). *Finite element methods for fluids*. Wiley, Chichester.

[Pope, 2000] Pope, S. (2000). *Turbulent Flows*. Cambridge University Press, Cambridge.

[Popinet and Zaleski, 1999] Popinet, S. and Zaleski, S. (1999). A front-tracking algorithm for accurate representation of surface tension. *International Journal for Numerical Methods in Fluids*, 30:777–793.

[Pralong and Funk, 2004] Pralong, A. and Funk, M. (2004). A level-set method for modeling the evolution of glacier geometry. *Journal of Glaciology*, 50(171):485–491.

[Quaini and Quarteroni, 2007] Quaini, A. and Quarteroni, A. (2007). A semi-implicit approach for fluid-structure interaction based on an algebraic fractional step method. *Mathematical Models and Methods in Applied Sciences*, 17(6):957–983.

[Quarteroni and Formaggia, 2003] Quarteroni, A. and Formaggia, L. (2003). *Mathematical modelling and numerical simulation of the cardiovascular system*, chapter Modelling of living systems. Handbook of Numerical Analysis. Elsevier.

[Ramakrishnan and Collis, 2004] Ramakrishnan, S. and Collis, S. (2004). Variational multiscale modeling for turbulence control. *AIAA Journal*, 42:745–753.

[Randles and Libertsky, 1996] Randles, R. and Libertsky, L. (1996). Smoothed particle hydrodynamics, some recent improvements and applications. *Computer Methods in Applied Mechanics and Engineering*, 139:375–408.

[Rappaz and Glowinski, 2003] Rappaz, J. and Glowinski, R. (2003). Approximation of a nonlinear elliptic problem arising in a non-Newtonian fluid flow model in glaciology. *Mathematical Modelling and Numerical Analysis*, 37:175–186.

[Rappaz and Reist, 2005] Rappaz, J. and Reist, A. (2005). Mathematical and numerical analysis of a three-dimensional fluid flow model in glaciology. *Mathematical Models and Methods in Applied Sciences*, 15(1):37–52.

[Rappaz et al., 2000] Rappaz, M., Desbiolles, J., Gandin, C., Henry, S., Semoroz, A., and Thevoz, P. (2000). Modelling of solidification microstructures. *Material Science Forum*, 329(3):389–396.

[Renardy et al., 2003] Renardy, Y., Popinet, S., Duchemin, L., Renardy, M., Zaleski, S., Josserand, C., Drumright-Clarke, M. A., Richard, D., Clanet, C., and Quéré, D. (2003). Pyramidal and toroidal water drops after impact on a solid surface. *Journal of Fluid Mechanics*, 484:69–83.

[Renardy et al., 2004] Renardy, Y., Renardy, M., Chinyoka, T., Khismatullin, D., and Li, J. (2004). A viscoelastic VOF-PROST code for the study of drop deformation. In *Proceedings of the ASME Heat Transfer/Fluids Engineering Summer Conference, Charlotte, North Carolina*.

[Richtmeyer and Morton, 1967] Richtmeyer, R. and Morton, K. (1967). *Difference methods for initial value problems*. Interscience Publishers, New York.

[Rider and Kothe, 1998] Rider, W. and Kothe, D. (1998). Reconstructing volume tracking. *Journal of Computational Physics*, 141:112–152.

[Robinson and Monaghan, 2008] Robinson, M. and Monaghan, J. (2008). Forced two-dimensional wall-bounded turbulence using SPH. In *Proceedings of the 3th SPHERIC ERCOFTAC Workshop on SPH Simulations, Lausanne, Switzerland, June 4-6*.

[Rodi, 1982] Rodi, W. (1982). Examples of turbulence models for incompressible flows. *AIAA Journal*, 20:872–889.

[Rodi et al., 1997] Rodi, W., Ferziger, J., Breuer, M., and Pourquié, M. (1997). Status of large-eddy simulation: results of a workshop. *Journal of Fluids Engineering*, 119:248–262.

[Roe, 1981] Roe, P. (1981). Approximate Riemann solvers, parameters, vectors and difference schemes. *Journal of Computational Physics*, 43:357–372.

[Rogé and Martin, 2008] Rogé, G. and Martin, L. (2008). Goal-oriented anisotropic grid adaptation - adaptation de maillage anisotrope orienté objectif. *Comptes Rendus à l'Académie des Sciences, Mathématiques*, 346(19-20):1109–1112.

[Romerio et al., 2005] Romerio, M., Lozinski, A., and Rappaz, J. (2005). A new modelling for simulating bubble motion in a smelter. In *Light Metals*, pages 547–552.

[Rothenbühler, 2000] Rothenbühler, C. (2000). Erfassung und Darstellung der Geomorphologie im Gebiet Bernina (GR) mit Hilfe von GIS. Master's thesis, University of Zurich, Switzerland.

[Saad, 2003] Saad, Y. (2003). *Iterative Methods for Sparse Linear Systems*. SIAM, Philadelpha, PA, 2nd edition.

[Sagaut, 2001] Sagaut, P. (2001). *Large-eddy simulation for compressible flows*. Springer.

[Salvetti et al., 2007] Salvetti, M., Koobus, B., Camarri, S., and Dervieux, A. (2007). Simulation of bluff-body flows through a hybrid RANS/VMS-LES model. In *Proceedings of the IUTAM Symposium on Unsteady Separated Flows and their Control, Corfu, Grece, June 18-22*.

[Sarkis and Koobus, 2000] Sarkis, M. and Koobus, B. (2000). A scaled and minimum overlap restricted additive Schwarz method with application on aerodynamics. *Computer Methods in Applied Mechanics and Engineering*, 184:391–400.

[Scardovelli and Zaleski, 1999] Scardovelli, R. and Zaleski, S. (1999). Direct numerical simulation of free-surface and interfacial flow. *Annual Review of Fluid Mechanics*, 31(7):567–603.

[Schall et al., 2004] Schall, E., Leservoisier, D., Dervieux, A., and Koobus, B. (2004). Mesh adaptation as a tool for certified computational aerodynamics. *International Journal for Numerical Methods in Fluids*, 45:179–196.

[Schewe, 1983] Schewe, J. (1983). On the forces acting on a circular cylinder in cross flow from subcritical up to transcritical Reynolds numbers. *Journal of Fluid Mechanics*, 133:265–285.

[Schmidt and Thiele, 2002] Schmidt, S. and Thiele, F. (2002). Comparison of numerical methods applied to the flow over wall-mounted cubes. *International Journal of Heat and Fluid Flow*, 23:330–339.

[Schofield et al., 2009] Schofield, S., Garimella, R., François, M., and Loubère, R. (2009). A second-order accurate material-order-independent interface reconstruction technique for multi-material flow simulations. *Journal of Computational Physics*, 228(3):731–745.

[Schoof, 2005] Schoof, C. (2005). The effect of cavitation on glacier sliding. *Royal Society of London Proceedings Series A*, 461:609–627.

[Schoof, 2010] Schoof, C. (2010). Coulomb friction and other sliding laws in a higher order glacier flow model. *Mathematical Models and Methods in Applied Sciences*, 20(1):157–189.

[Selmin and Formaggia, 1998] Selmin, V. and Formaggia, L. (1998). Unified construction of finite element and finite volume discretizations for compressible flows. *International Journal for Numerical Methods in Engineering*, 39(1):1–32.

[Sethian, 1999] Sethian, J. (1999). Fast marching methods. *SIAM Review*, 41(2):199–235.

[Shaskov and Wendroff, 2004] Shaskov, M. and Wendroff, B. (2004). The repair paradigm and application to conservation laws. *Journal of Computational Physics*, 198(1):265–277.

[Shelley et al., 1997] Shelley, M., Tian, F.-R., and Wlodarski, K. (1997). Hele-Shaw flow and pattern formation in a time-dependent gap. *Nonlinearity*, 10(6):1471–1495.

[Shepard, 1968] Shepard, D. (1968). A two-dimensional interpolation function for irregularly spaced points. In *Proceedings of A.C.M. National Conference*, pages 517–524.

[Shin and Juric, 2002] Shin, S. and Juric, D. (2002). Modeling three-dimensional multiphase flow using a level contour reconstruction method for front tracking without connectivity. *Journal of Computational Physics*, 180:427–470.

[Shu and Osher, 1988] Shu, C. and Osher, S. (1988). Efficient implementation of essential non-oscillatory shock capturing schemes. *Journal of Computational Physics*, 77:439–471.

[Shyue, 1999a] Shyue, K.-M. (1999a). A fluid-mixture type algorithm for compressible multicomponent flow with van der Waals equation of state. *Journal of Computational Physics*, 156:43–88.

[Shyue, 1999b] Shyue, K.-M. (1999b). A volume-of-fluid type algorithm for compressible two-phase flows. *International Series of Numerical Mathematics*, 130:895–904.

[Sidilkover, 1994] Sidilkover, D. (1994). A genuinely multidimensional upwind scheme and efficient multigrid solver for the compressible Euler equations. Research Report 94-84, ICASE.

[Sirnivas et al., 2006] Sirnivas, S., Wornom, S., Dervieux, A., Koobus, B., and Allain, O. (2006). A study of LES models for the simulation of a turbulent flow around a truss spar geometry. In *Proceedings of 25rd International Conference on Offshore and Arctic Engineering (OMAE'06)*.

[Smagorinsky, 1963] Smagorinsky, J. (1963). General circulation experiments with the primitive equations. *Monthly Weather Review*, 91(3):99–164.

[Sohankar et al., 2000] Sohankar, A., Davidson, L., and Norberg, C. (2000). Large-eddy simulation of flow past a square cylinder: comparison of different subgrid scale models. *Journal of Fluids Engineering*, 122:39–47.

[Spalart et al., 1997] Spalart, P., Jou, W., Strelets, M., and Allmaras, S. (1997). *Advances in DNS/LES*, chapter Comments on the feasibility of LES for wings and on a hybrid RANS/LES approach. Columbus (OH).

[Speziale, 1998] Speziale, C. (1998). A combined large-eddy simulation and time-dependent RANS capability for high-speed compressible flow. *Journal of Scientific Computing*, 13(3):253–274.

[Steger and Warming, 1981a] Steger, J. and Warming, R. (1981a). Flux vector splitting for the inviscid gas dynamic equations with applications to the finite difference methods. *Journal of Computational Physics*, 40(2):263–293.

[Steger and Warming, 1981b] Steger, J. and Warming, R. (1981b). Flux vector splitting of the inviscid gasdynamic equations with application to finite-difference methods. *Journal of Computational Physics*, 40:263–293.

[Stoufflet et al., 1996] Stoufflet, B., Periaux, J., Fezoui, L., and Dervieux, A. (1996). 3D hypersonic Euler numerical simulation around space vehicles using adapted finite elements. *AIAA*, 86-0560.

[Stroeven, 1996] Stroeven, A. (1996). The robustness of one-dimensional, time-dependent, ice-flow models: a case study from Storglaciären, Northern Sweden. *Geogr. Ann.*, 78A(2-3):133–146.

[Sussman et al., 1999] Sussman, M., Almgren, A., Bell, J., Colella, P., Howell, L., and Welcome, M. (1999). An adaptive level set approach for incompressible two-phase flows. *Journal of Computational Physics*, 148:81–124.

[Sussman et al., 1998] Sussman, M., Fatemi, E., Smereka, P., and Osher, S. (1998). An improved level set method for incompressible two-phase flows. *Computers and Fluids*, 27(5-6):663–680.

[Sussman and Puckett, 2000] Sussman, M. and Puckett, E. (2000). A coupled level set and volume-of-fluid method for computing 3D and axisymmetric incompressible two-phase flows. *Journal of Computational Physics*, 162:301–337.

[Sussman et al., 2007] Sussman, M., Smith, K., Hussaini, M., Ohta, M., and Zhi-Wei, R. (2007). A sharp interface method for incompressible two-phase flows. *Journal of Computational Physics*, 221(2):469–505.

[Sweby, 1984] Sweby, P. (1984). High resolution schemes using limiters for hyperbolic conservation laws. *SIAM Journal in Numerical Analysis*, 21:995–1011.

[Tait, 1888] Tait, P. (1888). Report on some of the physical properties of fresh water and sea water. *Physical Chemistry*, 2.

[Tam et al., 2000] Tam, A., Ait-Ali-Yahia, D., Robichaud, M., Moore, M., Kozel, V., and Habashi, W. (2000). Anisotropic mesh adaptation for 3D flows on structured and unstructured grids. *Computer Methods in Applied Mechanics and Engineering*, 189:1205–1230.

[Teleaga and Struckmeier, 2008] Teleaga, D. and Struckmeier, J. (2008). A finite-volume particle method for conservation laws on moving domains. *International Journal for Numerical Methods in Fluids*, 58:945–967.

[Tennekes and Lumley, 1972] Tennekes, H. and Lumley, J. (1972). *A first course in turbulence*. MIT Press, Cambridge, MA.

[Thomas and Lombard, 1979] Thomas, P. and Lombard, C. (1979). Geometric conservation law and its application to flow computations on moving grids. *AIAA Journal*, 17:1030–1037.

[Thompson, 1986] Thompson, E. (1986). Use of pseudo-concentrations to follow creeping viscous flows during transient analysis. *International Journal for Numerical Methods in Fluids*, 6:749–761.

[Torres and Brackbill, 2000] Torres, D. and Brackbill, J. (2000). The point-set method: front-tracking without connectivity. *Journal of Computational Physics*, 165:620–644.

[Travin et al., 1999] Travin, A., Shur, M., Strelets, M., and Spalart, P. (1999). Detached-eddy simulations past a circular cylinder. *Flow, Turbulence and Combustion*, 63:293–313.

[Tryggvason et al., 2001] Tryggvason, G., Bunner, B., Esmaeeli, A., Juric, D., Al-Rawahi, N., Tauber, W., Nas, J., and Jan, Y.-J. (2001). A front tracking method for the computations of multiphase flow. *Journal of Computational Physics*, 169:708–759.

[Tsubota et al., 2006] Tsubota, K., Wada, S., and Yamaguchi, T. (2006). Simulation study on effects of hematocrit on blood flow properties using particle method. *Journal of Biomechanical Engineering*, 1(1):159–170.

[Turkel, 1993] Turkel, E. (1993). Review of preconditioning methods of fluid dynamics. *Applied Numerical Mathematics*, 12:257–284.

[Unverdi and Tryggvason, 1992] Unverdi, S. and Tryggvason, G. (1992). Computations of multi-fluid flows. *Physica D*, 60:70–83.

[Van Leer, 1977a] Van Leer, B. (1977a). Towards the ultimate conservative difference scheme. IV. A new approach to numerical convection. *Journal of Computational Physics*, 23:276–299.

[Van Leer, 1977b] Van Leer, B. (1977b). Towards the ultimate conservative scheme. IV. A new approach to numerical convection. *Journal of Computational Physics*, 23:276–299.

[Van Leer, 1979] Van Leer, B. (1979). Towards the ultimate conservative difference scheme. V. A second order sequel to Godunov's method. *Journal of Computational Physics*, 32:101–136.

[Vassilevski and Agouzal, 2005] Vassilevski, Y. and Agouzal, A. (2005). Unified asymptotic analysis of interpolations errors for optimal meshes. *Doklady Mathematics*, 72(3):295–298.

[Vàzquez et al., 2004] Vàzquez, M., Koobus, B., and Dervieux, A. (2004). Multilevel optimisation of a supersonic aircraft. *Finite Element in Analysis and Design*, 40:2101–2124.

[Vàzquez et al., 2002] Vàzquez, M., Koobus, B., Dervieux, A., and Farhat, C. (2002). Spatial discretization issues for the energy conservation in flow problems on moving grids. Research Report RR-4742, INRIA.

[Venditti and Darmofal, 2002] Venditti, D. and Darmofal, D. (2002). Grid adaptation for functional outputs: application to two-dimensional inviscid flows. *Journal of Computational Physics*, 176(1):40–69.

[Venditti and Darmofal, 2003] Venditti, D. and Darmofal, D. (2003). Anisotropic grid adaptation for functional outputs: application to two-dimensional viscous flows. *Journal of Computational Physics*, 187(1):22–46.

[Vengadesan and Nithiarasu, 2007] Vengadesan, S. and Nithiarasu, P. (2007). Hybrid LES: review and assessment. *Sadhana*, 32:501–511.

[Venkatakrishnan, 1996] Venkatakrishnan, V. (1996). A perspective on unstructured grid flow solvers. *AIAA Journal*, 34:533–547.

[Venkatakrisnan, 1998] Venkatakrisnan, V. (1998). Barriers and challenges in CFD. In *Proceedings of the ICASE Workshop*, volume 6 of *ICASE/LaRC Interdisciplinary Series*, pages 299–313. Kluwer Academic Publishers.

[Verfürth, 1996] Verfürth, R. (1996). *A review of a posteriori error estimation and adaptative mesh-refinement techniques*. Wiley Teubner Mathematics, New York.

[Versteeg and Malalasekera, 1995] Versteeg, H. and Malalasekera, W. (1995). *An introduction to computational fluid dynamics: the finite volume method*. Prentice-Hall.

[Vieli et al., 2000] Vieli, A., Funk, M., and Blatter, H. (2000). Tidewater glaciers: frontal flow acceleration and basal sliding. *Annals of Glaciology*, 31(5):217–221.

[Vila, 1999] Vila, J. (1999). On particle weighted method and smoothed particle hydrodynamics. *Mathematical Models and Method in Applied Science*, 9:161–209.

[Vila, 2005] Vila, J. (2005). SPH renormalized hybrid methods for conservation laws: applications to free surface flows. In *Meshfree Methods for Partial Differential Equations II*, volume 43 of *Lecture Notes in Computational Science and Engineering*. Springer.

[Violeau, 2004] Violeau, D. (2004). One and two-equations turbulent closures for smoothed particle hydrodynamics. In Liong, Phoon, and Babovic, editors, *Proceedings of the 6th International Conference on Hydroinformatics, Singapore*. World Scientific Publishing Company.

[Violeau et al., 2002] Violeau, D., Piccon, S., and Chabard, J. (2002). Two attempts of turbulence modelling in smoothed particle hydrodynamics. In *Advances in Fluid Modelling and Turbulence Measurements*. World Scientific.

[Vreman, 2003] Vreman, A. (2003). The filtering analog of the variational multiscale method in large-eddy simulation. *Physics of Fluids*, 15(8):61–64.

[Vreman, 2004a] Vreman, A. (2004a). An eddy-viscosity subgrid-scale model for turbulent shear flow: algebraic theory and application. *Physics of Fluids*, 16:3670–3681.

[Vreman, 2004b] Vreman, A. (2004b). An eddy-viscosity subgrid-scale model for turbulent shear flow: algebraic theory and application. *Physics of Fluids*, 16:3670–3681.

[Wang et al., 2009] Wang, T., Pan, T., Xing, Z., and Glowinski, R. (2009). Numerical simulation of rheology of red blood cell rouleaux in microchannels. *Physical Review E*, 79:041916.

[Warnecke, 2005] Warnecke, G., editor (2005). *Meshless methods for conservation laws, analysis and numerics for conservation laws*. Springer.

[Wendland, 1995] Wendland, H. (1995). Piecewise polynomial, positive definite and compactly supported radial functions of minimal degree. *Advances in Computational Mathematics*, 4:389–396.

[Whitaker et al., 1989] Whitaker, D., Grossman, B., and Lohner, R. (1989). Two-dimensional Euler computations on a triangular mesh using an upwind finite-volume scheme. *AIAA*, 89-0365.

[Wilcox, 2006] Wilcox, D. (2006). *A first course in turbulence modeling for CFD*. DCW Industries, USA, 3rd edition.

[Williams et al., 1999] Williams, M., Kothe, D., and Puckett, E. (1999). Accuracy and convergence of continuum surface tension models. In *Proceedings of Fluid Dynamics at Interfaces*, pages 294–305. Cambridge University Press, Cambridge. Gainesville, FL, 1998.

[Wu and Wang, 1995] Wu, H. and Wang, L. (1995). Non-existence of third order accurate semi-discrete MUSCL-type schemes for nonlinear conservation laws and unified construction of high accurate ENO schemes. In *Proceedings of the Sixth International Symposium on Computational Fluid Dynamics*. Lake Tahoe, NV.

[Xiao and Ikebata, 2003] Xiao, F. and Ikebata, A. (2003). An efficient method for capturing free boundaries in multi-fluid simulations. *International Journal for Numerical Methods in Fluids*, 42:187–210.

[Yakhot and Orszag, 1986] Yakhot, V. and Orszag, S. A. (1986). Renormalization group analysis of turbulence. I. Basic theory. *Journal of Scientific Computing*, 1(1):3–51.

[Yates, 1987] Yates, E. (1987). AGARD standard aeroelastic configuration for dynamic response, candidate configuration. I. Wing 445.6. Research Report TM-100492, NASA.

Editor Biography

Frédéric Magoulès is Professor at École Centrale Paris in France, leading the High Performance Computing research team in the Applied Mathematics and Systems Department.

Dr. Magoulès received his B.Sc. in Engineering Sciences (1993), M.Sc. in Applied Mathematics (1994), M.Sc. in Numerical Analysis (1995), and Ph.D. in Applied Mathematics (2000) from Université Pierre et Marie Curie, where he then did post-doctoral work and taught for a year as Assistant Professor of Numerical Analysis. In 2000, he joined Université Henri Poincaré as Assistant Professor of Applied Mathematics and Engineering, later becoming Associate Professor. He received his HDR (Habilitation à Diriger des Recherches, "accredation to supervise research") in 2005 from Université Pierre et Marie Curie. In 2006, he joined École Centrale Paris as Professor of Applied Mathematics and Professor, by courtesy, of Computer Science.

With backgrounds in computational science and engineering, applied mathematics, computer science, and consulting experience with industry and national laboratories, Dr. Magoulès works at the algorithmic interface between parallel computing and the numerical analysis of partial differential equations and algebraic differential equations. He and his research team design, analyze, develop, and validate mathematical models and computational methods for the high-performance simulation of multidisciplinary scientific and engineering problems.

Author or co-author of over eighty refereed publications in computational science and engineering, applied mathematics, and computer science, Dr. Magoulès has authored two books and edited seven, and co-edited thirteen special issues of journals. He has also delivered over forty invited presentations at universities, laboratories, and industrial research centers worldwide.

List of Contributors

Frédéric Alauzet
Institut National de Recherche en Informatique et Automatique
Le Chesnay, France

Jérôme Boudet
Laboratoire de Mécanique des Fluides et d'Acoustique
École Centrale de Lyon
Écully, France

Alexandre Caboussat
Department of Mathematics
University of Houston
Houston, Texas, USA

Simone Camarri
Dipartimento di Ingegneria Aerospaziale
Università di Pisa
Pisa, Italy

Alain Dervieux
Institut National de Recherche en Informatique et Automatique
Sophia-Antipolis, France

Fabien Godeferd
Laboratoire de Mécanique des Fluides et d'Acoustique
École Centrale de Lyon
Écully, France

Guillaume Jouvet
Institute of Analysis and Scientific Computing
École Polytechnique Fédérale de Lausanne
Lausanne, Switzerland

Bruno Koobus
Département de Mathématiques
Université Montpellier 2
Montpellier, France

Francis Leboeuf
Laboratoire de Mécanique des Fluides et d'Acoustique
École Centrale de Lyon
Écully, France

Frédéric Magoulès
Applied Mathematics and Systems Laboratory
École Centrale Paris
Châtenay-Malabry, France

Jean-Christophe Marongiu
ANDRITZ HYDRO
Vevey, Switzerland

Hilde Ouvrard
Département de Mathématiques
Université Montpellier 2
Montpellier, France

Marco Picasso
Institute of Analysis and Scientific Computing
École Polytechnique Fédérale de Lausanne
Lausanne, Switzerland

Jacques Rappaz
Institute of Analysis and Scientific Computing
École Polytechnique Fédérale de Lausanne
Lausanne, Switzerland

Maria-Vittoria Salvetti
Dipartimento di Ingegneria Aerospaziale
Università di Pisa
Pisa, Italy

Stephen Wornom
Les Algorithmes
Société Lemma
Biot, France

Index